# HISTORY OF ANALYTIC GEOMETRY

# History of Analytic Geometry

*by*

## CARL B. BOYER

*The Scholar's Bookshelf*

Princeton Junction, New Jersey

Published by The Scholar's Bookshelf
51 Everett Drive, P.O. Box 179
Princeton Junction, N.J. 08550


Copyright © 1956 by Yeshiva University

Originally published as Numbers Six
and Seven of The Scripta Mathematica Studies

First Scholar's Bookshelf printing, 1988

ISBN  0-945726-11-2
ISBN  0-945726-12-0  (pbk.)


Printed in the United States of America

# Table of Contents

Preface . . . . . . . . . . . . . . . . . vii–ix

I. THE EARLIEST CONTRIBUTIONS . . 1–20

II. THE ALEXANDRIAN AGE . . . . . 21–39

III. THE MEDIEVAL PERIOD . . . . . 40–53

IV. THE EARLY MODERN PRELUDE . . 54–73

V. FERMAT AND DESCARTES . . . . . 74–102

VI. THE AGE OF COMMENTARIES . . . 103–137

VII. FROM NEWTON TO EULER . . . . 138–191

VIII. THE DEFINITIVE FORMULATION . . 192–224

IX. THE GOLDEN AGE . . . . . . . . 225–268

Analytical Bibliography . . . . . . . . 269–284

Index . . . . . . . . . . . . . . . . 285–291

# Preface

THE history of analytic geometry is by no means an uncharted sea. Every history of mathematics touches upon it to some extent; and numerous scholarly papers have been devoted to special aspects of the subject. What is chiefly wanting is an integrated survey of the historical development of analytic geometry as a whole. The closest approach to such a treatment is found in two articles by Gino Loria. One of these is in Italian and appeared in 1924 in a periodical (the *Memorie* of the *Reale Accademia dei Lincei* for 1923) which is not easily accessible; the other is in French and was published, in several installments, from 1942 to 1945, in a Roumanian journal (*Mathematica*) which is still less readily available. These two articles together constitute perhaps the most extensive and dependable account of the history of analytic geometry. Somewhat less inclusive treatments are found in German as parts of the works of Heinrich Wieleitner (*Geschichte der Mathematik*, part II, vol. 2, 1921) and Johannes Tropfke *Geschichte der Elementar-Mathematik*, 2nd ed., vol. VI, 1924). Had convenient translations of any of the above works been at hand—or had J. L. Coolidge collated and amplified those portions of his admirable *History of Geometric Methods* (1940) which pertain to analytic geometry—the present work might never have been written. As it was, there seemed to be room for an historical volume of modest size devoted solely to coordinate geometry. It soon became evident that, in view of the amount of material available, some limitations would have to be imposed if the work were to remain within reasonable proportions. Not all relevant details could be included, for H. G. Zeuthen (in *Die Lehre von den Kegelschnitte im Altertum*, 1886) had devoted more than 500 pages to one specific aspect of a single chronological subdivision. It was therefore decided that the present history should cover only such parts of analytic geometry as might reasonably be included in an elementary general college course. Consequently developments of the last hundred years or so are largely omitted, for they are of a more advanced and highly specialized nature. Even within this self-imposed limitation, the account is not intended to be exhaustive with respect to detail. Factual information is presented largely to the extent to which it is suggestive of the general development of ideas. Biographical details in the main have been overlooked, not for want of attractiveness, but because often they have little bearing upon the growth of concepts. For similar reasons peculiarities of terminology and notation have been accorded very limited space. Some attention has been given to the status of analytic geometry *vis-à-vis* other branches of mathematics; but the impact of the wider intellectual milieu has been referred to only where it was regarded as of particular significance. It is of interest to note in this connection that the development of coordinate geometry was not to any great extent bound up with general philosophical problems. The discoveries of Descartes and Fermat in particular are relatively free of any metaphysical background. Indeed, *La géométrie* was in many respects an isolated episode in the career of Descartes—one suggested by a classical problem of Greek geometry. It was the natural outcome of historical tendencies; and had Descartes not lived, mathematical history—in sharp contrast to philosophical—probably would have been much the same, by virtue of Fermat's simultaneous discovery. The work of Fermat is practically devoid of philosophical interest, his discoveries being the result of a close study of the achievements of his predecessors. Perhaps nowhere does one find a better example of the value of historical knowledge for mathematicians than in the case of Fermat, for it is safe to say that, had he not been intimately acquainted with the geometry of Apollonius and Viète, he would not have invented analytic geometry.

It is frequently held that mathematics develops most effectively when it is closely associated with the world of practical affairs—when scholars and artisans work together. However, to this general rule there seem to be more exceptions than there are instances of it; and the discovery of analytic geometry certainly seems to be one of the exceptions. For this reason the sociological background has, in the present account, gone unemphasized. On the other hand, bibliographical references to the source material have been granted a place of some prominence in order to enable the reader to pursue the subject further in directions which he may find especially attractive. Not all works cited in the footnotes are included in the "Analytical Bibliography," with which the volume closes. It is felt that the usefulness of the bibliography is enhanced by limiting it to items which are directly pertinent to the history of algebraic geometry and by incorporating in each case a very brief indication of the nature of the material.

A conscientious effort has been made to see that the information presented is substantially correct in detail, although perfect accuracy in this respect is rarely achieved. However, it is the broad general picture which represents the principal object of the book.  Here there undoubtedly are further points which the reader could wish to have seen included, but it is hoped that there are few portions of the work which he would prefer deleted.  For both inspiration and information the author is heavily indebted to the works of Loria and Wieleitner, and a special measure of credit is due also to the books of Coolidge and Tropfke.  To all the scholars whose studies have served as the basis for the present volume, the author would express his appreciation.

The manuscript of this work was completed about half a dozen years ago, and major portions of it have appeared from time to time in *Scripta Mathematica*.  The bibliography contains a few items published since the completion of the manuscript, but in most cases it was not feasible to make use of these in the body of the work.  There have, however, been few recent developments which would lead one to alter materially judgments on the history of analytic geometry expressed some six years ago.

The appearance of this book is due to the suggestion and encouragement of Professor Jekuthiel Ginsburg of Yeshiva University, and to him, for continued inspiration and assistance in the completion of the project, we extend our warmest gratitude.

*January 3, 1956*                                      CARL B. BOYER

# CHAPTER I

# The Earliest Contributions

*Mighty are numbers, joined with art resistless.*
—EURIPIDES

MATHEMATICS originally was the science of number and magnitude. At first it was limited to the natural numbers and rectilinear configurations; but even from the early primitive stages mankind presumably was concerned with the problem from which analytic geometry arose—the correlation of number with geometrical magnitude. The beginnings of the association of numerical relationships with spatial configurations are prehistoric, as are also the first connections between number and time. The *harpedonaptae* ("rope-stretchers" or surveyors) of Egypt and the astronomers of Chaldea bear witness to the early concern of mathematics with such associations. The very oldest written documents from Mesopotamia, Egypt, China, and India give evidence of the concern with mensuration. Pre-Hellenic papyri and cuneiform texts abound in elaborate problems involving the concepts of length, area, and volume.[1] So highly developed was this aspect of the Egyptian and Babylonian civilizations that one finds there, among other things, the correct result for the volume of the frustum of a pyramid with a square base.

It is indeed possible to have an analytic geometry of points and straight lines alone, a direction toward which ancient mensuration pointed; but historically the subject arose instead from the comparison of curvilinear with rectilinear magnitudes. Here also the Egyptians and Babylonians, in their geometry of the circle, took the first steps. The former made a remarkably accurate estimate of the ratio of the area of the circle to the area of the square on the diameter, taking this ratio to be $(1-\frac{1}{9})^2$, equivalent to taking a value of about 3.16 for $\pi$. The Babylonians adopted the cruder approximation 3 for $\pi$ (although an instance is known in which the value is taken as $3\frac{1}{8}$),

[1] A full account of this work is found in Otto Neugebauer, *Vorlesungen über Geschichte der antiken mathematischen Wissenschaften*, v.I, *Vorgriechische Mathematik* (Berlin, 1934). An excellent bibliography of Egyptian and Babylonian mathematics is found in A. B. Chace, L. S. Bull, H. P. Manning, and R. C. Archibald, *The Rhind Mathematical Papyrus* (2 vols., Oberlin, 1927–1929). For other contributions, including reliable popular expositions, see the "Literature List and Notes" in Archibald, *Outline of the History of Mathematics* (6th ed., Mathematical Association of America, 1949).

but their geometry of the circle nevertheless surpassed that of the contemporary Egyptians. They recognized that the angle inscribed in a semicircle is right, anticipating Thales by well over a thousand years. Moreover, they were familiar at about the same time with the Pythagorean theorem. Combining these two famous propositions, they found—for a given circle of radius $r$—the relationship between the length of a chord $c$ and its sagitta $s$. This property, when symbolically expressed in such a form as $4r^2 = c^2 + 4(r-s)^2$, may in a sense be regarded as an equation of the circle in terms of the rectangular coordinates $c$ and $s$. The Babylonians never reached this point of view, for such essential elements of analytic geometry as coordinates and equations of curves arose considerably later; but it is well to bear in mind how closely certain aspects of ancient mathematics approach their modern counterparts. Primitive systems of coordinates were used by Nilotic surveyors as early as 1400 B.C., and probably also by Mesopotamian star-gazers;[2] but there is no evidence that Egyptian or Babylonian geometers ever explicitly developed a formal geometric coordinate system.

The nascent state of the idea of coordinates was not the only difficulty in the way of the development of analytic geometry. Deficiencies in arithmetic were possibly just as serious. The systems of numeration used in the Nile and Mesopotamian valleys were not so well adapted to calculation as is ours. The hieratic script of the Egyptians made use of the principle of cipherization in connection with the ten-scale, but did not apply the idea of local value or position; the Babylonian sexagesimal notation, on the other hand, employed the positional principle, but cipherization was impracticable in conjunction with a base or radix of that size. Granted that these systems of numeration were imperfect, it is nevertheless open to doubt that difficulties in methods of computation operated as seriously to obstruct the growth of algebra as did other factors. After all, the Babylonians calculated the diagonal of a square to the equivalent of half a dozen decimal figures! The shortcomings were probably more in number *concepts* than in number *symbols*. Algebra calls for a higher degree of abstraction than does geometry, and this element seems to have been lacking in pre-Hellenic mathematics. Number referred essentially to concrete whole numbers, and the idea of general fractions was missing in Egyptian writings. Much time was spent in finding ways of avoiding all but unit fractions, so that the ratio of 2 to 43 would be written as $^1/_{24} + ^1/_{258} + ^1/_{1032}$ or as $^1/_{42} + ^1/_{86} + ^1/_{129} + ^1/_{301}$. Whether the Babylon-

[2] See E. W. Woolard, "The Historical Development of Celestial Co-ordinate Systems," *Publication of the Astronomical Society of the Pacific*, v. LIV (1942), p. 77–90.

ians achieved the concept of general rational number is open to question because of ambiguities in the interpretation of tables, the use of which was greatly emphasized. Elaborate tables give pairs of numbers the product of which is unity (or a power of sixty?). Presumably these tables of reciprocals served then as decimal fractions do today—as a means of avoiding general common fractions.

There are striking examples of high levels of attainment in Babylonian algebra of several thousand years ago. Numerous cases are given of quadratic equations which evidently were solved by the equivalent of the now customary "completion of the square" or its formula analogue (using the positive sign before the radical); and cubic equations are solved by the use of tables of cubes. Some work indicates a rough equivalent of logarithms, and there are instances of the use of negative numbers. Recent disclosures[3] indicate that the Babylonians possessed some rudiments of an abstract theory of numbers, including a rule for determining Pythagorean triads. They may also have been familiar with the ideas of arithmetic, geometric, and harmonic mean. Such a level of algebraic technique is in itself truly wonderful; but it is difficult to determine the extent to which such work definitively determined the developments in Greece, where the next steps toward analytic geometry were taken in a different manner and spirit.

The pre-Hellenic civilizations bequeathed to their successors a large body of knowledge in both arithmetic and geometry; but the association of these two fields which later characterized algebraic geometry was the outgrowth of an abstract generalization which the Egyptians and Babylonians failed to achieve. The earliest discoveries of numerical and spatial relationships followed from the empirical investigation of concrete cases, and were extended by a rough process of induction to include other similar cases. The results arrived at by this method may have been conceived of in general terms, but they invariably are stated in specific numerical terms rather than as universal theorems. Moreover, extant evidence indicates that formal deductive reasoning was not used by pre-Hellenic peoples. For these reasons the Greeks[4] ordinarily are regarded as the founders of mathematics in the strict sense of the word, for they emphasized the value of abstract generalizations (of which analytic geometry is a striking example) and the deductive elaboration of these. Just how or why this momentous

---

[3] See O. Neugebauer and A. Sachs, *Mathematical Cuneiform Texts* (American Oriental Series, v. XXIX), New Haven, Conn. (1945). Cf. also Neugebauer, *The Exact Sciences in Antiquity*, Princeton University Press, 1952.

[4] For the best general account of the Greek contribution see T. L. Heath, *A History of Greek Mathematics*, 2 vols., Oxford, 1921 (or his later but briefer *Manual of Greek Mathematics*, Oxford, 1931). See also B. L. van der Waerden, *Science Awakening* (transl. by Arnold Dresden), Groningen, 1954.

change took place has been a favorite topic of speculation from which no categorical conclusion has been derived. It is of interest to note, however, that this early intellectual revolution occurred at about the time of a distinct geographical shift in the centers of civilization. The focal points previously had been river valleys, such as that of the Nile, or of the Tigris and Euphrates; but by the middle of the eighth century B.C. these ancient *potamic* civilizations were confronted with a vigorous young *thallasic* civilization established about the Mediterranean Sea.

Thales (ca. 640–546 B.C.) and Pythagoras (ca. 572–501 B.C.) were largely responsible for, or at least typical of, the intellectual climate in Greece during the sixth century B.C. from which mathematics properly so-called arose; but their contributions lay more in their abstract point of view and in their deductive arrangement of material than in any novelty of subject matter. The theorems of Thales and Pythagoras are misnamed as far as original discovery is concerned, but these names are perhaps justifiable on the basis of the rational deduction of the theorems from other known relationships. The works of these men have not survived, but later accounts—especially by Pappus and Proclus—agree in ascribing the use of deduction to Thales of Miletus, "the first mathematician," and in attributing the rise of mathematics to the status of an independent and abstract discipline—a liberal art— to Pythagoras of Samos and Crotona, "the father of mathematics." In short, these two men—the first mathematicians to be known by name— were the founders of demonstrative geometry. Thales contributed especially to geometry. He seems to have added little to arithmetic or to the pre-Hellenic association of algebra and geometry; but Pythagoras and his disciples went much further in this direction. Earlier peoples had related time and space to number; but the Pythagoreans sought to explain all phenomena through the association of things with the properties of the natural numbers. Their well-known slogan, "All is number", served as the inspiration for much mathematics, both good and bad—elements of analytic geometry, as well as of numerology. As part of this program, the Pythagoreans[5] continued the pre-Hellenic problems in length, area, and volume, confident that number could in all cases be associated with geometrical magnitude. They made the plausible assumption implicit in earlier work, that the relationships of line segments to one another (and similarly for areas and volumes) are expressible through ratios of integers; and hence the concept of ratio and proportion became basic in all Greek mathematics.

Simple proportions had been used in many aspects of pre-Hellenic

[5] See, e. g., Heinrich Vogt, "Die Geometrie des Pythagoras," *Bibliotheca Mathematica* (3), v. IX (1909), p. 15–54.

mathematics, especially in geometric problems of mensuration. Clear indications of the ratio concept are found in the Ahmes papyrus of about 1650 B.C., and in the earlier Moscow papyrus there is a term indicating the ratio of the larger to the smaller side in a right triangle. The Babylonians of this period made use of proportions in connection with linear interpolation within the tables of lunar phases, and they were acquainted also with simple geometric progressions. But there seems to have been no abstract study of ratio and proportion before the Hellenic era.

The lack of the general fraction concept in ancient thought played a powerful role in science and mathematics, for it led to a domination of thought, lasting for two thousand years, by the idea of proportionality instead of the more general notion of function. For the modern word "ratio" the Greeks had two expressions:[6] *diastema*, which meant literally "interval," and *logos*, which meant "word," especially in the sense of conveying meaning or insight. The latter term generally was used in mathematics, pointing to the Pythagorean idea that ratios express the intrinsic nature of things. The language and theory of ratios were developed largely from musical theory, in connection with which Pythagoras discovered the oldest law of mathematical physics— the essence of harmony lies in the fact that the lengths of vibrating strings should be to each other as certain ratios of simple whole numbers. The Greek expression for proportionality or the equality of ratios, was *analogia*, which meant, literally, having "the same ratio." This was somewhat equivalent to the modern use of equations as expressions of functional relationships, although far more restricted, and for two millenia it served as the chief algebraic tool of geometry.

In the days of Thales and the early Pythagoreans, the realm of number included only the positive integers; the only curves recognized in geometry were still the straight line and the circle. Had this situation continued, there would have been little real need for either analytic geometry or the calculus. However, toward the middle of the fifth century B.C. there occurred a crisis which rocked the very foundations of Pythagorean philosophy and its association of number and configuration. This second intellectual revolution—the one which ultimately paved the way for elementary analysis—centered about figures narrowly concentrated in time but widely scattered throughout the Mediterranean world: Zeno of Elea (born ca. 496 B.C.), Hippasus of Metapontum (fl. 445 B.C.), Democritus of Abdera (ca. 460–357 B.C.), Hippocrates of Chios (born ca. 460 B.C.), and Hippias of Elis (born

[6] For a thorough scholarly account in this connection, see Kurt von Fritz, "The Discovery of Incommensurability by Hippasus of Metapontum," *Ann. Math.*, 2nd series, v. XLVI (1945), p. 242–64.

ca. 460 B.C.).  It is of interest to note that in each case the contributions of these men were not the outcome of problems in natural
science or technology, but they were motivated instead by purely
philosophical or theoretical difficulties.  Contrary to a widely held
belief, important developments in mathematics are not necessarily
related to the world's work or to man's material needs.

The Greek search for essences had led the Pythagoreans to picture
the universe as a multitude of mathematical points completely subject
to the laws of number—a sort of arithmetic geometry, but not at all an
analytic geometry.  The rival Eleatic philosophy of Parmenides up-
held the essential "oneness" of the universe and the impossibility of
analyzing it in terms of the "many."  Zeno of Elea sought dialectically
to defend his master's doctrine by demolishing the Pythagorean association of multiplicity with number and magnitude.[7]

Zeno proposed four paradoxes on motion, of which the first two—the
*dichotomy* and the *Achilles*—are directed against the infinite divisibility
of space and time, and the last two—the *arrow* and the *stade*—refute
the finite divisibility of space and time into ultimate countable elements, indivisibles, or monads.  The paradoxes, as one sees now, involve such notions as infinite sequence, limit, and continuity, concepts
for which neither Zeno nor any of the ancients gave precise definition.
They represented a confusion of sense and reason, and hence at that
time were not answerable; but their influence was profound.  The
Greeks banned from their mathematics any thought of an arithmetic
continuum or of an algebraic variable, ideas which might have led to
analytic geometry; and they refused to place any confidence in infinite
processes, the methods which would have resulted in the calculus.
Whereas the Pythagoreans had envisioned a union of arithmetic and
geometry, Greek mathematicians after Zeno saw only the mutual in-
compatibility of the two fields.

The work of Hippasus, roughly contemporary with that of Zeno, was
perhaps even more obstructive to the development of analytic methods.
The Pythagoreans had continued the pre-Hellenic study of length, area,
and volume, confident that number always could be associated with
geometrical magnitude; but not long after 450 B.C. Hippasus (or

---

[7] Accounts of the Zeno paradoxes are well-nigh numberless.  Two of the most thoroughgoing of these—and also the most appropriate from the point of view of mathematics—are
by Florian Cajori: "History of Zeno's Arguments on Motion," *Am. Math. Monthly*, v.
XXII (1915), p. 1–6, 39–47, 77–82, 109–115, 143–149, 179–186, 215–220, 253–258, 292–297;
and "The Purpose of Zeno's Arguments on Motion," *Isis*, v. III (1920), p. 7–20.  See also
Adolph Grünbaum, "A Consistent Conception of the Extended Linear Continuum as an
Aggregate of Unextended Elements," *Philosophy of Science*, v. XIX (1952), p. 288–306.

possibly someone else) blasted this doctrine by the discovery[8] that there exist simple cases of line segments which are mutually incommensurable. The ratio of the diagonal of a square to its side, for example, cannot be expressed in terms of integers. Just how this was discovered or proved cannot be determined with certainty. It may have resulted from the recognition of the non-termination of the geometrical equivalent of the process of finding the greatest common divisor; or it may have originated in the method given by Aristotle— the demonstration that the existence of such a ratio leads to the contradiction that an integer can be at once both even and odd.

That the discovery of the incommensurability of lines made a strong impression on Greek thought is indicated by the story that Hippasus suffered death by shipwreck as a penalty for his disclosure. It is demonstrated more reliably by the prominence given to the theory of irrationals by Plato and his school. The crisis which incommensurability caused in Pythagorean philosophy and Greek mathematics might have been met by the introduction of infinite processes and irrational numbers, but the paradoxes of Zeno blocked this path. Hence the Greeks were led by Zeno and Hippasus to abandon the pursuit of a full arithmetization of geometry, and the path was not resumed until after analytic geometry had reached maturity through more roundabout channels. Throughout Greek history there was no such thing as algebraic analysis. Geometry was the domain of continuous magnitude, arithmetic was concerned with the discrete set of integers; and the two fields were irreconcilable. Length, area, and volume were not numbers attached to a given configuration; they were

---

[8] There is considerable doubt about the time and circumstances of this disclosure, and much has been written on the subject. One of the most recent and convincing accounts is that by von Fritz (*op. cit.*). Another critical discussion is that by Heinrich Vogt: "Haben die alten Inder den Pythagoreischen Lehrsatz und das Irrationale gekannt," *Bibliotheca Mathematica* (3), v. VII (1906–1907), p. 6–23; "Die Entdeckungsgeschichte des Irrationalen nach Plato und anderen Quellen des 4. Jahrhunderts," *Ibid.*, (3), v. X (1910), p. 97–155; and "Zur Entdeckungsgeschichte des Irrationalen," *Ibid.*, (3), v. XIV (1914), p. 9–29. See also Siegmund Guenther, "Die quadratischen Irrationalitäten der Alten und deren Entwickelungsmethoden," *Abhandlungen zur Geschichte der Mathematik*, v. IV (1882), p. 1–134.

There appears to be no satisfactory evidence of an anticipation of the discovery in India. H. G. Zeuthen has pointed out ("Sur l'origine historique de la connaissance des quantites irrationelles," *Oversigt over det Kongelige Danske Videnskabernes Selskabs. Forhandlinger* (1915), p. 333–362) that the discovery must have been preceded by a clear distinction between exact and approximate values, but this apparently was not made by the Hindus. Vogt (*op. cit.*) and Heath (*Euclid*, v. I, p. 364) point out that three stages are implied by the discovery: (1) all values based upon direct measurement must be recognized as inaccurate, (2) a conviction must prevail that it is impossible to arrive at an accurate arithmetical expression of the value, and (3) the impossibility must be proved. Heath adds that there is no real evidence that at the date in question [the time of the Sulvasutras], the Hindus had reached even the first of these stages.

undefined geometric concepts.   Greek "algebra" was a geometry of lines instead of an algorithm of numbers; and classical problems called for the *construction* of lines—a sort of equivalent of modern existence theorems in analysis—for they had no independent algebraic formulas. Greek mathematicians, for example, always considered the ratio of *two* lines rather than the length of *one*.   The quadrature of the circle called for the construction of a square, not the determination of a number.

An enlightening example of the Greek attitude toward arithmetic and geometry is seen in the classical treatment of quadratic equations. The Babylonians of a thousand years before had reduced geometrical problems in mensuration to quadratic equations and then solved these numerically, using algebraic symbolism, much as is done nowadays. Greek geometers, on the other hand, made no such easy transition from the one field to the other.[9]   For them an equation arising from a geometrical problem represented an equality of lines, areas, or volumes, and hence the solution of quadratic equations was a sort of translation of Babylonian methods into the language of geometrical construction.[10]   The method by which this was accomplished, known as the application of areas, is given systematically in Euclid but may well go back to the Pythagoreans.   An area was said to be *applied* to a straight line (segment) when an equal area was described upon this line as base, or, more generally, when one side of the area was thought of as lying along the line, even if the side exceeded the line or fell short of it.   In its simplest form the application of areas amounted to finding the line which, together with a given line, determines a rectangle of given area—that is, it corresponded to the division of a given product by one of its factors.   In more general form it amounted to an algebra of factoring, used in solving quadratic equations.   As an illustration[11] of its use, let it be required to solve $x^2+c^2=bx$ (where all terms are positive and $b>2c$)—or, in Greek terminology, to apply to a straight line segment $b$ a rectangle equal to a given square $c^2$ and falling short (of the end of the segment) by a square figure.   Draw $AB=b$ and let this be bisected at the point $C$.   (See Fig. 1.)   Draw $CO=c$ perpendicular to $AB$.   With $O$ as center and $b/2$ as radius, draw an arc cutting $AB$ in $D$.   Then $BD=x$ is the required line.   ($APQD$ is the rectangle applied to the segment $b$, and $DBRQ$ is the square by which

[9] See, for example, Federigo Enriques, *L'evolution des idées géométriques dans la pensée grecque* (translated by Maurice Solovine, Paris, 1927), and H. G. Zeuthen, "Sur les rapports entre les anciens et les modernes principes de la géométrie," *Atti del IV Congresso Internazionale dei Mathematici*, v. III (1909), p. 422–427.

[10] Neugebauer, "Zur geometrischen Algebra," *Quellen und Studien zur Geschichte der Mathematik...*, Abt. B, Studien, v. III (1934–1936), p. 245–259.

[11] For this method, as well as other aspects of Euclidean and pre-Euclidean mathematics, see *The Thirteen Books of the Elements of Euclid*, edited by T. L. Heath, 3 vols., Cambridge 1908.

*APQD* falls short of the end of the segment.)   By similar procedures the equations $x^2+bx=c^2$ and $x^2=bx+c^2$ (the only other quadratics with positive roots) were solved geometrically.[12]   Such solutions show

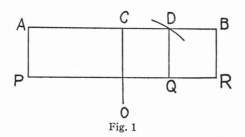

Fig. 1

that Greek algebra—as distinct from arithmetic and logistic—was wholly dependent upon geometry.   Probably one of the chief reasons that Greece did not develop an *algebraic geometry* is that they were bound by a *geometrical algebra*.   After all, one cannot raise himself by his own boot straps.

During the critical years when Zeno and Hippias were confuting the best efforts of mathematicians in the mensuration of figures, there arose three famous challenges within this very area.[13]   Had men of the time realized that all three of these classical problems—the squaring of the circle, the trisection of the angle, and the duplication of the cube—were unsolvable, the whole history of mathematics undoubtedly would have been quite different.   This is particularly true of analytic geometry, for the search for new loci was the direct outgrowth of these questions.   The origin of the three problems is not known, but it is said that Anaxagoras (ca. 499–427 B.C.), the teacher of Pericles, worked on the first one while in prison, presumably without success. So far as we know, the earliest exact results on curvilinear mensuration were due to his younger contemporaries, Democritus and Hippocrates.

The middle of the fifth century B.C. saw the rise of one of the greatest scientific theories of all times—that of physical atomism.

---

[12] On the history of the solution of equations, the following two works are especially valuable: A. Favaro, "Notizie storico-critiche sulla costruzione delle equazioni," *Memorie della Regia Accademia di Scienze, Lettere ed Arti in Modena*, v. XVIII (1878), p. 127–330; and Ludwig Matthiessen, *Grundzüge der antiken und modernen Algebra der litteralen Gleichungen* (Leipzig, 1878).

[13] For a good general account of the problems, see Felix Klein, *Famous Problems in Geometry*, translated by Beman and Smith with notes by Archibald (2nd ed., New York, 1930). Cf. also Arthur Mitzcherling, *Das Problem der Kreisteilung. Ein Beitrag zur Geschichte seiner Entwicklung* (Leipzig and Berlin, 1913), which includes also valuable comments on the Delian problem.   On this problem in particular see also J. H. Weaver, "The Duplication Problem," *Am. Math. Monthly*, v. XXIII (1916), p. 106–113.

Democritus, one of the founders of the atomic doctrine, was also a mathematician, and to him Archimedes ascribed the determination or demonstration of the volume of the pyramid and the cone. This work is significant not only as an extension to three dimensions of Pythagorean mensurational efforts, but also for the bold use of infinite processes. Democritus composed numerous works bearing on critical

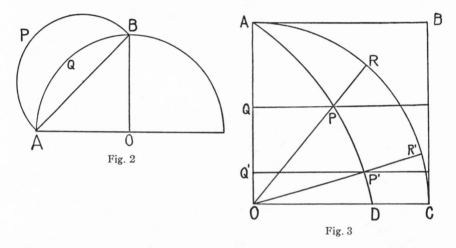

Fig. 2

Fig. 3

aspects of the principles of geometry, but virtually all of what he said has been lost. It is consequently difficult to reconstruct his thought; but it seems clear that to him is largely due the introduction of the infinitesimal in geometry. This mathematical atomism became, even in Greek days, a powerful heuristic device, and in the seventeenth century it was the motivating force which led to the calculus. However, the use of the infinitely small in antiquity could not be made rigorous because the algebraic notion of a continuous variable had not been developed. The Greeks therefore searched for, and later found, a meticulous but circuitous geometrical procedure by which to establish their theorems on curvilinear mensuration. This device, known as the method of exhaustion, was formulated by Eudoxus of Cnidus, but there is reason to believe that it goes back to Hippocrates of Chios, the contemporary of Democritus. That Hippocrates was familiar with the attempts to unify arithmetic and geometry through measurement is indicated by the report that he was for a while a Pythagorean. (The story has it that he was expelled from the school because he accepted a much-needed fee from a student of geometry.) The method of exhaustion, the Greek equivalent of the integral calculus, was based on the so-called axiom of Archimedes which assumes that continuous magnitudes by successive bisection can be reduced to elements as small as desired. The argument proceeded much as it would in the case of the modern method of limits except that the point of view was geometric instead of numerical. Inasmuch as length, area, and volume were not defined numerically as limits, the procedure was supplemented by a *reductio ad absurdum* argument.

The method of exhaustion made it possible to prove that the areas of circles are to each other as the squares on their diameters. This theorem has been ascribed to Hippocrates, who made it the basis for the earliest exact quadrature of a curvilinear area. His proof that the lune $APBQ$ (Fig. 2) bounded by circular arcs, was exactly equal to the triangle $ABO$ led him to believe, mistakenly, that the exact quadrature of the whole circle was possible. Interest in the three classical problems was thus intensified.

The circle and the straight line had possessed for the Greeks a peculiar fascination, and upon these alone they had sought to build all of their science and mathematics. The apotheosis of the straight-edge and compasses has played an enormous role in the development of mathematics; but it favored synthetic geometry at the expense of analysis. Fortunately, however, the three famous problems are unsolvable under the classical restriction, a fact which motivated the search for, and discovery of, other curves.

It is reported that Hippias the Sophist (ca. 425 B.C.) invented the first curve other than the circle and straight line, through the intrusion into geometry of the notion of a mechanical movement. If a horizontal bar or line-segment AB moves downward with a uniform motion of translation to the position $OC$ in the same time that an equal vertical bar or segment $OA$ rotates about $O$ to the position $OC$, the intersection $P$ of these bars or line segments will trace out a curve known as the quadratrix of Hippias (see Fig. 3). This curve was used by the Greeks to resolve two of the three classical problems. It easily "solved" all multisection questions, including that of trisection. To trisect the angle $COR$, for example, one first trisected the segment $OQ$ by the point $Q'$, then found the corresponding point $P'$ on the quadratrix, and finally extended $OP'$ to intersect the circle in $R'$, the desired point trisecting the arc $CR$. Moreover, it was shown later by Dinostratus (fl. ca. 350 B.C.) that, once the quadratrix is constructed, the circle can be squared through the fact that the side of the square $AB$ is the mean proportional between the length of the quarter circular arc $ARC$ and the linear segment $OD$. Thus, of the three classical problems, only the duplication problem remained "unsolved."

The quadratrix is of importance not only as a new curve, but also as heralding one of the basic ideas of analytic geometry—that of a locus. This idea is implicit in the definition of the circle, but the dynamic point of view seems not previously to have been appreciated.   However, the plotting of the quadratrix presented practical difficulties. So long as one has no apparatus for describing the curve by continuous motion, a pointwise construction is necessary even though the language of the definition is kinematic.   The distinction between curves defined geometrically and those described mechanically by a continuous motion was not made clear, and one does not know which point of view Hippias adopted.   It is not even known whether he and Dinostratus regarded the curve as furnishing solutions of the classical problems in a strict theoretical sense.   Unfortunately, the limitless possibilities, in the idea of a locus, for the definition of new curves seems not to have been appreciated by Hippias and his contemporaries.

Of the three famous problems of geometry, the duplication of the cube was the one which played the greatest role in the development of analytic geometry; and it evidently was one which fired the imagination of the ancient Greeks, if we are to believe the legend relating to it. The story goes that the people of Athens appealed to the oracle at Delos to relieve them from a devastating plague.   Upon being told to double the altar of Apollo (presumably making use only of an unmarked straight-edge and compasses), the Athenians ingenuously increased each dimension twofold.   The plague continued; and when complaint was lodged with the oracle, the people were reminded that they had increased the volume of the altar eightfold—i.e., they had solved geometrically the equation $x^3 = 8$ instead of the equation $x^3 = 2$. The plague finally abated; but attempts to duplicate the cube continued.   Not until some two thousand years later did it become clear that the oracle sardonically had proposed an unsolvable problem— henceforth known as the Delian problem.

Following unsuccessful efforts to duplicate the cube according to the rules, the Greeks turned to other devices.   The first "solutions" of the Delian problem differed considerably from those of the other two classical problems.   Hippocrates of Chios made some progress toward the duplication of the cube in showing that if two mean proportionals $x$ and $y$ can be determined so as to satisfy the continued proportion $a:x = x:y = y:2a$, then the proportional $x$ will be the side of the cube desired—i.e., it will satisfy the equation $x^3 = 2a^3$.   The problem thus called for the construction, through geometric methods, of such a proportional.   The first one to cut the Gordian knot in this case seems to have been the Pythagorean scholar Archytas (ca. 428–347 B.C.).

He is reputed to have determined the required mean proportional through a remarkable construction calling for the intersection of three surfaces of revolution: a cone, a cylinder, and a tore.

His construction is now easily verified by analytic methods: letting the equations of the three surfaces be

$$b^2(x^2 + y^2 + z^2) = a^2x^2$$
$$x^2 + y^2 = ax$$
$$x^2 + y^2 + z^2 = a\sqrt{x^2+y^2}$$

it is a simple matter to arrange these equations in the form of the continued proportion

$$\frac{a}{\sqrt{x^2+y^2+z^2}} = \frac{\sqrt{x^2 + y^2 + z^2}}{\sqrt{x^2+y^2}} = \frac{\sqrt{x^2 + y^2}}{b}.$$

For $a = 2b$ these equations obviously lead to the solution of the Delian problem. But such an anachronistic application of modern analysis fails to do justice to the ingeniousness of Archytas in inventing this solution with the aid of synthetic solid geometry alone. In his day surfaces were not defined by means of equations but by the revolution of known curves, such as the line and the circle.

Following the Peloponnesian war, the center of mathematical activity shifted to Athens, although of the leading mathematicians there, only Plato (ca. 427–347 B.C.) was a native. Here Plato, the friend of Archytas, established the famous Academy. Plato exerted a powerful influence on mathematics by his enthusiasm for the subject, but his interests did not lie in the direction of analytic geometry. Archytas is said to have devised an organic solution for the duplication problem, and Plato is reported to have devised another mechanical locus for this purpose. But it seems that in general Plato condemned the use of mechanical contrivances in geometry on the grounds that these tend to materialize a subject which he felt belonged to the realm of eternal and incorporeal ideas. He realized that mathematics does not deal with things of the senses, such as the figures which are drawn, but with the ideals which they resemble. He seems to have been one of the first men to recognize the status of the premises of the subject as pure hypotheses and hence to see the need for stating carefully the assumptions made. Inasmuch as the straight-edge and compasses are in a real sense mechanical contrivances, it is difficult to see why Plato felt that a gulf lay between the straight line and the circle on the one hand and all remaining curves on the other. It may have been the ease with which the line and circle are described, or possibly the perfection of these curves from the point of view of symmetry; but in any case, tradition holds him largely responsible for the canonization of the ruler and compasses in geometry.[14]

Plato's rejection of curves other than the line and circle undoubtedly inhibited the development of analytic geometry, yet to him is ascribed (by Proclus and Diogenes Laertius) one of the fundamental aspects of the subject—the use of the analytic method.   In the broad sense of a preliminary investigation, analysis is not to be ascribed to any one individual, for it undoubtedly has been used since the beginnings of mathematics.   Even in the more technical sense it may antedate the time of Plato.   If incommensurability was first proved in the manner described later by Aristotle—i.e., by showing that if such a ratio of integers exists, it must be at the same time both odd and even—then at least one type of analytical reasoning, the argument by a *reductio ad absurdum*, was in existence probably before Plato's birth.   Plato, however, paid particular attention to the principles and methods of mathematics, and so it is likely that he formalized and pointed out the limitations of the analytical procedure.   It is reported (in the "Eudemian Summary" of Proclus) that Eudoxus, an associate of the school of Plato, made use of the method of geometrical analysis.   As Plato seems to have used the term, analysis meant the method in which one assumes as true the thing to be proved and then reasons from it until one arrives at propositions previously established or at an acknowledged principle.   By reversing the order of the steps (if possible), one obtains a demonstration of the theorem which was to have been proved. That is, analysis is a systematic process of discovering *necessary* conditions for the theorem to hold, and if by synthesis these conditions are then shown to be *sufficient*, the theorem is thereby established.   It should be noted, however, that it is not primarily by virtue of this order of steps in the reasoning process that coordinate geometry now is known as analytic geometry.   The signification of the word analysis has changed with circumstances, and today this term has several more or less distinct meanings.[15]   The more recent applications of the word differ from the original Platonic use especially in an increased emphasis upon symbolic techniques.   In Plato's day there was no

---

[14] The correctness of this tradition, however, may be questioned. See D. A. Steele, "Ueber die Rolle von Zirkel & Lineal in der grieschischen Mathematik," *Quellen und Studien zur Geschichte der Mathematik...*, Abt. B, Studien, v. III (1934–1936), p. 287–369, for the view that Plutarch's statement in this connection has been misinterpreted by modern historians.   Cf. also, by the same author, "A Mathematical Reappraisal of the Corpus Platonicum," *Scripta Mathematica*, v. XVII (1951), p. 173–189.

[15] See Paul Tannery, "Du sens des mots analyse et synthèse chez les grecs et de leur algèbre géométrique," *Notions historique*, in Jules Tannery, *Notions de mathématiques* (Paris, 1903), p. 327–333.

formal algebra; but when, almost two thousand years later, the analytic method of Plato came to be applied to primitive forms of algebraic geometry, then the invention of analytic geometry quickly followed.

Plato appreciated keenly the problems which were raised in mathematics by the paradoxes of Zeno and the discovery of the incommensurable, and although he did not solve these, he suggested a possible method of attack. He regarded the continuum as generated by the flowing of an abstract unbounded infinite, rather than as made up by an aggregation of indivisibles. Such a point of view is somewhat analogous to the effective heuristic use by Leibniz and Newton of generative infinitesimals and fluxions, but it is in essence a *petitio principio* through the lack of definition of terms. Indeed, in pursuing the thought of Plato on motion and continuity, his successors took two diametrically opposite points of view. The members of the Academy rejected the physical atomism of Democritus, yet they attempted to develop the idea of indivisibles or fixed infinitesimals in mathematics. In this they were vigorously opposed by two of Plato's outstanding students, Aristotle (384–322 B.C.) and Eudoxus (ca. 408–355 B.C.), who were more inclined toward the natural sciences. The decisions of Aristotle on the indivisible, the infinite, and the continuous were those dictated by immediate common sense. He denied categorically the existence of minimal indivisible line segments and of an actual or completed infinite. The essence of continuity he found in that which is divisible into divisibles that are infinitely divisible. Emphasizing the gap in Greek days between geometry and arithmetic, he denied that number can produce a continuum, inasmuch as there is no contact in numbers. His study of motion was concerned with qualitative explanation rather than quantitative description, and so precluded a satisfactory science of dynamics. However, it was the study of Aristotle's works which, during the late Middle Ages, inspired those attempts to build a mathematical theory of motion and the continuum which brought medieval mathematicians very close to analytic geometry.

The mathematical counterpart of the philosophical views of Aristotle is seen in the work of Eudoxus, most of which is known only indirectly through other sources. It is concerned largely with the early equivalent of the integral calculus, inspired presumably by some problems in stereometry suggested to him by Plato. Democritus had given the volumes of pyramids and cones, but the demonstration of these is ascribed to Eudoxus. His proof, by the method of exhaustion, may have been original with him—inasmuch as Archimedes ascribed the basic postulate to Eudoxus—or it may have been adopted from Hippoc-rates of Chios.  The method of Eudoxus is somewhat equivalent in type of procedure to that involving limits, but it was quite different in point of view.  The Greek method, dealing as it did with con-tinuous magnitude, was wholly geometrical, for there was at the time no knowledge of the continuum of numbers; the present method of limits, on the other hand, is essentially arithmetical.

By the method of exhaustion Eudoxus (and perhaps Hippocrates before him) was enabled to compare, through the theory of ratio and proportion, curvilinear geometric magnitudes with rectilinear.  How-ever, the discovery of incommensurability had shown that ratios, even of rectilinear figures, can not always be defined in terms of integers. The Greeks of the time of Theaetetus (fl. ca. 375 B.C.) seem therefore to have made use of a modified definition (due possibly to Hippocrates of Chios) in terms of the process of finding the greatest common divisor: Magnitudes have the same ratio if they have the same mutual successive subtraction, analogous to the process of successive division in the Euclid algorithm for the highest common factor of two quanti-ties.  For example, whether the bases of two rectangles of equal altitude are commensurable or incommensurable, the ratio of the bases is equal to the ratio of the areas inasmuch as the successive application of the lines in the one case of mutual subtraction corresponds directly to the application of areas in the other.  But the usefulness of the new definition was limited, and so Eudoxus (about 370 B.C.) developed another which he found was needed in his proofs by exhaustion: Magnitudes are said to be in the same ratio, the first to the second and the third to the fourth, when, if any equimultiples whatever be taken of the first and third, and any equimultiples whatever of the second and fourth, the former equimultiples alike exceed, are alike equal to, or alike fall short of, the latter equimultiples respectively taken in corre-sponding order.  This definition might have served as a basis for a general definition of real number, whether rational or irrational.[16] As used by the Greeks, however, it served to avoid all reference to numbers other than natural numbers or integers.  All four of the en-tities involved in the definition might be geometrical, as in the propo-

---

[16] For a profound study of pertinent aspects of the work of Eudoxus see Oskar Becker, "Eudoxos-Studien," *Quellen und Studien zur Geschichte der Mathematik, Astronomie und Physik*, Part B, *Studien*, v. II (1933), p. 311–333, 369–387; v. III (1936), p. 236–244, 370–410.   For a good general account of the relation of Eudoxus' theory of proportion to ancient and modern thought, see Coolidge, *A History of Geometrical Methods* (Oxford, 1940), p. 29–34.   Bell has pointed out ("Sixes and Sevens," SCRIPTA MATHEMATICA, v. XI (1945), see p. 153) that Eudoxus' theory of proportion still depends on infinities inasmuch as the phrase "any equimultiples" implies an infinite number of possibilities; but the Greeks seem to have overlooked this fact.

sition that the areas of circles are to each other as squares on the diameters of the circles. The ancient equivalents of the integral calculus were thus concerned with ratios of geometrical figures, such as circles and squares, rather than with analytic functions of continuous variables, such as $A = \pi r^2$. This emphasis made it difficult to recognize or establish a general analytic theory of curves or the algorithmic procedures which form the basis of modern analysis. The emphasis in the method of exhaustion was on the synthetic form of exposition; it was not an analytic instrument of discovery. It represented a conventional type of demonstration, but the Greeks never developed it into a concise and well-recognized operation with a characteristic notation. In fact they did not take the first step in this direction: they did not formulate the principle of the method as a general proposition, reference to which might serve to abbreviate subsequent demonstrations. However, the Greek theory of proportions had survived the crisis caused by incommensurability, and it continued, even in geometric form, to be the chief algebraic tool of Greek mathematics.

Eudoxus was without doubt the most brilliant mathematician of his century, but it remained for one of his pupils, Menaechmus (ca. 360 B.C.)—brother of Dinostratus and tutor of Alexander the Great—to make the most spectacular contribution of the time to the development of analytic geometry. This work seems to have been inspired by the Delian problem for which Archytas had given his amazing triad of surfaces. Had Archytas studied carefully the sections of his cylinder by a plane, he would have discovered a new curve with striking properties. The ellipse may indeed first have entered Greek geometry as a section of a cylinder, or in some way not now known. Of all curves the ellipse is, with the exception of the straight line, the one most commonly seen in routine experience. Wheels and other circular objects, when viewed obliquely, appear as ellipses, and the shadows cast by circles generally are elliptical. Did not the earlier mathematicians see these? Democritus, in his infinitesimal geometry, is known to have studied the circular sections of a cone. Did he not, in this connection, note the more general elliptic sections? There appears to be no evidence that he did. It is reported by Proclus and Eutocius that the ellipse, hyperbola, and parabola all were discovered[17]

[17] For extensive historical accounts of the conic sections see H. G. Zeuthen, *Die Lehre von den Kegelschnitten im Altertum* (Kopenhagen, 1886); J. L. Coolidge, *A History of the Conic Sections and Quadric Surfaces* (Oxford, 1945); the prolegomena in Charles Taylor, *An Introduction to the Ancient and Modern Geometry of Conics* (Cambridge, 1881); the introductions to Apollonius of Perga, *A Treatise on Conic Sections* (translated by T. L. Heath, Cambridge, 1896), and *Les coniques* (translated by Paul Ver Eecke, Bruges, 1925). For further historical notes refer also to the section on "Coniques" in the *Encyclopédie des Sciences Mathématiques*, v. III (3), p. 1–256, by F. Dingeldey.

by Menaechmus toward the middle of the fourth century B.C., and that accordingly these curves were at first called the "Menaechmian triads." Menaechmus seems to have been led to them by following the very path which Archytas had suggested—that is, he sought to solve the Delian problem by a consideration of sections of geometrical solids. Taking three right circular cones, one acute-angled, one right-angled, and one obtuse-angled, he cut each of these by a plane perpendicular to an element.    This disclosed to Greek geometers for the first time a whole family of curves which, unlike lines, circles, and quadratrices, differ in shape as well as in size.    By means of these conic sections the cube is easily duplicated, either through the determination of the intersection of the two parabolas $x^2 = ay$ and $y^2 = 2ax$, or through the intersection of the first of these with the hyperbola $xy = 2a^2$.

If Menaechmus duplicated the cube with these curves, as is reported, he must have known the geometric equivalent of the equation of the equilateral hyperbola referred to its asymptotes as axes.    Zeuthen, Heath, and Coolidge[18] suppose that this equation was derived from the form with respect to a vertex—$y^2 = 2ax - x^2$—by translating axes to obtain the central equation $x^2 - y^2 = a^2$, and then rotating axes through half a right angle to get $2xy = a^2$.

The probable achievement of Menaechmus is given above in analytic form and so the account fails to do justice to his ingenuity. He was a trail-blazer in discovering the most useful and intriguing family of curves in all science and mathematics, and the path was made difficult by the lack of algebraic ideas and symbolism.    The conic sections now are defined as loci in a plane—the locus of points for which the distance from a fixed point (focus) is to the distance from a fixed line (directrix) in a fixed ratio (eccentricity).    It is a relatively simple matter to translate this defining property into analytical language by means of modern algebraic notations and technique.    Trigonometric symbolism and formulas now enable one easily to transform equations under a rotation of axes, passing readily from the equation of the hyperbola with respect to its axis to the equation with reference to its asymptotes.    An extraordinary degree of originality would be required for Menaechmus to conceive of the equivalent of all of this in geometric form, but evidence seems to indicate that he did so.    What he wrote has been almost completely lost,[19] and even the names he attached to his curves are unknown.    His thought must be reconstructed, as best one can, from bits of information supplied by later commentators.

[18] See Coolidge, *History of the Conic Sections*, p. 5.
[19] See M. C. P. Schmidt, "Die Fragmente des Mathematikers Menaechmus," *Philologus*, v. XLII (1884), p. 72–81.    However, these fragments, in Greek with German commentary, are quite inadequate to give a picture of his mathematics.

To begin with, even the manner in which the conics were discovered is open to some doubt. Did the Greeks stumble upon the new curves or was the discovery the reward of a systematic search? It has been suggested[20] that Menaechmus may first have looked upon them as plane loci, constructed kinematically in a manner similar to that adopted by Hippias for the quadratrix. This view is in harmony with the fact that Hippocrates had shown the duplication problem to be solvable if the relationships in a continued proportion $a/x = x/y = y/b$ can be constructed, from which, by analysis, the properties of the desired loci were clear. It would remain merely to draw curves—the parabolas $x^2 = ay$ and $y^2 = bx$ and the hyperbola $xy = ab$—having these properties, after which it would not be difficult to recognize the curves as sections of cones. In fact, the continued proportion above would be satisfied by the intersection of the circle $x^2 + y^2 - bx - ay = 0$ with any *one* of the new curves. Against this view, however, is the fact that such a procedure is contrary to Greek custom. Had they used such a method here, why would they have hesitated to draw curves with other desired (algebraic) properties? Any problem could easily be solved in the same manner by sketching curves suitably defined. Such a cutting of the Gordian knot would hardly be in keeping with the strong Greek feeling for the appropriateness of methods of solution, and it was not, in fact, used in other connections.[21] Then, too, the designation the Greeks invariably adopted for problems and loci determined by conic sections—*solid* problems and loci, as distinguished from the *plane* problems and loci which are constructed by line and circle—would seem to point to a three-dimensional origin of the conics along the lines suggested above.

Whatever the original source of the conics—planimetric or stereometric—the amazing part of the discovery by Menaechmus is not so much the curves themselves as the fact that he apparently was able to go from the one aspect of them to others. The sections of cones were shown to have fundamental properties as plane loci; and from these basic "geometric equations," countless other plane properties of the curves were deduced. It is this aspect of the early work on conic sections which has led a number of historians, notably Zeuthen[22] and Coolidge,[23] to claim for the Greeks the invention of analytic geometry. The thesis of the latter is "that the essence of analytic

[20] See, e.g., Taylor, *Geometry of Conics*, p. xxxi f.
[21] See Loria, "Aperçu sur le développement historique de la théorie des courbes planes," *Verhandlungen des ersten internationalen Mathematiker-Kongress in Zürich*, 1897 (Leipzig, 1898), p. 289–98.
[22] "Sur l'usage des coordonnées dans l'antiquité," *Kongelige Danske Videnskabernes Selskabs. Forhandlinger* (1888–1890).
[23] Coolidge, *History of Geometric Methods*, p. 117–119.

geometry is the study of loci by means of their equations, and that this was known to the Greeks and was the basis of their study of the conic sections. The original discoverer seems to have been Menaechmus."

It is a great pity that the works of Menaechmus have been lost, so that reconstructions of his work are largely conjectural.[24] Under the circumstances it may be well to postpone a further consideration of the methods he might have used until one is on more solid ground in the published works of Apollonius. The opening portions of the *Conics* of Apollonius presumably are representative of the thought of Menaechmus. Before passing on to the period of Greek mathematical maturity, however, it should be pointed out that although Menaechmus probably did study the conics more or less analytically, under the limitations imposed by the geometric character of Greek algebra, there is no hint, either in Menaechmus or in later works of antiquity, of a *general* analytic geometry in the sense of a mutual correspondence between curves and equations. Had the Greeks possessed such an analytic geometry, it is doubtful that the conics would have undergone an apotheosis second only to that of the line and circle. As it was, they never developed a theory of curves in general. In fact, they did not discover more than half a dozen new curves in all of their enormous mathematical activity, and these were not systematically classified. It was to be about two thousand years before Descartes undertook to bring order into higher plane curves, and in so doing he invented analytic geometry in a sense far more general than that of Menaechmus. To carry out his program, however, it will be seen that Descartes found it necessary to substitute for the theory of proportion and the application of areas a symbolic algebra of which the Greeks did not develop even the rudiments. The incommensurable had left so deep an impression upon Greek thought that they carefully distinguished between cases in which the magnitudes were rational and those in which they were irrational. Only in modern times was the ancient doctrine of proportion given arithmetic freedom by permitting the quantities in an equation or proportion to be indifferently either rational or irrational.

[24] Neugebauer has made the interesting suggestion that the discovery of the conic sections may have resulted from a study of the shadows cast by sundials. See "The Astronomical Origin of the Theory of Conic Sections," *Proceedings of the American Philosophical Society*, v. XCII (1948), p. 136–138.

# CHAPTER II

# The Alexandrian Age

*Alexander, the king of the Macedonians, began like a wretch to learn geometry that he might know how little the earth was whereof he had possessed very little.*

—SENECA

GREEK history customarily is divided into two periods, the Hellenic and the Hellenistic. In the history of mathematics the division is strikingly illustrated by the fact that of the earlier period no mathematical treatises have survived, whereas representative works from even the earliest part of the later period have been preserved. The golden age in art and literature fell during the first period, but the great age in mathematics belonged to the early part of the second. Yet in mathematics, as in science, the earlier Hellenic period may be characterized as the heroic age, for it was during that time that the fundamental attitudes and principles were established. The Hellenistic age added tremendously to geometrical knowledge, but it did so along the lines laid down before the time of Alexander the Great. The three greatest extant mathematical works of antiquity were composed—all of them during the period of one hundred years following Alexander—by Euclid (fl. ca. 300 B.C.), Archimedes (287–212 B.C.), and Apollonius (fl. ca. 225 B.C.); but these represent the superstructure erected upon the foundation begun by Pythagoras, Eudoxus, and Menaechmus.

The *Elements* of Euclid[1] undoubtedly exerted a wider influence than any other mathematical work ever written, but this was not primarily in the direction of analytic geometry. The work represented the definitive formulation of earlier achievements, and so its significance lay in its characteristic form of logical exposition rather than in suggestions for future developments in methodology. Nevertheless, it is important to note that the *Elements* contains not only the elementary pure geometry of present-day high-school courses, but also extensive sections on

[1] See *The Thirteen Books of Euclid's Elements*, edited by T. L. Heath, 3 vols., Cambridge (1908).

21

the application of geometry to what now would be known as algebra. Ratio and proportion are now topics in generalized arithmetic, but Euclid systematized the Eudoxian theory in geometrical form—the nearest approach in his day to the ideas of function and equation. Euclid followed Eudoxus also in the use of the analytic method in connection with the division of a line in mean and extreme ratio, but such a hint of a general point of view was overshadowed by an emphasis upon special geometric constructions. The solution of quadratic equations, for example, is carried out by Euclid in a geometric manner contrasting sharply with the algebraic methods of the Babylonians. Because of the lack of a general concept of negative number, quadratic equations were divided by the Greeks into three types equivalent to the following: $x^2 + bx = c^2$, $x^2 = bx + c^2$, and $x^2 + c^2 = bx$, where $b$ and $c$ are real and positive. The Euclidean geometrical solution of such equations was based on the method of the application of areas, described above, which the Pythagoreans seem to have devised. It strikes a modern reader as far removed from the graphical solution of analytic geometry, obtained by plotting a polynomial on a coordinate system; and yet the modern method arose through gradual modifications of such ancient geometrical constructions.

Euclid contributed more directly to the growth of analytic geometry through other treatises which have long been lost. One of these was a work on conic sections, much of the material of which was later incorporated into the *Conics* of Apollonius. Another was a treatise on porisms which has been partially reconstructed a dozen or more times on the basis of information from later commentators. Pappus (*Mathematical Collection*, VII) says that a porism is intermediate between a theorem, in which something is proposed for demonstration, and a problem, in which something is proposed for construction. We are told that the Greeks divided geometrical propositions into three types— theorems, problems, and porisms—according as one was asked to demonstrate, to construct, or to find something. Chasles, referring to Simson, says that a porism is a proposition in which one announces the possibility of determining, and where one determines, certain things which have an indicated relation with those fixed and known and with others variable, establishing a law of variation. This description would cover the study of loci, and hence the loss of Euclid's *Porisms* is the more to be regretted in that it leaves an unfortunate gap in the history of analytic geometry.[2] As Chasles puts it, porisms were in a

[2] For a full account see Michele Chasles, *Les trois livres de porismes d'Euclide*, Paris (1860); or see his *Aperçu historique sur l'origine et le développement des méthodes en géométrie*, Paris (1875), p. 12–13, 274–84. Cf. also his excellent article, "Sur la doctrine des porismes d'Euclide," *Correspondance Mathématique et Physique*, v. X (new series IV, 1838), p. 1–20, which includes an excellent bibliography.

sense equations of curves, and the doctrine of porisms was the "analytic geometry of the ancients," differing from ours chiefly in the lack of symbols and algebraic processes. Not possessing the idea of the equation of a curve, Euclid seems to have used porisms to substitute for a geometric expression of a locus another geometric expression of the same locus, thus relating one construction to another equivalent one—a sort of transformation of coordinates, or the equivalent of the modern use of formulas for points, lines, and circles. Chasles gives, as an instance appropriate to the Euclidean use of porisms, the following: to find a point for which the sum of the squares of its distances from two fixed points is a given constant (area). It is today a simple matter to show that the result is a circle with center midway between the given points; but the ancients, lacking uniform procedures, evidently experienced far more difficulty in reconciling the conditions of the locus with the definition of the circle.

The *Porisms* was included in an ancient collection of works by Euclid, Aristaeus, and Apollonius which was known as the "Treasury of Analysis" and which Pappus describes as a special body of doctrine for those who, after going through the usual elements, wish to obtain power to solve problems involving curves. The "Treasury," which undoubtedly comprised much of what now goes to make up analytic geometry, included another lost work by Euclid, one on surface loci. The content of this treatise is quite conjectural. It may have been a study of surfaces known to the ancient Greeks—the sphere, cone, cylinder, tore, ellipsoid of revolution, paraboloid of revolution, and hyperboloid of revolution of two sheets—or it may have concerned twisted curves drawn on these surfaces. Chasles believed that it covered surfaces of revolution of second degree.[3] Tannery ascribes the discovery of the conoids and spheroids of revolution to Archimedes,[4] but Heath[5] thinks it possible that Euclid's "Surface Loci" may have included these. In any case, this Euclidean work, did we but have it, would be an important link in the history of solid analytic geometry.

The titles alone of the lost works of Euclid show that ancient geometry is comparable in breadth of interest with that of the seventeenth century. The difference between ancient and modern geometry is one of method and manner of expression rather than of content. This impression is strongly confirmed by a study of the *Conics* of Apollonius, where one finds the closest approach in antiquity to coordinate geometry. Pappus reports that Aristaeus the Elder (living between the

[3] See Chasles, *Apercu*, p. 273, note II.
[4] Paul Tannery, "Pour l'histoire des lignes et surfaces courbes dans l'antiquité," *Mémoires scientifiques*, v. II (1912), p. 1–47.
[5] See *The Works of Archimedes* (edited by T. L. Heath, Cambridge (1897)), p. lxi ff.

time of Menaechmus and that of Euclid) wrote a treatise in five books entitled "Solid Loci." This work, perhaps the first textbook on the conic sections, has been lost, probably irretrievably, along with the four books on conics by Euclid. Presumably, however, the first four books of the Conics of Apollonius[6] include most of the material given by Aristaeus and Euclid. The author himself says that the first four books coordinate and complete previous knowledge, but he emphasized the originality of his mode of treatment; the other four books, of which the last has been lost, are presumed to be substantially original in content. Apollonius made numerous changes in the method of handling the conic sections, especially in the direction of generalization. Menaechmus and his early followers had used right sections only. Archimedes undoubtedly knew (as perhaps Euclid did also) that a single cone with circular sections would suffice, by varying the angle of the cutting plane, to produce all three types of conics; but he seems not to have taken advantage of this fact and he continued to use right sections of right cones. Archimedes therefore retained the old names ascribed by Pappus to Aristaeus—section of an acute-angled cone, section of a right-angled cone, and section of an obtuse-angled cone. Apollonius, on the other hand, started at once with the most general section of an oblique circular cone. (He noticed, incidentally, not only that sections parallel to the base are circles, but that there is also a secondary set of circular sections.) His predecessors appear to have noticed the opposite branch in the case of the hyperbola, but Apollonius first investigated the general properties of the hyperbola as a double-branched central conic. Moreover, whereas earlier geometers had emphasized the differences between the three fundamental types, Apollonius developed the tendency to consider the curves as a single general family rather than as triads.

The names ellipse, parabola, and hyperbola were attached to the conic sections by Apollonius in introducing a further significant change in the manner of treatment. Archimedes had preferred to characterize the curves in terms of the Euclidean theory of proportions. Thus the parabola has the property that abscissas, measured along the axis from the vertex, are to each other as the squares of the corresponding ordinates. Apollonius, on the other hand, made use of the Pythagorean idea and language of the application of areas, so that the basic name-property of the parabola would be described by the fact that (for a

[6] The English edition by Heath (Apollonius of Perga, *Treatise on Conic Sections*, Cambridge (1896)) is altogether excellent. Some further information, especially on the provenance of Apollonian manuscripts, is found in the introduction to the French edition by Paul Ver Eecke, *Les coniques d'Apollonius de Perge*, Bruges (1923).

parabola referred to its vertex as origin and its axis as line of abscissas) the rectangle formed with the parameter by any abscissa is equal to the square on the ordinate corresponding to that abscissa. The names ellipse and hyperbola referred to the fact that for these two curves (referred to a vertex as origin and major or transverse axis as line of abscissas) the squares on an ordinate respectively fell short of, or exceeded, the rectangle formed by the corresponding abscissa and the parameter.[7] It is possible that this property was known before Apollonius, but it was he that made it basic. The Archimedean equality of ratios is better adapted to the theory of proportions, as used in medieval and early modern times, while the Apollonian equality of areas is more in harmony with the later symbolic algebra of equations and with the association of curves and equations in analytic geometry. It is a simple matter to convert the Apollonian propositions on the name-property of the ellipse, parabola, and hyperbola into the corresponding algebraic equations with respect to the vertex, so widely used in the seventeenth and eighteenth centuries—$y^2 = lx - (lx^2/t)$, $y^2 = lx$, and $y^2 = lx + (l\,x^2/t)$, where $l$ is the latus rectum and $t$ the major or transverse axis.

The Apollonian treatment of conic sections approaches the modern view likewise in its emphasis on planimetric study. His predecessors already had made progress in this direction, but Apollonius went further and used the stereometric origin of the curves only so far as necessary to derive a fundamental plane property for each. Thus, in the case of the parabola, Apollonius began with an oblique cone with vertex $A$ and a circular base with diameter $BC$. Through any point $P$ on the element $AB$ he passed a plane cutting the base in a chord $ED$ perpendicular to $BC$. (See Fig. 4.) The section of the cone by this plane will be the parabola $EPD$. To derive the basic property of the curve, let $Q$ be any point on it and pass a plane through $Q$ parallel to the base, cutting the cone in the circle $HQK$. Now in this circle $QV^2 = HV \cdot VK$. But from similar triangles $HV = \dfrac{PV \cdot BC}{AC}$ and $VK = \dfrac{PA \cdot BC}{BA}$. Hence $QV^2 = PV \left( \dfrac{PA \cdot BC^2}{AC \cdot BA} \right)$; i. e., the square of the ordinate $QV$ is to the abscissa $PV$ in a constant ratio. For the ellipse Apollonius similarly derived the basic property $QV^2 = PV \cdot VP' \left( \dfrac{BF \cdot CF}{AF^2} \right)$; i. e., the

[7] Eutocius said that the names given to the conic sections referred to the fact that the vertex angle of the cone was less than, equal to, or greater than a right angle; or that the cutting plane fell short of, ran along with, or cut into the second nappe of the cone. These explanations, plausible enough on first glance, have been often repeated; but Apollonius clearly gives the application of areas as the reason for the names.

square of the ordinate is to the product of the segments, *PV* and *VP'*, of the axis in a constant ratio. A similar property was found for the hyperbola. From such planimetric relationships Apollonius then proceeded to derive innumerable properties of the curves, including those of the foci, asymptotes, and tangent and normal lines. The foci of the

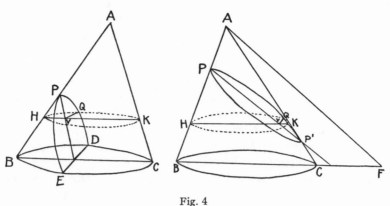

Fig. 4

ellipse furnish another instance of the application (or parabolism) of areas, for he refers to these as points determined by "an application"— i. e., as the points which divide the major axis into two segments the rectangle on which is equal in area to the square on the semi-minor axis. Since this method of application fails for the parabola, Apollonius here overlooked the focus.

The work of Apollonius in many respects approaches so closely to the modern form of treatment that it not infrequently has been regarded as constituting analytic geometry. A decision on this point depends on precise definitions of such terms as coordinates and equations, and upon one's view of the essential features of analytic geometry. The use of reference lines by Apollonius, as previously by Archimedes and perhaps also by Menaechmus, does indeed resemble the modern application of rectangular and oblique Cartesian coordinates. A diameter of a conic corresponds to the axis of abscissas, where the abscissa is measured from the point of intersection with the conic; the tangent line to the conic at this point serves as a second axis. Ordinates, then, are represented by segments, from points on the axis of abscissas to points on the conic, of lines drawn parallel to the second axis (or conjugate to the first axis). The relationships, verbally expressed, between these ordinates and the corresponding abscissas, are tantamount to the equations of the curves. However, the Greek use of auxiliary lines differs in several particulars from the modern applications of coordinates.[8] In the first place, the geometrical algebra of antiquity did not provide for negative quantities or lines. More important than this, however, is the fact that the system of reference lines was in every case simply an auxiliary construction superimposed *a posteriori* on a given curve in order to study its properties. Hence the curve always passed through what would now be called the origin. There appear to be no cases in ancient geometry in which a coordinate frame of reference was constructed *a priori* for purposes of graphical representation or in order to solve a given problem.

In modern courses in analytic geometry the equations of the ellipse and hyperbola almost always are expressed in terms of a coordinate system in which the origin coincides with the center of the conic and in which the coordinate axes generally coincide with the axes of the conic; but in antiquity the curves more frequently were studied with reference to other auxiliary lines, notably a tangent to the curve and the diameter of the conic which is terminated at the point of tangency. This meant that the Greek systems of auxiliary lines usually constituted the equivalent of an *oblique* coordinate reference frame; and in this respect the ancient predilections were in agreement with those of Descartes many centuries later. The absence of rigid conventions in the ancient auxiliary-line geometry meant that there were no formulas for the transformation of axes; but classical geometers nevertheless

[8] For an extensive discussion on this point see Siegmund Günther, "Le origini ed i gradi di sviluppo del principio delle coordinate," *Bullettino di Bibliografia e di Storia delle Scienze Matematiche*, X (1877), p. 363–406. This article originally appeared as "Die Anfange und Entwickelungsstadien des Coordinatenprincipies," *Abhandlungen der Naturforschenden Gesellschaft zu Nürnberg*, v. 6. So nearly does the Greek work resemble the modern use of coordinates that Günther and Zeuthen, both unusually competent historians of mathematics, disagreed sharply, the former insisting on the differences, the latter emphasizing the similarities. See H. G. Zeuthen, "Sur l'usage des coordonnées dans l'antiquité," *Kongelige Danske Videnskabernes Selskabs, Forhandlinger* (1888), p. 127–144. Of more recent historians, the majority tend to agree with Günther, although Heath and Coolidge are sympathetic to the view of Zeuthen. Heath holds that "His [Apollonius'] method does not essentially differ from that of modern analytic geometry except that in Apollonius geometric operations take the place of algebraical calculations." (See his *Apollonius*, p. xcvi f.) L. C. Karpinski, in "Is There Progress in Mathematical Discovery and Did the Greeks Have Analytic Geometry," *Isis*, v. XXVII (1937), p. 46–52, rejects this view. An excellent summary of the situation is given by Gino Loria:

"In truth, whoever studies thoroughly the treatise of Apollonius on *Conics* must confess the profound analogy it bears to an exposition of the properties of the curves of second degree by means of Cartesian coordinates; not only do the fundamental properties employed by the Greek geometer to distinguish the three curves one from the other translate into the canonical equations of the same in Descartes' method, but many of the reasonings given, when translated into the ordinary language of algebra, answer to elimination, solution of equations, transformation of coordinates, and the like. What we would however seek in vain in the Greek geometer is the concept of a system of axes, given *a priori*, independent of the figure to be studied." (In "Sketch of the Origin and Development of Geometry Prior to 1850," translated by G. B. Halsted, *Monist*, v. XIII (1902–1903), p. 80–102, 218–34. See especially p. 94, note.)

were adept in problems involving the equivalent of coordinate transformations.  In the *Conics* of Apollonius there are numerous instances
of propositions in which it is proved that a property derived with respect to a *given* tangent and the related diameter holds as well for *any*
tangent and the corresponding diameter.  Apollonius frequently first
assumed that the angle between the tangent and the diameter is a right
angle, and then he reduced to this case the more general situation in
which the angle is oblique.

The use of coordinates in a broad sense is without doubt prehistoric.
Coordinates are simply magnitudes associated with objects in order to
locate them with respect to other objects taken as a frame of reference.
Instructions given, or diagrams drawn, to indicate relative positions are
in this sense applications of the coordinate principle.   Primitive Chal-
dean star charts are specific and systematized instances of the use of co-

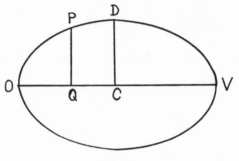

Fig. 5

ordinates.   Systems of coordinates in this sense were used also in the
early Egyptian cadastral surveys, and again later in Greek astronomy
and geography;  but the graphical representation of variable quantities
was not associated with coordinates until the period of the middle ages.
It is to be doubted that the coordinate systems used by Hipparchus, for
example, directly influenced Greek or medieval geometry—or, for that
matter, the early modern invention of analytic geometry.   Greek geometers did not seek to reduce the number of unknown quantities or
lines in a figure to one or two, as is done in astronomy or in plane
Cartesian coordinates, but they sought instead to make as simple as
possible the relationships in terms of areas.   For example, in modern
notation the equation of an ellipse with respect to a vertex is written as
$\dfrac{(x-a)^2}{a^2} + \dfrac{y^2}{b^2} = 1$, where $x$ and $y$ are the two unknown lines $OQ$ and
$PQ$.  (See Fig. 5.)   Apollonius would express this same property in rhe-

torical language equivalent to the abbreviated notation $\dfrac{PQ^2}{CD^2} = \dfrac{OQ \cdot QV}{OC \cdot CV}$ where $C$ is the center and $OV$ the major axis), a form involving a third unknown line, $QV$. That is, his equations for the ellipse and hyperbola are in terms of an ordinate and *two* abscissas.

An important aspect of the Greek use of lines as coordinates arises in connection with the study of the conic sections by means of their fundamental plane properties. In the *Conics* of Apollonius such properties, known as *symptomae*, play much the same role as do the equations of curves in modern analytic geometry.[9] Inasmuch as the Greeks lacked symbolic algebra, a statement by Apollonius of a given property might run to as much as half a page in length, which compares unfavorably with concise modern terminology. This fact has led Tannery to say that what the Greeks lacked was not so much the methods of analytic geometry as the formulas appropriate to the methods. This is largely, but not entirely, true. The essence of an equation does not, indeed, lie in its brevity; but on the other hand, out of symbolic notation there arose much later the concepts of algebraic variable and equation of a curve. Philosophical difficulties precluded the Greek acceptance of the former concept, but it was probably deficiencies in symbolic algebra which prevented Apollonius from developing the latter. Of Greek geometry one may say that equations are determined by curves, but not that curves are defined by equations. Coordinates, variables, and equations were subsidiary notions derived from a given geometrical situation. Presumably it was not sufficient to define curves abstractly as loci of points satisfying certain given conditions. Hence finding a conic meant to Apollonius localizing it in a cone or finding a cone of which the required curve is a section. To guarantee that a locus was really a curve, Greek geometers had to exhibit it stereometrically as sections of a solid or to describe it kinematically by means of a mechanical construction. Perhaps nowhere is the gap between Apollonius and modern analytic geometry brought out more clearly than in the fact that at no time did he begin with a coordinate system, two unknowns or variables, and a relationship or equation between the variables, and then plot the curve corresponding to the equation as the locus of points whose coordinates are values of the variables which satisfy the given equation. And yet so thorough was the Apollonian treatment of conics that he would have recognized that a geometrical relationship equivalent to $xy + ax + by + c = 0$ is representative of an equilateral hyperbola. Such recognition is equiva-

[9] See H. G. Zeuthen, "Sur l'origine de l'algèbre," *Det Kongelige Danske Videnskabernes Selskab. Mathematisk-fysiske Meddelelser*, v. II, p. 4.

lent to the modern transformation of coordinates, the differences being more in emphasis and point of view than in substance. Apollonius knew also the equivalent of other forms of the equations of conics, including the constancy of the sum or difference of focal radii for the ellipse or hyperbola. Tangents and polars he constructed by using the properties of harmonic division; and he showed that normals are the shortest lines to a conic, and that the subnormal is constant for the parabola. He found that in general as many as four normals could be drawn to a conic from a given point; and by noting the locus of points from which fewer normals could be constructed, he found and studied evolutes of conics, even though he did not regard such a locus as defining a curve in the strict sense of the word. In this connection he worked with the geometrical equivalent of equations of sixth degree in two unknowns. One might almost say that the Greeks had an analytic geometry of conic sections but not an analytic geometry in the general sense. There are two sides to the fundamental principle of analytic geometry: the one is that with a given known plane curve there is associated, with respect to any coordinate system in its plane, an equation in two indeterminate quantities. With this fact Apollonius seems to have been throughly familiar, at least as far as the conic sections are concerned. The converse is that an equation in two indeterminate quantities determines, with respect to a given plane coordinate system, a curve in this plane. This the Greeks did not know, except for special cases. The differences, however, are more in emphasis and point of view than in substance.[10]

The material in the *Conics* of Apollonius is remarkably extensive and includes numerous well-known properties of the curves. It is presumed that he, and perhaps also Aristaeus and Euclid, was aware of the familiar focus-directrix ratio definition of these curves, but this is not even mentioned in his *Conics*. There is no reference to the focus of the parabola, but from a number of theorems it would appear that he knew of this important point (although not necessarily of the directrix).[11] He seems to have known how to determine the conic through five points, but this also is omitted from the *Conics*, perhaps because it was given elsewhere in one of the many lost works of Apollonius. The books of the second half of the *Conics* include a study of normals approaching the modern theory of evolutes, work on the similitude of conic sections and their segments, and the properties of conjugate diam-

---

[10] H. G. Zeuthen, *Historie des mathématiques dans l'antiquité et le moyen âge* (translated by Jean Mascart, Paris (1902)) treats Apollonius' *Conics* in modern notation, and hence he is led to say that the treatment corresponds exactly to ours. See, e. g., p. 177.

[11] See Neugebauer, "Apollonius-Studien," *Quellen und Studien zur Geschichte der Mathematik . . .* , Abt. B, Studien, v. II (1932–1933), p. 215–54.

eters, topics beyond the scope of the now-customary course in elementary analytic geometry.

The *Conics* was not the only contribution of Apollonius to analytic geometry. In fact, it was probably his work on loci which, a long time later, chiefly influenced Fermat in the formation of coordinate geometry. Unfortunately, the *Plane Loci* of Apollonius has been lost, but the nature of its contents is to some extent known through comments of Pappus, Eutocius, and others. Two of the loci are now quite familiar in synthetic geometry: the locus of points the difference of the squares of whose distances from two fixed points is constant is a straight line perpendicular to the line joining the points; and the locus of points the ratio of whose distances from two fixed points is constant is either a circle or a line.[12] The latter locus is now known as the "Circle of Apollonius," but this is a misnomer inasmuch as it was known to Aristotle who used it to give a mathematical justification of the semicircular form of the rainbow. In modern times the numerous efforts made to restore the *Plane Loci* were influential in the development of analytic geometry. Attempted restorations have been made also of his *Determinate Section*, which seems to have been a sort of algebraic analysis.

The work of Archimedes on conics was somewhat earlier than that of Apollonius, but it has not been reviewed in detail inasmuch as it centered about problems of the calculus rather than those of analytic geometry. Archimedes probably did not write a separate book on the conic sections, but his treatises include an extensive study of them, with special reference to mensuration.[13] Attempts to square the ellipse seem to have been made before his time, but the first successful quadrature of a conic section was that given by Archimedes for the segment of a parabola. This was followed by other quadratures and cubatures of conics, conoids (solids obtained by revolving a segment of a parabola or a branch of an hyperbola about its axis), and spheroids (ellipsoids of revolution), as well as by determinations of centers of gravity. In connection with this work he indicated the construction of the ellipse in terms of two concentric circles and the eccentric angle.[14] The results of Archimedes were so spectacular that they seem to have overshadowed those of Apollonius. When in the early modern period the works of these two great geometers were again eagerly studied, the in-

---

[12] See notes by R. C. Archibald in *American Mathematical Monthly*, v. XXIII (1916), p. 159–61.

[13] See Heath's *The Works of Archimedes*. Cf. also Heath's *History* and *Manual* of Greek mathematics; J. L. Heiberg, "Die Kentnisse des Archimedes über Kegelschnitte," *Zeitschrift für Mathematik und Physik*, v. XXV (1880); Zeuthen, *Geschichte der Kegelschnitte*, Coolidge, *History of Conic Sections*.

[14] See Dingeldey, *op. cit.*, and *Works of Archimedes*, "On Conoids and Spheroids," V.

terest in infinitesimal mensurations exceeded that in so-called pure geometry, so that Cartesian geometry at first made slow progress in comparison with that of the calculus. However, so closely related is the subject matter of the two fields that advances in the one were directly related to achievements in the other.

Archimedes contributed to analytic geometry not only indirectly through his infinitesimal methods but also through the introduction of a new curve. The spiral of Archimedes is defined as the locus of a point which moves at a uniform rate along a ray which revolves uniformly about its end-point, always lying in the same plane. As the Greek study of conics approaches closely to the modern application of rectangular coordinates, so also does the work on the spiral resemble the use of polar coordinates. The definition is, in fact, equivalent to the polar equation $r = k\theta$. The attempt sometimes made to read into Archimedes the concept of polar coordinates can be carried further, if one wishes, to cover the pre-Hellenic geometry of the circle, for the definition of the circle leads immediately to the equation $r = k$. Precisely where one places the origin of the idea of coordinates, whether polar or rectangular, is largely a matter of taste and judgment; but it should be noted that not until modern times were curves derived from polar or Cartesian equations by plotting them on coordinate systems. The classical Greek geometers divided curves into three ranks or orders: the highest place was reserved for the only perfect curves, the line and the circle. These were called the plane loci. Second place was granted to the Menaechmian conics which, probably on account of their original mode of definition, were known as solid loci.[15] All other curves, whether algebraic or transcendental, were grouped together under the heading linear loci. Pappus described this last category as made up of those curves "the origin of which is more complicated and less natural [than that of the plane and solid loci], as they are generated from more irregular surfaces and intricate movements." In this description we see the two types of curve definition which the Greeks recognized—the kinematic and the stereometric.

The most important of the Greek linear loci were the quadratrix of Hippias and the spiral of Archimedes, both transcendental. The original definition of these was kinematic, although Pappus later studied them as projections of curves of double curvature traced on surfaces. Other curves defined kinematically were the cissoid of Diocles (a cubic)

[15] Heath, *Works of Archimedes*, p. cxl–cxli, suggests that the phrases "plane loci" and "solid loci" were used at first to express the fact that the former sufficed to solve quadratic equations—i. e., problems involving the comparison of areas—whereas the latter were necessary in the solution of cubic equations—i. e., in questions concerning relationships between volumes. The more extensive classification given later by Descartes was simply a generalization of this idea.

and the conchoid of Nichomedes (a quartic), both introduced during the earlier part of the second century B.C. All four of these curves, as well as the conic sections, served to furnish solutions of the three famous problems. Centuries later Proclus (ca. 412–85) applied the kinematic approach to the conics, showing that if a line segment moves so that its extremities slide along two perpendicular straight lines, then each point of the segment describes an ellipse. Similar constructions were revived and varied during the seventeenth and eighteenth centuries when the "organic description" of conics loomed large as a part of analytic geometry.

The most elaborate ancient attempt, apart from the study of the conics, to extend the number of curves by the stereometric means of definition was made by Perseus, probably during the period between Euclid and Apollonius.[16] He first determined solid figures, called spires, generated by a circle revolving about an axis such that the plane of the circle passes through the axis. There are three general cases, according as the radius of the circle is less than, equal to, or greater than the distance from the center of the circle to the axis of rotation. Perseus then cut these surfaces by planes, obtaining certain quartic oval-shaped curves (of which the lemniscate of Bernoulli is a special case) which he called spiric sections. He appears to have studied these in a manner comparable to the treatment of conic sections by Apollonius.

There are evidences that a few further curves, not all plane, were known to the Greeks. Apollonius may have squared the circle through the cylindrical helix, a curve known later to Geminus, Pappus, and Proclus. Eudoxus apparently had sought to represent the motions of the planets by what was known as the hippopede, a curve perhaps derived through Archytas as the intersection of a sphere with a cylinder. A little later, however, the motions of the planets were represented in the classic manner, by Apollonius and others, as combinations of circular motions. Such representations are tantamount to constructions of epicyclic curves, but the interest of Greek astronomers and geometers seems to have been focused on the uniform circular motions rather than on the geometrical curves resulting from these superimposed movements. Even the cycloid appears to have escaped the attention of ancient geometers, although this may have been the "line of double motion" which, Iamblichus tells us, Karpis of Antioch constructed in order to square the circle.

A comparison of the Greek definition and study of curves with the

---

[16] D. E. Smith, however, places Perseus in the middle of the second century. See *History of Mathematics*, v. I, p. 118.

flexibility and extent of the modern treatment shows surprising narrowness of the ancient point of view.    Inspired by the Pythagoreans, they had found number everywhere in nature, but they overlooked much of the geometric beauty which natural phenomena afford. Aesthetically one of the most gifted people of all times, the only curves which they found in the heavens and on the earth were circles and straight lines.    Even in theoretical geometry they restricted themselves for the most part to the dyads of Plato and the triads of Menaechmus. The Greeks did not effectively exploit even the two means of definition which they possessed.    The kinematic approach and the method of the section of surfaces are capable of far-reaching generalizations, yet scarcely a dozen curves were familar to the ancients.    The failure of Greek mathematicians to develop analytic geometry may well have resulted from the lack of a theory of curves.    General methods are not necessary where problems concern always one of a limited number of particular curves.

With the exception of the spiric sections, curves were not sought and studied in and for themselves, but only in so far as they possessed properties useful for the solution of problems arising in other connections. The determination of volumes of spherical segments, for example, had led Archimedes to the equivalent of the cubic $x^3 + a^2 b = cx^2$; and this equation, Eutocius tells us (ca. 520), he solved by finding the intersection of the parabola $cx^2 = a^2 y$ with the hyperbola $cy = xy + bc$. Whereas now the tendency is to find the intersection points of curves through algebraic elimination and solution, the Greek habit was to solve an algebraic equation—such as $x^3 = 2$—by reducing the problem to the geometric determination of the intersections of curves.    That is, the Greeks used curves, especially the conics, as a sort of *ersatz* algebra in the solution of certain equations.    The application of areas of Euclid and the Pythagoreans showed that the line and circle sufficed in the case of quadratics, but the cubics of Hippocrates and Archimedes required the use of conic sections.[17]    With an important exception to be noted in the case of Pappus, Greek geometry did not in general lead to equations beyond the third degree, but Descartes long afterward was led to the invention of analytic geometry largely by the desire to extend and systematize the traditional geometrical solution of equations of degree greater than four through the intersection of curves of order higher than two.

The centuries immediately following the time of Euclid, Archimedes, and Apollonius added little to the history of analytic geometry.    The

---

[17] See H. G. Zeuthen, "Note sur la résolution géométrique d'une équation du 3ᵉ degré par Archimède", *Bibliotheca Mathematica* (2), v. VII (1893), p. 97–104.

period from Hipparchus (fl. 150 B.C.) to Ptolemy (fl. 150 A.D.) was significant largely for the rise of applied mathematics, where one finds coordinates widely used in astronomy and geography. The use of rectangular coordinates in laying out plans for cities was, in fact, one of the few portions of geometry adopted by the Romans. The *Metrica* of Heron illustrates clearly the use of coordinate methods (both plane and spherical) in mensuration during the later Alexandrian Age; and some of his methods are tantamount to the use of coordinates in three-space.[18] However, the use of coordinate methods in astronomy, geography, and technology seems not to have exerted any clearly definable influence on the later rise of analytic geometry. Of more significance, paradoxically, was the tendency in Heron to dissociate arithmetic and algebra from geometry. Whereas Euclid and Apollonius would have refused to add an area to a line, Heron occasionally did so. This practice may well have resulted from the influence of the Babylonians who in a great number of problems had added magnitudes of unequal dimensionality.[19] This early arithmetization of mathematics appeared in more pronounced form a century later in the *Arithmetic* of Diophantus (ca. 250 A.D.). In his work the classical graphical solution of equations was replaced by the older Babylonian non-geometrical methods. Diophantus used letters and abbreviations to represent the powers of an unknown, and his symbols were no longer thought of as lines; they were numbers. For this reason it was easy for him to go beyond cubic equations and to consider powers of the unknown up to the sixth, which he called a "cube-cube."[20] The tendency to widen the gap between algebra and geometry was accentuated by his strong interest in the theory of numbers. The paradoxes of Zeno had left a profound impression on Greek thought causing the ideas of change and variability to be relegated to metaphysics. Geometric magnitudes were static and continuous; algebraic quantities were discrete constants. The symbols of indeterminate quantities in the *Arithmetic* of Diophantus therefore represent unknown *numbers* rather than *variables* in the sense of analytic geometry. Modern higher analysis emphasizes the Weierstrassian static nature of variables and the theory of aggregates, but historically both analytic geometry and the calculus arose from the idea of variability. From this point of view Diophantine analysis may be regarded as having constituted a hindrance to the development of coordinate geometry; but in another respect it was a

---

[18] See Max Dehn, "Historische Übersicht," in A. Schoenflies and M. Dehn, *Einführung in die analytische Geometrie der Ebene und des Raumes* (2nd ed., Berlin (1931)), p. 379–393.

[19] See, e. g., O. Neugebauer and A. Sachs, "Mathematical Cuneiform Texts" (*American Oriental Series*, v. XXIX), New Haven, Conn. (1945), *passim*.

[20] See T. L. Heath, *Diophantus of Alexandria*, Cambridge (1910), p. 129.

step forward.  The chief deficiency of Greek mathematics was the
lack of an independent symbolic algebra, and the work of Diophantus
is one link in the algebraic chain from the Babylonians to Descartes.

The century of Diophantus and Pappus (fl. ca. 300) represents a
sufficient revival in mathematical interest to be called the silver age of
Greek mathematics.  However, whereas the work of the former betrays
a strong Babylonian-Egyptian influence, the work of the latter represents a return to the classical interest of the golden age.  Pappus made
no striking new advance in method, his role being rather that of organizer and commentator.  He rivals Euclid as an expositor, and Proclus as a preserver, of the knowledge of antiquity.  Most ancient
treatises on advanced geometry have been lost.  For Apollonius alone
the lost works include his *Proportional Section*, his *Determinate Section*,
his *Tangencies*, his *Inclinations*, and his *Plane Loci*.  In many cases the
*Mathematical Collection* of Pappus is now the chief source of information concerning these lost works.  For example, the focus-directrix-
eccentricity property of the conic sections, including that for the parabola, is given here for the first time, although for the central conics the
general focal properties were known to Apollonius.[21]  It is surprising
that this property was so frequently overlooked.  It is said that even
in modern times it remained unfamiliar to some geometers who first
read of it in the *Principia* of Newton (I, 14).  Pappus solved the
problem of finding a conic through five points, a construction apparently also known to Apollonius but subsequently lost.  Pappus gave a
definitive formulation of the Greek division of curves or loci into three
categories, as indicated above; and he gave a clear-cut statement of the
classical ideas on the curves appropriate to the solution of a given problem—solid and linear loci should not be used where plane loci suffice,
and linear loci are not to be applied where the problem can be solved by
means of solid loci.  This important notion of the appropriateness of
solutions (equivalent to the idea, arising much later, of the irreduci-
bility of equations) was a natural extension of the Platonic glorification
of the line and circle, and one which probably had arisen long before
the time of Pappus.  Archimedes, for example, had made a practice of
avoiding the use of two geometric means (the determination of which is
a solid problem) where arithmetic means (found by plane loci) suffice,
even when the latter are less convenient.[22]  Long afterwards, the insistence on the use of the simplest possible means appropriate to a
given geometrical problem was strongly emphasized in the analytic
geometry of Descartes and his successors.

[21] J. H. Weaver, "On Foci of Conics," *Bulletin, American Mathematical Society*, v. XXIII
(1916–1917), p. 357–365.
[22] See *The Works of Archimedes* (Heath), p. lxvii.

To Pappus one owes also the clearest statement in antiquity on the nature of analysis. Greek geometry was divided into three parts: the elements (as found in Euclid); practical geometry or geodesy (represented by Heron); and higher geometry (illustrated by Apollonius and Archimedes). The loss of many ancient works in the third category makes the *Mathematical Collection* of Pappus (itself incompletely preserved) an indispensable work on the elements of higher geometry. Pappus says that in the last division two methods are used—synthesis, or composition, and analysis, or decomposition. These words do not refer to the *method* used, but to the *order* of demonstration. Kant had this in mind when, long afterward, he referred to the former as "progressive" and the latter as "regressive." That is, analysis is used by Pappus in the Platonic sense of assuming as known the thing to be found or proved. One finds a little of this in the Euclidean proofs by the indirect method, and more in Archimedes' treatise "On the Sphere and Cylinder," in which one reasons with unknown quantities as though they were known. The Diophantine use of unknowns is a clear example of analysis, but this was not related to geometry.

It was through Pappus that Descartes became acquainted with the problem—the locus to three or four lines—which inspired him to invent his coordinate geometry. The problem of Pappus[23] calls for the locus of a point such that if line-segments are drawn from the point to meet three or four given lines at given angles, the product of two of these segments shall be proportional to the product of the other two (if there are four lines) or to the square of the third (if there are three lines). Euclid evidently had constructed the locus only for some special cases, but evidence indicates that Apollonius, in works now lost, probably had given a complete solution of the problem.[24] Pappus, however, gave the impression that Greek geometers had failed in their attempts at a general solution, and that it was he who first showed the locus in all cases to be a conic section. Pappus then went on to consider the equivalent problem for more than four lines.[25] For six lines he recognized that a curve is determined such that for any point on it the solid contained by the distances from three of the lines is proportional to the solid contained by the distances from the other three.

[23] Throughout this book the phrase "problem of Pappus" will be used in this sense. It should be pointed out, however, that the phrase is often used for another problem—through a point on a bisector of a given angle to pass a line segment of given length with extremities on the sides of the angle. See A. Maroger, *Le problème de Pappus et ses cent premières solutions*, Paris (1925).

[24] See J. J. Milne, *An Elementary Treatise on Cross-Ratio Geometry, with Historical Notes*, Cambridge (1911), p. 146–49, for a good account of this problem in the history of geometry.

[25] The definitive edition of the *Mathematical Collection* of Pappus is the Latin one by Hultsch (3 vols., Berlin (1876–1878)). The French translation by Ver Eecke (2 vols., Bruges (1933)) is also very useful.

Pappus hesitated to go on to cases involving more than six lines since "there is not anything contained by more than three dimensions"; but, he continued, "men a little before our time have allowed themselves to interpret such things, signifying nothing at all comprehensible, speaking of the product of the content of such and such lines by the square of this or the content of those. These things might however be stated and shown generally by means of compounded proportions."[26] These unnamed predecessors of Pappus evidently were attempting to take a highly important step in analytic geometry—the consideration of lines not directly as such but only in terms of numerical measures of their length—one which would have led to a truly algebraic treatment. Had Pappus pursued the matter further, he might conceivably have been the inventor of analytic geometry, for his observations pointed toward the practicability of a general classification and theory of curves and loci far beyond the scope of the classical distinction between plane, solid, and linear loci. His recognition that, no matter what the number of lines in the Pappus problem, a specific curve is determined, is the most general observation on loci in all ancient geometry. Greek geometrical algebra would indeed have been far from ideal as a tool with which to develop the theory of higher plane curves, but it would have sufficed. But Pappus either did not perceive the possibilities lying in this direction or else he regarded the locus problem as inadequate for the definition of new curves; and hence he did not follow the fruitful lead which was here suggested. He was satisfied to point out that the discussion of the locus to three or four lines can be generalized to show that for any number of lines, whether odd or even, the point will lie on a curve given in position. This observation is equivalent to saying that the equations

$$\frac{x_1 \cdot x_3 \dots x_{n-1}}{x_2 \cdot x_4 \dots x_n} = k \text{ and } \frac{x_1 \cdot x_3 \dots x_n}{x_2 \cdot x_4 \dots a} = k$$

(where the $x$'s are the variable segments drawn to the fixed lines at given angles, $a$ is a given line segment, and $k$ is a given constant) define curves; and no one again so nearly anticipated the invention of analytic geometry before modern times. Pappus remarked with surprise that no one had made a synthesis of this problem for any case beyond that of four lines, and hence one was unable to recognize the loci; but he himself made no further study of these loci "of which one has no further knowledge and which are simply called curves."[27]

[26] Quoted from Charles Taylor, *An Introduction to the Ancient and Modern Geometry of Conics*, p. xlvi. Cf. also Ivor Thomas, *Selections Illustrating the History of Greek Mathematics* (2 vols., Cambridge, Mass. (1939–1941)), v. II, p. 601 f.
[27] *La collection mathematique* (edited by Ver Eecke), v. II, p. 508–510.

It is customary to hold that Greek algebra, because of its geometrical bonds, was limited to the first three powers or dimensions, but work of Diophantus and Pappus makes such a view open to question. It should be noted, however, that Descartes finally was led to the invention of his geometry by precisely the problem which Pappus here touched upon but did not develop—the study of higher plane curves determined as loci for the case of more than four lines. The problem of Pappus long after served as a challenge to the vain-glorious Descartes to surpass the best efforts of antiquity. So well did Descartes succeed in his efforts that he boasted, with some justification, that his method was to ancient geometry as the rhetoric of Cicero was to the ABC of children. Between Pappus and Descartes, however, there is a gap of some thirteen hundred years. This interval witnessed the development of a number of significant achievements bearing on the history of analytic geometry, but of these the most important was the rise of symbolic algebra—the tool which Apollonius and Pappus wanted above all else.

# CHAPTER III

# The Medieval Period

*Neglect of mathematics works injury to all knowledge, since he who is ignorant of it cannot kwow the other sciences or the things of this world.*

—ROGER BACON

THE medieval age is conveniently divided, with respect to the history of mathematics, into an earlier and a later period. The first part covers the long interval between Boethius (†524) and Fibonacci (ca. 1170–1250), and during this time interest in mathematics was found chiefly in the Hindu, Byzantine, and Arabic civilizations. Among the Hindus the computational side of the subject overshadowed the logical and speculative aspects, so that one finds there little material related to the rise of analytic geometry. As David Eugene Smith wrote, "To the Orient we may look for early progress in algebra, trigonometry, and the creation of a remarkable number system, but not for any geometry whatever until relatively modern times."[1] Among the Hindus, for example, so little concern was shown for the theory of curves that there is hardly a trace even of the conic sections. Applications of arithmetic to geometry were limited largely to questions of measure. It is true that the Hindus made use of negative numbers, and this may have had some influence upon the later generalization of coordinates to include negative values, but such a conjecture is open to doubt in view of the fact that negative coordinates were rarely used before the eighteenth century. There appears to be still less likelihood that the Hindu acceptance of zero and the irrationals as numbers had any significant effect upon the Cartesian association of numbers and lines in analytic geometry. One frequently reads that the Hindus *began* the practice, so important in algebraic geometry, of substituting letters for magnitudes and of operating with literal symbolisms. This view is fundamentally erroneous,

[1] "The Geometry of the Hindus," *Isis*, v. I (1913), p. 197–204. See also F. W. Kokomoor, "The Status of Mathematics in India and Arabia during the 'Dark Ages' of Europe," *Mathematics Teacher*, v. XXIX (1936), p. 224–231.

as a study of Aristotle's symbolism of letters in the *Physica*, for example, shows. But Hindu arithmetic did develop further a syncopated calculation with symbols and letters which Diophantus had used, and so it represents a step from the Greeks toward Viète. Moreover, the introduction of the so-called Hindu numerals (which may, however, have been invented by the Greeks or the Egyptians or some other people) served further to simplify algebraic calculations and so to make attractive later the substitution of algorithmic devices for geometric constructions. On the whole, however, the Hindus were indifferent to fundamental principles and logical order, and they displayed little concern for mathematical method.[2]

In the Byzantine empire mathematical works continued to be written in Greek from Pappus and Proculus down to the dawn of the Renaissance, but they did not rise above the level of undistinguished commentaries. In the sixth century Anthemius of Trales (†534), architect of St. Sophia, described the use of the parabola in connection with burning mirrors, and he gave also the "gardner's construction" in the laying out of ellipses; but both of these basic properties had been known to Apollonius. Isidore of Miletus, an associate, seems to have known the string-and-ruler construction of the parabola, but this is simply a mechanization of a property familiar to Pappus. The commentaries of the same period by Isidore's pupil, Eutocius (ca. 560), on the works of Apollonius and Archimedes preserved some knowledge of earlier work which might otherwise have been lost; and the commentaries of Simplicius (fl. 529) on the *Physica* of Aristotle inspired later discussions which bordered more on the calculus than on analytic geometry.

Arabic mathematics in general characteristics was a compromise between the arithmetic of the Hindus and the geometry of the Byzantines, with additional Greek, Babylonian, and Egyptian elements. Irrational numbers were retained but negatives were dropped. Algebra followed mainly the Babylonian and Hindu pattern, and remained largely independent of geometry; but Al-Khowarizmi's solution of quadratic equations nevertheless betrays Greek influence in that the process of completing the square is illustrated in terms of geometrical areas. It is to be noted, however, that this solution, in spite of its geometrical background, represents, as far as the development of analytic geometry is concerned, a retrogression from classical Greek work. It did not call for a strict construction in conformity to specifically stated postulates (as had the Pythagorean-

2 G. R. Kaye, "Some Notes on Hindu Mathematical Methods," *Bibliotheca Mathematica* (3), v. XI (1911), p. 289–299.

Euclidean solutions), nor did it make use of the intersections of new curves (as in the solutions of cubics by Menaechmus and Archimedes); and these were the very principles upon which Descartes later based his work.

Diophantus is sometimes called the father of algebra, but this title belongs more properly to Al-Khowarizmi (ca. 825) in that the work of the latter emphasized the Babylonian solution of determinate equations rather than the Diophantine analysis which is characteristic of higher arithmetic. The Hindus had been much attracted by the Diophantine problems and had extended the development of syncopated forms of expression. It is unfortunate that the Arabs did not adapt the Hindu syncopation to the solution of equations but reverted instead to rhetorical forms. Specific forms of notation in themselves are not always of great significance, but frequently the use of symbols exerts a decisive influence on the subsequent development of concepts. This is aptly illustrated by the relationship of algebra to analytic geometry. It is customary to recognize three stages in the development of algebra: the rhetorical, the syncopated, and the symbolic. Diophantus had risen to the second stage but Al-Khowarizmi slipped back to the first. The ideas of algebraic variable and equation of a curve arose only after the last stage had been reached, and hence the Arabic beginnings in algebra had to be developed further in the sixteenth and seventeenth centuries to serve as a basis for modern algebraic geometry.

Muslim mathematicians were much interested in the geometry of the Greek golden age, translating into Arabic the works of Euclid, Archimedes, and Apollonius. Had it not been for their translations, some portions of Archimedes (especially his trigonometry) would now be unknown, and all of the last four books of Apollonius would be lost, instead of only the last one. The conic sections, overlooked by the Hindus, did attract the Arabs, and the latter made some further contributions in this connection, the optical properties of the parabola being given a place of prominence. Archimedes had studied the solid obtained by revolving a parabola about its axis, and Alhazen (Al-Haitham, ca. 1000) generalized this by revolving parabolic segments about arbitrary diameters and ordinates, both perpendicular and oblique, determining the cubatures of the various figures.[3] Such work, however, was more closely related to the calculus than to analytic geometry, for surfaces at that time were regarded as boundaries of solids

---

[3] See "Die Abhandlung über die Ausmessung des Paraboloides von el-Hasan b. el Hasan b. el-Haitham," translated with commentary by Heinrich Suter, *Bibliotheca Mathematica* (3), v. XII (1911–1912), p. 289–332.

and were not studied in terms of coordinates. Similarly, the "problem of Alhazen"—to find the point on a given circle at which light issuing from a given point as source will be reflected toward a second fixed point—was not associated with coordinate geometry until the seventeenth century.

One of the high points of Arabic mathematics is the geometrical solution of cubic equations in the traditional manner of Menaechmus. Alhazen here, too, followed Archimedes, solving the cubic $x^3 + a^2b = cx^2$ through the intersection of $x^2 = ay$ and $y(c - x) = ab$. His work was continued on a broader scale by Omar Khayyam (ca. 1100) who, although he believed numerical solutions for general cubic equations to be impossible, nevertheless asserted that by means of the intersections of conic sections he could give a solution for each type.[4] This claim is illustrated for various cases: those of the form $x^3 + b^2x = b^2c$ were solved by using the parabola $x^2 = by$ and the circle $x^2 + y^2 = cx$; those of the form $x^3 + ax^2 = c^3$ by using the hyperbola $xy = c^2$ and the parabola $y^2 = cx + ac$; those of the form $x^3 \pm ax^2 + b^2x = b^2c$ through the hyperbolas $x(b \pm y) = bc$ and the circles $y^2 = (x \pm a)(c - x)$. As in the case of quadratic equations, cubics having no positive roots were omitted from consideration. It is significant that Omar felt that even where algebraic solutions were possible, it was necessary to supplement and verify these by means of geometric constructions, a view which makes him an important link between the Greeks and the geometry of Descartes.

While the Hindus, Byzantines, and Arabs were maintaining an interest in mathematics, Latin Europe had been struggling through the dark ages. Little of any consequence was added to the history of analytic geometry, unless one excepts rough graphical representations of the courses of the planets through the zodiacal constellations. Using thirty subdivisions in longitude and a dozen in latitude, the paths of the planets, as given in Pliny's *Natural History*, were plotted during the tenth or eleventh century on a crude system of rectangular coordinates.[5]

The Latin world was largely unfamiliar with the ancient mathematical treatises until, in the twelfth century, Latin translations were made from Arabic, Hebrew, Syriac, and Greek manuscripts. Even after the Greek classics in geometry and the Arabic works in algebra

[4] See D. S. Kasir, *The Algebra of Omar Khayyam* (New York, 1931), chaps. IV–VIII. Cf. also Matthiessen, *op. cit.*

[5] See Harriet Lattin, "The Eleventh Century MS Munich 14436: its contribution to the history of coordinates, of logic, of German studies in France," *Isis*, v. XXXVIII (1948), p. 205–225, and H. G. Funkhouser, "Historical Development of the Graphical Representation of Statistical Data," *Osiris*, v. III (1937), p. 269–404. See also Funkhouser, "A Note on a Tenth Century Graph," *Ibid.*, v. I (1936), p. 260–262.

became available, western European scholars at first displayed little interest in them, occupied as they were with questions in theology and metaphysics.  Nevertheless, the thirteenth century opened auspiciously with works on algebra and geometry by Fibonacci (Leonardo of Pisa, ca. 1170–1250).  His *Liber abaci* in 1202 did more than popularize the Hindu-Arabic numerals, for it emphasized the interrelationships of arithmetic and geometry.  This work opens with the assertion that the complete doctrine of number cannot be presented without looking to geometry inasmuch as many of the demonstrations are in terms of geometric figures.  Conversely, in his *Practica geometriae* of 1220 he solved many questions "secundum algebram."[6]  Such an association of algebra and geometry does not in itself constitute analytic geometry; but three and four centuries later it was the reappearance and rapid spread of similar work which paved the way for coordinate geometry.

The interval from Fibonacci to Chuquet (fl. 1477), known as the later medieval age, was noteworthy more for certain abortive efforts in new directions than for achievement in the traditional subject matter. This is especially true in connection with the application of curves to the study of dynamics.  The Greeks had built up an elaborate mathematical theory, but they had applied only the most elementary portions of it to science; the scholastic philosophers of the fourteenth century, on the other hand, possessed the most elementary mathematical tools but sought ambitiously to make an elaborate quantitative study of science.  Archimedes had not tackled the simple problems of dynamics, even though he possessed rudiments of the calculus.  The Greek sciences of astronomy, optics, and mechanical statics had been elaborated geometrically, but there was no such representation of variable physical phenomena.  The scholastics, however, essayed, in what they called the latitude of forms, a broad study of physical variation.  The word form referred to any quality which admits of variation, and the latitude of a form was the degree to which it possessed this quality. In general the discussion centered about the *intensio* and *remissio* of the form, or the rate of change of the quality.  Aristotle had distinguished between uniform and non-uniform velocity, but the scholastics at Oxford and Paris went further.  They applied these ideas to acceleration, intensity of illumination or thermal content, and density; and they distinguished not only between uniform and non-uniform rates of change, but subdivided the latter according as the rate of change of the rate of change was constant or not.

[6] See Ettore Bortolotti, *Lezioni di geometria analitica* (2 vols., Bologna, 1923), "Introduzione storica," v. I, p. ix–xxxix.

The origin of the study of the latitude of forms is far from clear. Duns Scotus appears to have been among the first to consider the increase and decrease of forms, and shortly thereafter treatises on the latitude of forms were composed by James of Forli, Walter Burley, Albert of Saxony, and Richard Suiseth (ca. 1345). The work of the last named, better known as Calculator, was a leading model which exerted a wide influence for several centuries. In this the author discussed at length the average intensity of a form representing thermal intensities and concluded that if the rate of change over an interval is uniform, then the average intensity is the mean of the first and last intensities. The rigorous proof of this requires the use of the limit concept, but Calculator based his reasoning on the crude experience of rate of change. He argued at great length that if the greater intensity is allowed to decrease uniformly to the mean while the lesser is increased at the same rate to this mean, then the whole is neither increased nor decreased. For example, if the intensity of a form increases uniformly from four to eight, or if for the first half of the time it is four and for the last half it is eight, then the effect is that which would result from a uniform intensity of six operating throughout the whole time.

The latitude of forms illustrates more clearly than had any earlier mathematical work the idea of one quantity varying as a function of another; nor was this notion limited to cases of direct proportionalities. The *Liber calculationum* of Suiseth includes problems such as the following: if throughout half of a given time interval a variable quantity has a certain constant intensity, and if throughout the next quarter of the interval it continues at double the initial intensity, and throughout the next eighth of the interval at triple the initial intensity, and so on ad infinitum; then the average intensity for the whole interval will be the intensity of the variable during the second subinterval. This is equivalent to saying that the sum of the infinite series $\frac{1}{2} + \frac{2}{4} + \frac{3}{8} + \frac{4}{16} + \ldots + \frac{n}{2^n} + \ldots$ is 2. The study of infinite series is important in the calculus, but the example of Calculator is significant in analytic geometry as another instance of a verbally defined functional relationship—one which has discontinuities and in which the dependent variable increases indefinitely.

The proofs given by Suiseth are tediously long because of the complete lack of algebraic or geometric symbolism, other than the use of letters for quantities and of the customary abbreviations in manuscript writing. However, as the center of the new mathematical science shifted from the logicians of Oxford to the Ockhamites of Paris, a striking change took place. At the University of Paris, Nicole

Oresme (ca. 1323–1382) felt that the study of the latitude of forms could be clarified through reference to geometrical figures. Although his work is clearly an outgrowth of earlier Scholastic philosophy, Oresme seems to have had no predecessors in the matter of graphical representation.[7] The manner in which this is to be accomplished is described at length in a manuscript work, *Tractatus de Figuratione potentiarum et mensurarum*, written probably before 1361,[8] and also in the briefer printed work, the *Tractatus de latitudinibus formarum*. Following the Greek tradition which regarded number as discrete and geometrical magnitude as continuous, he said that measurable quantities *other than numbers* can be represented by points, lines, and surfaces. To carry this out, intensities are to be represented by lines drawn perpendicular to the interval or region under consideration. If, for example, the velocity of an object is to be represented as a function of time, then time is measured along a horizontal straight line (Oresme calls it a "longitude") and the intensities of the velocity are then drawn perpendicular to it (as representations of "latitude"). This does not mark the first use of coordinates, but it appears to be the earliest use of the graphical representation of functions on a coordinate system. The work of Apollonius may be interpreted as the first stage in the mathematical development of coordinates—in which one introduces as coordinate axes certain auxiliary lines determined by a figure or curve previously given. The work of Oresme represents a second stage in which the coordinate system is laid down first and the points of a curve are then determined with respect to this, subject to given conditions verbally expressed.

Günther[9] would recognize three stages in the development of the idea of coordinates: (1) the introduction of two axes on the surface to be studied; (2) the plotting of a curve by constructing ordinates for given abscissas and then connecting the end points; (3) the use of equations permitting one to go from abscissas to ordinates, or vice versa. He correctly points out that Oresme clearly belongs to the second stage; but there seems to be room for doubt that this stage is a necessary preliminary to step three, as he holds. Historically, the

---

[7] See Adolf Krazer, *Zur Geschichte der graphischen Darstellung von Funktionen* (Karlsruhe, 1915). This Festschrift of 31 pages gives a cogent summary of Oresme's work.

[8] This work is known also by other titles, such as *De uniformitate et difformitate intentionum* and *De configuratione qualitatum*. For an extensive account of this work see Heinrich Wieleitner, "Ueber den Funktionsbegriff und die graphische Darstellung bei Oresme," *Bibliotheca Mathematica* (3), v. XIV (1914), p. 193–243. Commonly ascribed to Oresme is another smaller work, the *Tractatus de latitudinibus formarum*, which is probably a student's notes on Oresme's longer treatise. A full description of this shorter work is given by Wieleitner in an article, "Der 'Tractatus de latitudinibus formarum' des Oresme," *Bibliotheca Mathematica* (3), v. XIII (1912–1913), p. 115–145.

[9] *Op cit.*

three stages are easily discernible, but it appears likely that the third step developed from the first step, possibly uninfluenced by the intermediary stage.

Oresme used the method of graphical representation to give a simple proof of Calculator's proposition on average intensities where the rate of change of velocity is uniform. If a body, starting from rest, moves with uniformly increasing speed, the lines representing the intensities or velocities will form a surface area in the shape of a right triangle. (See Fig. 6.) Now the area of the triangle $ABC$ is exactly equal to the area of the rectangle $ABGF$ (where $F$ bisects $AC$), and this rectangle is the graphical representation of a motion for the same time interval but having a uniform speed equal to the mean of the initial and final speeds for the former case. Oresme did not state explicitly the fact, demonstrated in the integral calculus, that the areas $ABGF$ and $ABC$ represent in each case the distance covered; but this seems to have been his interpretation inasmuch as he derived the equality of the distances from the congruence of the triangles $CFE$ and $EBG$.

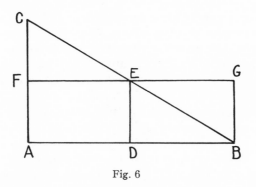

Fig. 6

Oresme studied graphically other instances of functional relationship, including the discontinuous function of Calculator described above. This he then modified in various ways such as the following: let the body move during the first half of the time interval with a given uniform velocity; then for the next quarter of the time let the velocity increase uniformly from the given velocity to double this speed, for the next eighth of the interval with the uniform speed attained at the end of the previous sub-interval, for the next sixteenth with uniform acceleration until the velocity again is doubled, and so on ad infinitum. Oresme found that in this case the total distance covered is to that of the first half of the time as seven is to two.

The work of Oresme has been hailed by some historians as equiva-

lent to Cartesian geometry.   Duhem, a competent but over-enthusi-
astic authority, has asserted that Oresme "gives the equation of the
right line, and thus forestalls Descartes in the invention of analytical
geometry."[10]   It is true that Oresme knew that a constant rate of
change or a rate of change which is proportional to time can be represented on a coordinate system by means of a rectilinear configuration,
and that other types of variation are associated with characteristic
diagrams.   However, the facility Oresme displayed in handling linear
and broken-line graphs could not be extended to curvilinear figures,
although his discussion in connection with the latter sometimes included significant observations.   In connection with a form represented
graphically by a semicircle, it is pointed out that the rate of change of
an intensity (such as velocity) is least at the point corresponding to the
maximum intensity.   Nevertheless, he was prevented from taking full
advantage of his novel idea by deficiencies in geometrical knowledge
and algebraic technique.   Consequently, there is in the work of Oresme
no systematic association of algebra and geometry in which an equation
in two variables determines a specific curve, and conversely.   As a
matter of fact, his graphs are associations of (physical) variables with
geometric representation from which number is specifically excluded.
His treatise *De continuitate intentionis* opens with the statement,
"Anything measurable, except for number, is imaginable in the manner
of continuous quantity."   The Greek idea of the ratio between two
geometrical magnitudes still was dominant.

Even apart from the lack of the fundamental principle that algebra
and geometry can be associated, Oresme's graphical representation
differs in several respects from the later point of view.   Wieleitner
stresses the fact that Oresme lacked a clear idea of an origin in his coordinate system and that his longitudinal axis was a finite time interval
and not infinite in extent; but these differences are of lesser importance,
as is also the fact that there is no reference to negative coordinates.
More important, it appears, is the fact that he looked upon the figures
formed by the latitudinal lines as plane figures the *areas* of which were
his chief concern.   A form with uniform intensity was not a horizontal
line, but a rectangle; or again, as Oresme explains, "Any uniformly
difform quality terminating in zero intensity is imagined as a *right
triangle*"—not as a straight line.[11]   The upper boundary line played

[10] See his article, "Oresme," in the *Catholic Encyclopedia*.   Duhem here similarly exaggerates the role of Oresme as a precursor of Copernicus: "and the whole of his argument in
favor of the earth's motion is both more explicit and much clearer than that given by
Copernicus."   A more reliable estimate of Oresme's views on the motion of the earth is
given by Lynn Thorndike, *History of Magic and Experimental Science* (6 vols., New York,
1923–1941).

[11] Cf. Krazer, *op. cit.*

only a secondary role, whereas in analytic geometry properly understood it is precisely the locus of the *end points* of the lines of latitude which is associated with the equation or law of variation. The "form" or function in problems of uniformly increasing velocity was not velocity but distance, the latitudes of which were the vertical lines representing the intensities of the rate of change of distance with respect to time. Oresme pointed out in connection with such graphs (see Fig. 6) the constant-slope property: $CF:ED = AD:DB$; and it is largely upon this that Duhem bases his argument that "analytic geometry of two dimensions was created by Oresme," for he reads into it the two-point form of the equation of the straight line.[12] It should be noted, however, that the slope of the line here represented the rate of change of the rate of change (the second derivative rather than the first) of the function with respect to the independent variable. That is, what would now be regarded as a velocity-time graph was to Oresme a distance-time graph. The same point of view dominated his suggestion with respect to solid analytic geometry—the *intensio* or function in the case of a *qualitas superficialis* (two independent variables) was not represented by a surface but by the volume made up of all the ordinates erected upon the portion of the reference plane. Incidentally, even long after the time of Descartes surfaces were looked upon generally as bounding volumes rather than as loci satisfying equations in three variables. Oresme boldly sought to extend consistently his idea of graphical conception to a fourth dimension. For a *qualitas corporalis* (involving three independent variables) one takes not a reference line or plane, but a reference body or volume. With each point of this volume there is associated a line indicating the degree of intensity, and the (four-dimensional) totality of these lines represents the intensity (i. e., function) of the quality or form.[13] Obviously pictorial representation failed in this case. What was needed here was an *algebraic* geometry which Oresme did not possess.

Oresme's graphical method made possible the diagrammatic representation of any type of variation of one variable in terms of one or two others. That is, curves and surfaces might have been defined by analytic means or by the equivalent of differential equations, for his work foreshadowed the very important notions of variable, function, and rate of change. Smith[14] has written: "The real idea of functionality as shown by the use of coordinates was first clearly and publicly

[12] Pierre Duhem, *Études sur Léonard de Vinci* (3 vols., Paris, 1906–1913), v. III, p. 386.
[13] Wieleitner, "Zur Frühgeschichte der Räume von mehr als drei Dimensionen," *Isis*, v. VII (1925), p. 486–489. For later history of this subject see Cajori, "Origins of Fourth Dimension Concepts," *American Mathematical Monthly*, v. XXXIII (1926), p. 397–406.
[14] *History of Mathematics*, v. I, p. 376.

expressed by Descartes." It would appear more appropriate, however, to ascribe the graphical representation of *functions* (as distinct
from algebraic expressions) to Oresme, and to attribute to Descartes
(and to Fermat) the use of coordinates in connection with algebraic
curves and equations.

As the failure to use empirical methods hampered the development
of medieval science, so also the weakness of mathematical technique
precluded the construction at that time of an effective analytic geometry or a theory of curves and surfaces. The fourteenth century was,
in fact, scarcely familiar with the conic sections or of other curves beyond the line or circle. It is said that the limaçon had been recognized
by Johannes Campanus (fl. ca. 1260) and Jordanus Nemorarius in the
previous century, four hundred years before its rediscovery by
Étienne Pascal, but this discovery was not related to the latitude of
forms or to analytic geometry. Moreover, Oresme's graphs were discussed more in terms of physical variation than of geometric definition
and significance.

The relationship of the latitude of forms to the later development of
analytic geometry is difficult to determine. The work of Suiseth and
Oresme was much admired by men of the fifteenth and sixteenth centuries, and several printed editions of the *Liber calculationum* and the
*Tractatus de latitudinibus formarum*, as well as commentaries on each,
appeared in the interval from 1477 to 1520. In some cases the latitude
of forms was required work in the universities. The traditions developed at Oxford and Paris were continued at Padua and other Italian
universities by an Averroistic interest in quantitative science,[15] and
there can be no doubt that Galileo was thoroughly familiar with the
Scholastic doctrines in dynamics. The widespread belief that Galileo
created the idea of acceleration is completely false, for the general
concept of acceleration is found in Aristotle and the specific idea of a
uniform rate of change of the rate of change of distance with respect to
time goes back at least as far as Calculator, a scholar to whom Galileo
specifically referred in his early work. In his *Two New Sciences* of 1638
Galileo reproduced with striking fidelity the diagram and argument of
Oresme for uniformly accelerated motion, outlined above. The influence of Oresme on modern dynamics is clear, but the case for geometry is less certain. Descartes' early interest in mathematics was connected with the laws of falling bodies, and here he made use of dia-

[15] For a good account of such intellectual currents see J. H. Randall, Jr., "The Development of Scientific Method in the School of Padua," *Journal of the History of Ideas*, v. I
(1940), p. 177–206. For a more extensive but less judicious history of the physical science
of the time see Pierre Duhem, *Études sur Léonard de Vinci*.

grams resembling those of Oresme and Galileo.[16] Descartes carefully avoided any reference to his predecessors and so one cannot say with assurance that he was familiar with the work of Oresme, but this seems quite probable. Nevertheless, it will be seen later that the differences (in motivation and purpose, as well as in substance) between his analytic geometry and the graphical representation of the latitude of forms are so great as to make questionable any decisive influence of Oresme on Descartes.[17] The same thing is true also in the case of Fermat, and inasmuch as his interests were more narrowly classical, any significant indebtedness to medieval learning is unlikely. It is more likely that Oresme's graphs and the scholastic use of indivisibles made themselves felt in geometry indirectly through the medium of infinitesimal analysis and the function concept at about the time of the first appearance of Cartesian geometry. Whatever his influence, however, in estimating his graphical method one should note that whereas the later inventors of analytic geometry had predecessors—notably Apollonius and Pappus—out of whose work theirs was developed, Oresme was here a trail-blazer with no appreciable mathematical background. Had he been guided by Greek geometry, the history of mathematics might have been radically changed.

There is another medieval contribution to mathematics which indirectly may have affected analytic geometry through the development of formal algebra. Although Europe in the fourteenth century overlooked much of the classical geometry of Greece, there was at the time a widespread interest in the theory of proportion which had been handed down from Pythagorean musical theory. The fifth book of Euclid remains a monument in the application of the theory to geometry, but the ancients made practically no attempt to carry the ideas over into physical science. Through the primitive works of Boethius and

---

[16] As late as 1654 one finds a reproduction of Oresme's diagram on uniformly accelerated motion in the work of Huygens. (*Oeuvres complètes*, v. XVI, p. 114–115.)

[17] There is a great difference of opinion on this subject. C. R. Wallner, "Entwickelungs-geschichtliche Momente bei Entstehung der Infinitesimalrechnung," *Bibliotheca Mathematica* (3), v. V (1904), p. 113–124, especially p. 120, sees not the least influence of Oresme on Descartes; Edward Stamm, "Tractatus de continuo von Thomas Bradwardina. Eine Handschrift aus dem XIV. Jahrhundert," *Isis*, v. XXVI (1936), p. 13–32, especially p. 24, says that the problem of the latitude of forms of Oresme was undoubtedly the most important influence on Descartes; Curtze, who rediscovered the work of Oresme, and Cantor, in his *Geschichte der Mathematik*, see a strong similarity between the geometry of Oresme and that of Descartes; Weileitner, in his article "Ueber den Funktionsbegriff" (cited above), p. 242, says that Descartes undoubtedly knew of Oresme's latitude of forms, but he is inclined to doubt that Descartes recognized the element their work had in common. Dingeldey, *op. cit.*, doubts that medieval graphical representation played even a minor role in the definitive introduction of coordinate geometry. Hankel's estimate of Oresme's work is low, but Gelcich calls it "epoch-making." (E. Gelcich, "Eine Studie über die Entdeckung der analytischen Geometrie mit Berücksichtigung eines Werkes des Marino Ghetaldi," *Abhandlungen zur Geschichte der Mathematik*, v. IV (1882), p. 191–231.)

through Arabic translations of Euclid, the Latin medieval world was acquainted with the theory of proportions at least by the time of Campanus; and about a century later one finds treatises devoted to the subject, such as the *Liber de proportionibus* of Thomas Bradwardine in 1328. In this book one finds a continuation of the work of Euclid, Iamblichus, and Boethius in the idea of fractional proportions, in which, for example, one quantity varies as the cube of the square root of another. Somewhat later in the century Oresme composed an *Algorismus proportionum* in which fractional "powers" are freely used. Here again one finds a medieval advance which *might* have had considerable influence, for Oresme introduced occasional symbolic representations of proportions, the forerunners of the Cartesian notation of exponents.[18]  Thus he used expressions such as $\dfrac{1p}{1 \cdot 2 \cdot 4}$ and $\dfrac{3p}{2 \cdot 4}$ to designate $4^{1/2}$. He did not, however, carry out such abbreviation systematically, and his statements of rules of "exponents"—equivalent to such expressions as $(a^m)^{p/q} = (a^{mp})^{1/q}$ or $a^m \cdot a^{1/n} = a^{m + (1/n)}$—generally are verbal rather than symbolic. Had he and his successors gone further in the direction of syncopation, the history of algebra would be quite different, for it was here that emphasis was placed in the early modern period. Unfortunately, the medieval development was cut short by weakness in algebraic knowledge and technique. The spirit of quantitative science was strong, but the necessary mathematical equipment was lacking.

The artist Leone Battista Alberti (1404–1472) also has been characterized as a precursor of Descartes because of his application of coordinate systems in perspective and architecture. We are told[19] that in one treatise (now lost) he seems to have been occupied with problems of analytic geometry; but even if this be granted, there is no evidence that his work influenced the later inventors of the subject. The use, in early works on the theory of perspective, of concentric circles and radiating lines in problems relating to deformations (anamorphoses), represents an anticipation of the idea of polar coordinates, much as latitude and longitude in astronomy and geography are considered forerunners of rectangular coordinates. However, there was no correlation between such systems and an analytic theory of curves. Here, again, the practical arts and sciences seem to have played a smaller part than one would have anticipated in the history of analytic geometry.

[18] For further details and references see my paper, "Fractional Indices, Exponents, and Powers," *National Mathematics Magazine*, XVIII (1943), p. 81–86.
[19] Georg Wolff, "Leone Battista Alberti als Mathematiker," *Scientia*, v. LX (1936), p. 353–359, and supp., p. 142–147.

The medieval latitude of forms and the theory of proportions continued to dominate thought for some time. In the fifteenth century one finds the physicist Marliani using them to study problems in dynamics, but he made little if any progress, for he failed to master the elements of proportion of Bradwardine. His century missed the opportunity of making outstanding progress in the development of science because it was not mathematically prepared. It was not equipped to fashion the theory of proportions into an algebra, or the latitude of forms into an algebraic geometry, because the scholars of that period had not first mastered the background which they had inherited from ancient Greece and medieval Arabia. It was largely during the sixteenth century that this situation was remedied through a renewed interest in, and a systematic coordination of, the classic results of geometry and algebra.

# CHAPTER IV

# The Early Modern Prelude

*In mathematics I can report no deficiency, except it be that men do not sufficiently understand the excellent use of the Pure Mathematics.*

—FRANCIS BACON

THE traditional distinction between the Middle Ages and the Renaissance is a convenient one, but it is also deceptive. For one thing, 'a noticeable revival in the fields of art and literature was not at first accompanied by marked advances in science and mathematics; and in the second place, there are definite connections linking the medieval with the modern period in algebra and geometry. No new and outstanding mathematical trend, comparable to the artistic, is found, for example, between the time of Petrarch (1304–1374) and that of Leonardo da Vinci (1452–1519). On the other hand, numerous editions of the works of Bradwardine and Oresme appeared in the late fifteenth and early sixteenth centuries, and treatises on proportion continued to betray medieval influence for still another hundred years. The end of the fifteenth century produced two notable works, both of which were based on earlier sources but which in some measure anticipated future lines of development. These two books— Chuquet's *Triparty en la science des nombres* of 1484 and Pacioli's *Summa de arithmetica* of 1494—may for convenience be taken to mark a reasonably clear line between medieval and modern mathematics.

From the titles of the works of Chuquet (†ca. 1500) and Pacioli (†ca. 1509) it is clear that the modern period opened with emphasis in a new direction—toward algebra. The *Triparty* represents a marked expansion of the timid symbolization of Oresme. Here for the first time integral powers of an unknown are clearly indicated by exponents. One finds expressions such as .5. and .6.$^2$ and .10.$^3$ to designate what now would appear as $5x$ and $6x^2$ and $10x^3$. In this remarkable work negative integers and zero also are used as exponents, for $9x^0$ is written as .9.$^0$, and one reads correctly that .72.$^1$ divided by .8.$^3$ is .9.$^{2.m}$ (i. e., $72x \div 8x^3 = 9x^{-2}$). Chuquet possessed also a brief notation for roots—such as $\mathbf{R}^2.7.$ for the square root of 7 and $\mathbf{R}^4.10.$ for the fourth root of 10, corresponding to our forms $\sqrt{7}$ and $\sqrt[4]{10}$. The particular forms of such abbreviations are not nearly so significant as is the whole tendency toward symbolic algebra which they represent. This tendency enabled mathematics more easily to transcend the limits of geometric visualization and to use powers beyond the cube. Chuquet himself referred to terms of fourth degree, using a phrase equivalent to "square-square," a terminology reminiscent of that of Diophantus.

It would be of interest to know more about the inspiration for the work of Chuquet—whether it was influenced by Greek sources; whether or not it came, directly or indirectly, from Oresme; and who, if any, were the intermediaries. There is some appearance of Italian influence in the *Triparty*, and it is not impossible that the author knew of Fibonacci's *Liber abaci* of 1202; but Chuquet mentions only two authors, Boethius and Campanus, whose lives are separated by a span of seven hundred years. Further evidence on the forces at work here may some day serve to qualify the general impression that algebra developed primarily from the work of the Hindus and the Arabs. Possibly more credit for originality should be given to the late medieval scholars in Europe.

Unfortunately, the *Triparty* went unpublished for very nearly four hundred years,[1] but part of its substance appeared as early as 1520 (again in 1538) in an *Arismétique* of Estienne de la Roche (born ca. 1480). Chuquet's symbolic methods may well have been known to others also, and so they may have paved the way for the notations of Descartes; but the chief source of European algebra seems to have been Italy rather than France. Here the influence of scholastic philosophers—such as Bradwardine and Oresme—had continued in the universities at Pavia, Bologna, and Padua; but there was also a strong tendency in non-academic circles toward commerical arithmetic, such as had been evident almost three centuries before in the *Liber abaci*. Hindu and Arabic algebra had tended to emphasize applied aspects of the subject at the expense of questions of logical fundamentals; but in Italy at the time of Leonardo da Vinci, this oriental tendency was balanced by two contrary forces—a lingering scholasticism and rising interest in the Greek classics of geometry.[2] Luca Pacioli inherited all

---

[1] It appeared in print for the first time, edited by Aristide Marre, in Boncompagni's *Bullettino di Bibliografia e di Storia delle Scienze Matematiche e Fisiche*, v. XIII (1880), p. 555–659, 693–814; v. XIV (1881), p. 413–460. Cf. Ch. Lambo, "Une algèbre française de 1484. Nicolas Chuquet," *Revue des Questions Scientifiques* (3), v. II (1902), p. 442–472.

[2] See E. W. Strong, *Procedures and Metaphysics. A Study in the philosophy of mathematical-physical science in the sixteenth and seventeenth centuries* (Berkeley, Cal., 1936).

three tendencies, for he contributed to bookkeeping, he wrote on the golden section, and he referred to the work of medieval mathematicians. The algebra in his *Summa de arithmetica* was not greatly advanced over that to be found in the *Liber abaci*, but it shows the same tendency toward symbolism[3] that was so evident in the *Triparty*. Moreover, its influence was widespread, for it linked mathematicians and technicians of the Low Countries with the Latinized learning of Italy. Furthermore, the *Summa* initiated a movement which continued for almost a hundred and fifty years, culminating in the end in Cartesian geometry. One section of the work is devoted to the "method of solving various cases of rectangular quadrilateral figures by the algebraic method." Such applications of algebra to geometry were to become a commonplace in the work of Viète, the greatest mathematician of the sixteenth century. But Pacioli and his early successors were handicapped by the fact that algebra itself was not free of geometry. As in Greek geometric algebra, he constructed equations geometrically, a custom which persisted for another three centuries.

One of the important influences on analytic geometry was a discovery made in Italy which Pacioli had not anticipated. In the *Summa*, echoing the pessimism of Omar Khayyam, Pacioli had compared the impossibility of the algebraic solution of the cubic equation with that of the squaring of the circle. Little did he realize that a few years later the seemingly impossible was to be accomplished by his fellow countrymen. The discovery of the solution of the cubic equation seems to have been due to del Ferro (1465–1526) about 1515, but it was publicized a generation later by the unpleasant controversy between Cardan (1501–1576) and Tartaglia (1506–1557). The solution of the cubic equation was of great importance in the history of mathematics for several reasons, among which was the impetus it gave to the development of algebra in general and the theory of equations in particular. Such a development was indispensable to the rise of analytic methods; but in one sense it served, for just about a century, to direct attention *away* from coordinate geometry. Menaechmus long before had very nearly invented analytic geometry when in solving the Delian problem he gave the first plane geometric solution of a cubic equation. To be sure, he had not thought of this as a problem in determinate equations, for Greek mathematicians sought to circumvent these through geometric devices. Nevertheless, since that time determinate cubic equations and the curves given by indeter-

[3] For the development in algebra at this period see H. G. Zeuthen, "Sur l'origine de l'algèbre," *Det Kongelige Danske Videnskabernes Selskab. Mathematisk-fysiske meddelelser*, v. II, p. 4; Ettore Bortolotti, *Studi e ricerche sulla storia della matematica in Italia nei secoli XVI e XVII* (Bologna, 1928).

minate equations of second degree had been closely connected. The work of Ferro, Cardan, and Tartaglia on the cubic, and of Ferrari (1522–ca. 1560) on the quartic, temporarily broke down this association and caused the study of indeterminate equations in two unknowns to be relegated to the Diophantine theory of numbers. Cubics and quartics now were solvable by calculation instead of by intersecting conics. For about a century the relationship between algebra and geometry was to be simply a pact of mutual assistance in the solution of determinate problems—not an association of curves and indeterminate equations. But the dependence was no longer to be one-sided, as it had been in Greece, for there now had arisen a greater confidence in the operations of algebra, independent of any geometrical significance. As Stevin (1548–1620) was to express it, what can be done in geometry can also be done by arithmetic. The development of operations, notations and concepts in arithmetic and algebra was perhaps the chief contribution of the sixteenth century to the history of analytic geometry. The ancient Greeks had had a sort of algebraic analysis in geometric form in which the solution of determinate equations was avoided through the reduction of problems to questions of finding intersections of known curves. The Arabs had continued this point of view with respect to cubic equations. But the startling success of the early modern period in solving cubics and quartics by algebraic means led to the development of an elementary theory of equations.[4] Cardan was familiar with some of the simple relations between the roots of an equation and its coefficients, but the generalization of these required the formalization of algebraic quantities and the operations performed upon these. The *Algebra* of Bombelli (born ca. 1530) in 1572 and 1579 contributed greatly to this tendency, begun almost a century before by Chuquet, by the systematic use of letters and of abbreviations for operations and relationships. The idea of denoting quantities by letters was certainly not a new one,[5] for it had been found not only among the Hindus but also among the Greeks, at least as far back as Aristotle. However, the application of special signs and abbreviations for operations to literal

---

[4] In Matthiessen, *Grundzüge der antiken und modernen Algebra*, one will find more than 300 pages devoted to the methods used in solving these two equations. This is not a history in the strict sense, but it is a very useful volume of a thousand pages with a good index and copious bibliographical references. See also the valuable work of A. Favaro, "Notizie storico-critiche sulla costruzione delle equazioni," *Memorie della Regia Accademia di Scienze, Lettere ed Arti in Modena*, v. XVIII (1878), p. 127–330. This includes an exceptionally useful bibliography.

[5] J. M. Peirce, "References in Analytic Geometry," *Harvard University [Library] Bulletin*, v. I (1875–1879), p. 157–158, 246–250, 289–290, gives a generally excellent account of Viète's work (and an even better one of the work of Descartes), but repeats the erroneous idea that Viète was first to represent known quantities by letters.

symbols for quantities seems to be due in large part to Bombelli. Moreover, in the case of polynomial expressions he indicated powers in a manner somewhat similar to that of Chuquet; so that $x^2 + 2x - 3$ would be written as $1^2p.2^1m.3$. It will be noted, however, that the letters were simply abbreviations for the words which they replaced. The particular form of his notations[6] is of less significance here than is the idea of symbolic algebra, but both the form and idea seem to have had a wide influence. The notations of Stevin for decimal fractions and polynomials—he would have written $6789$ for $6.789$ and $1⓪ + 2① + 3② + 4③$ for $1+2x+3x^2+4x^3$—in all probability were inspired by those of Bombelli.

The *Algebra* of Bombelli is significant also for the anticipation of some later points of view: the use of algebraic proofs independent of geometric justification; the suggested application of rectangular coordinates in the location of a point in a plane; and the use of an arbitrary unit of length in geometrical constructions. But these ideas were largely overlooked by his successors. However, the work of Cardan and Bombelli on imaginary numbers was of immediate significance in that it showed the necessity of a serious consideration of these in real situations, even though geometric interpretation was lacking. One could dismiss the imaginary roots of a real quadratic equation with the simple statement that the equation is impossible of solution; but the role of imaginaries is quite different in connection with the cubic, for here, in the so-called "irreducible case," they lead in the end to real roots.

In a manuscript which did not form part of his *Algebra* and which was never published, Bombelli studied the constructions or graphical solutions of determinate problems in a manner somewhat analogous to that which Descartes made the core of his *Géométrie*.[7] About the same time, and at least by 1587, Paolo Bonasoni composed a somewhat similar work with the title *Algebra geometrica*. In it he sought to give a logical basis for algebra by basing it upon geometry, an idea going back to Fibonacci. All problems reducible to equations of second degree, Bonasoni showed, can be constructed with ruler and compasses. For such problems he gave various graphic constructions, including some by the application of areas. He did not use symbols of operations or Bombelli's exponential notation, but he did use letters to

---

[6] The best account of the rise of notations is that by Florian Cajori, *A History of Mathematical Notations* (2 vols., Chicago, 1928–1929).

[7] See Bortolotti, *Lezioni di geometria analitica*, v. I, p. xxxv.

represent both given and unknown quantities.[8] This represents an important anticipation of the notation of Viète, but unfortunately Bonasoni's work was never published and hence its influence is questionable.

The work of the Italian algebraists[9] encouraged the study of whole classes of equations, but there was at the time no satisfacory notation (with the possible exception of that of Bonasoni) for what would now be called a parameter. Quantities were either known numbers—in which case the Hindu-Arabic form would be available—or unknown numbers—in which case suitable abbreviations were invented. The problems which arose generally were particular instances leading to equations with specific numerical coefficients. In this respect the algebraists of Italy did not differ essentially from the arithmetical "cossists" of Germany. Cases of polynomials and of polynomial equations were familiar, but the notion of a *polynomial as such* seems not to have arisen. Perhaps further historical studies, especially of the German cossic works, will clarify this situation, but at the present time the credit for introducing the idea of a parameter seems to go largely to Viète (1540–1603). The importance of his contribution has been widely recognized and is perhaps not exaggerated in the statement of E. T. Bell that Viète was "the first mathematician of his age to think occasionally as mathematicians habitually think today."[10] His achievement was not so much a contribution to notation as it was to algebraic ideas. Algebra before his time was in general concerned with particular numerical equations, such as the cubic "cubus p. 6 rebus aequalis 20"—i.e., $x^3 + 6x = 20$—which Cardan gave. Viète, on the other hand, in the *De recognitione aequationum* studied the properties of equations of the form $A$ cub. $- B$ planum in $A$ aequatur $B$ plano in $Z$—i.e., $x^3 - b^2x = b^2c$. Using vowels to designate unknown quantities and consonants to represent quantities assumed to be known, Viète made it possible to distinguish not two, but three, types of magnitudes in algebra—specifically given numbers, parameters, and variables. Viète did not himself speak of parameters or variables, but his work prepared the way for these ideas. He was not the

---

[8] See Ettore Bortolotti, "Primordi della geometria analitica: l'algebra geometrica di Paolo Bonasoni," *Rend. delle Sessioni della R. Accademia delle Scienze dell'Istituto di Bologna* 1924–1925, *Classe di Scienze Fisiche, Sezione di Scienze Fisiche e Matematiche*. This is reprinted in his *Studi e ricerche sulla storia della matematica in Italia nei secoli XVI e XVII* (Bologna, 1928).

[9] See Ettore Bortolotti, "L'algebra nella storia e nella preistoria della scienza," *Osiris*, v. I (1936), p. 184–230.

[10] *The Development of Mathematics* (New York, 1940), p. 99.

first to use symbols in equations, for germs of a literal algebra are found in Bombelli[11]; but he seems to have originated the practice of using letters as coefficients of terms in an equation—i.e., of considering "affected" equations. It became possible to build up a general theory of equations—to study, not cubic equations, but *the* cubic equation. Viète realized the significance of this point of view,[12] for he contrasts the ordinary *logistica numerosa* with his "new" *logistica speciosa*. The former applied to calculations on numbers; the latter was concerned with "species" or "the forms of things," and this latter was made possible, he held, through his "alphabetic elements." These "things" might be incommensurable geometrical elements, the relationships among which are not expressible in terms of whole numbers. Here Viète came close to the idea of a real algebraic variable, one of the most important in the evolution of mathematics in general and of analytic geometry in particular. The invention of such variables is, in fact, frequently ascribed to him. Nevertheless it must be recalled that, on the one hand, there had been geometrical anticipations of the idea, notably in the medieval latitude of forms; and, on the other hand, Viète's vowels were not, strictly speaking, variables in the sense of symbols representing any of a whole class of values. The vowel-vs.-consonant notation, as applied to determinate equations, was not so much a distinction between *variable* and *fixed* magnitudes as it was between those constants which are taken to be *unknown* and those assumed to be *known*. It was only when such conventional notations were applied later to graphical representations of indeterminate equations that vowels came to be looked upon as variables rather than as fixed unknowns. But the transition from the one point of view to the other was a natural outgrowth of Viète's literal notation, and it was this transition which marked the beginning of analytic geometry in the strict sense of the word. As L. C. Karpinski has well said, it was Viète's algebraic literal notation which "gave a tongue to the analytical geometry of Descartes."[13]

Viète restricted himself to equations in a single unknown, and largely

[11] See Ettore Bortolotti, *L'algebra, opera di Rafael Bombelli da Bologna* (Bologna, 1929). I have not seen this book, but cite it on the basis of a review in *Scripta Mathematica*, v. IV (1936), p. 166–169.

[12] The importance of this viewpoint seems to have been realized by at least one contemporary, Adriaen Roomen (or Adrianus Romanus, 1561–1615), who in 1598 claimed as his own the distinction between an equation "numerosa" and an equation "figurata." Viète, however, seems to have anticipated him in this. See H. Bosmans, "Le fragment du Commentaire d'Adrien Romain sur l'algèbre de Mahumed ben Musa El-chowârezmi," *Annales de la Société Scientifique de Bruxelles*, v. XXX (1906), part 2, p. 266.

[13] See "The Origin of the Mathematics as Taught to Freshmen," SCRIPTA MATHEMATICA, v. VI (1939), p. 133–140. On this point consult also his paper, "Is There Progress in Mathematical Discovery and Did the Greeks Have Analytic Geometry?" *Isis*, v. XXVII (1937), p. 46–52.

for this reason he failed to invent analytic geometry; but he played a preparatory role in this direction above and beyond that of developing algebraic ideas. He was one of those who systematically applied algebra to the solution of geometric problems. In fact, his vowels and consonants generally referred to geometrical magnitudes, as is implied by the names by which they were designated. His distinction between parameters and unknowns is brought out in this terminology, as well as in the vowel-vs.-consonant convention: the first nine powers of a given constant quantity are known, respectively, as longitudo or latitudo (recalling the work of Oresme), planum, solidum, plano-planum, . . .solido-solido-solidum; the corresponding powers of an unknown magnitude are designated, respectively, by latus or radix, quadratum, cubus, quadrato-quadratum, . . .cubo-cubo-cubus.[14] It will be noted that Viète in his algebra did not hesitate to go beyond the third dimension, even though he continued to use a geometric nomenclature, reminiscent of that of Diophantus and Chuquet, for powers of quantities.

In the terminology of Viète for known and unknown quantities one sees a close connection between algebraic operations and geometric visualization, but this relationship in no way represented an anticipation of Cartesian geometry. As a matter of fact, it in one respect served to obscure the path toward the use of coordinates, for it encouraged the tendency (seen long before in Pappus) to visualize cubic equations in terms of stereometric representations, rather than graphically in two dimensions, as is required in algebraic geometry. That is, it would have been better to regard $A^3$ or $B^3$ as numerical magnitudes— or, better still, as linear quantities—rather than as geometrical cubes, for then the association of these quantities with lines on a coordinate diagram would have been facilitated. This fact shows how erroneous it is to speak of analytic geometry as nothing more than a combination of algebra and geometry. The association of geometry and algebra in the sense of Viète led inevitably to the notion that all equations must be homogeneous in terms of the variables and coefficients. This meant that the constants or parameters in a given expression, as well as the unknown magnitudes, possessed geometric dimensionality. The equation $x^2 + bx = c^2$, for example, would be interpreted as a proportion between the lines $x:c = c:(x+b)$. This idea of ratios between homogeneous magnitudes did not preclude the later development of analytic geometry, as the work of Fermat will testify; but it did misdirect attention, as Descartes to some extent realized.

It was not so much the geometric terminology in the algebra of

[14] See Viète, *Opera mathematica* (ed. by van Schooten, Lugduni batavorum, 1646).

Viète which prepared the way for Descartes: there was another aspect, not far removed, which formed a more direct link. This is the systematic application of syncopated algebra to the solution of geometric problems. The sixteenth century witnessed a rising interest in the classical problems of antiquity at a time when simplifications in arithmetic processes and advances in algebra were striking. The result was that men sought a royal road to geometry through numerical techniques. Viète was neither alone nor first in this respect, but he carried the idea further than did his predecessors. Vowels were substituted for unknown geometric lines, the construction of which was called for, and consonants for known lines; and these letters were then subjected to the appropriate algebraic operations. This was not the earliest use of unknowns in geometry, for ancient mathematics consisted largely of a rhetorical geometrical algebra of unknown magnitudes. The novelty lay rather in the application to geometrical problems of literal symbolism *followed by* mechanical methods of calculation—the translation of a problem from the field of geometry to that of algebra. This prepared the way for the free manipulation and simplification of the relevant algebraic expressions according to algorithmic rules. Where the geometrical problem was determinate, the result of the simplification invariably was an (irreducible) algebraic equation in one unknown, the roots of which gave the possible magnitudes of the original unknown lines.

An important part of Viète's work on the application of algebra to geometry, *Ad logisticam speciosam*, has been lost; but a simple example from the *Zeteticorum libri quinque* may serve to illustrate his general method of procedure: Given the area of a rectangle and the ratio of its sides, to find the sides of the rectangle. Viète takes the area as $B$ planum and the ratio of the sides as $S$ to $R$. Let the larger side be $A$. Then $S$ is to $R$ as $A$ is to $\dfrac{R \text{ times } A}{S}$. Therefore the smaller side will be $\dfrac{R \text{ times } A}{S}$. Hence $B$ planum is equal to $\dfrac{R \text{ times } A \text{ squared}}{S}$. Multiplying by $S$ one obtains the final equation, $R$ times $A$ squared equals $S$ times $B$. In this form the geometric construction of $A$ is easily indicated. Viète repeats this process to find the smaller unknown side $E$.[15]

The example above illustrates well the literal symbolism of Viète as applied to *quantities*, and its application to a simple geometric problem. It shows also the great need for further abbreviation through the application to the vowels and consonants of symbols for arithmetic

[15] *Opera mathematica*, p. 50.

*operations* and *relationships*. Viète did adopt the symbols + and −, but he went little beyond this. The change from syncopated to symbolic algebra and from the manipulation of proportions to the use of equation-forms took place largely in the interval between Viète and Descartes. Another example, chosen from a French edition (1630) of Viète's *Zetetic*, shows how gradual the transition was: to cut a given line into two parts so that a given proportion of the first part combined with another given proportion of the second part shall make a given sum. Let the line be $B$ and let the required proportion of the first part be $D$ to $B$ and let the given proportion of the second part be $F$ to $B$; and let the required sum be $H$. Let the desired portion of the first part of the line $B$ be $A$. Then the desired portion of the second part will be $H - A$. Then the first part will be $\dfrac{BA}{D}$ and the second part will be $\dfrac{BH - BA}{F}$ and the sum of the two parts, $\dfrac{BA}{D} + \dfrac{BH - BA}{F}$, will be equal to the whole line $B$. Multiplying this equality by $DF$, the result is $FBA + DBH - DBA$ equal to $BDF$. Dividing by $B$, one obtains $FA + DH - DA$ equal to $DF$. Transposing (assuming $D$ greater than $F$), we get $DH - FD$ equal to $DA - FA$. Dividing by $D - F$ results in $\dfrac{DH - FD}{D - F}$ equal to $A$. Then $D - F$ is to $H - F$ as $D$ is to $A$, by which proportionality $A$ is determined and can be constructed.[16]

Geometrical problems generally called for the construction of certain lines. In such cases it was necessary to show how to construct geometrically the roots of the resulting algebraic equation. The Euclidean geometrical algebra was adequate for equations of first and second degree, but it failed for those of higher degree. The early modern algebraic solutions of the cubic and quartic had not resolved the classical problems of antiquity inasmuch as such methods do not provide for the geometric constructibility of the roots. Through the use of intersecting conics Menaechmus, Archimedes, and Omar Khayyam had associated problems of construction with the study of loci, and a continuation in this direction would almost inevitably have led Viète to coordinate geometry, just as it did Descartes half a century later. As it was, Viète anticipated Descartes only to the extent of showing that algebra could be used to bring about some order in questions of constructibility, and that, conversely, the algebraic solution of deter-

[16] See *Les cinq livres des zetetiques de François Viète* (transl. by Vaulezard, Paris, 1630), p. 37–38.

minate equations could be given geometrical meaning by constructing the roots.

Analytic geometry frequently is described as the subject in which algebra is applied to geometry, and vice versa.   The inadequacy of such a statement becomes apparent with the realization that Viète's work clearly satisfies this description and yet it is not analytic geometry.   It is not this for a very simple reason—it is not coordinate geometry.   It already has been pointed out that Viète's application of algebra to geometry did not include locus problems.   Conversely, his application of geometry to algebra did not take the form of the graphical representation of equations or functions by plotting on a coordinate system.   It was, rather, a survival of the classical Greek geometrization of algebra, and the pattern here set was continued not only by Viète's immediate successors, but by geometers in general for over a century after the invention of analytic geometry.   In a canonical review of geometric constructions of algebraic operations (in the *Supplementum geometriae* of 1593) Viète pointed out that the representation of the roots of an irreducible cubic or biquadratic equation is equivalent either to the trisection of the angle or to the duplication of the cube, thus giving these problems a wider significance than they had had previously.   To solve these problems Viète proposed extending the Euclidean postulates to include constructions by instruments similar to the ancient mesolabe of Eratosthenes.   The work of Descartes was primarily an effort to extend such systematization to equations of higher degree, where he, too, suggested a bold liberalization of the usual postulates.   Descartes realized the power of Viète's analytic art as an algebraic tool, as contrasted with its limitations on the geometric side.   In seeking to remedy the latter weakness, Descartes was forced to seek new curves to effect the constructions, and it was for this reason that he, rather than Viète, invented analytic geometry.

It has been noted that in the work of Viète there is no analytic geometry properly so-called.   It is an algebraic geometry in the sense that algebra serves as a handmaiden to geometry, but it is not coordinate geometry.[17]   Menaechmus and Apollonius had used the equivalent of a coordinate system, but lacked symbolic algebra; Viète possessed the latter but overlooked the former.   Probably the explanation of Viète's oversight is to be found in the preponderant interest of his age in determinate geometrical problems.   In such cases the use of a coordinate frame may offer little advantage.   Prob-

[17] The claim that Viète anticipated analytic geometry by giving the relationship between the coordinates of points on a line has been refuted by G. Eneström, "Auf welche Weise hat Viète die analytische Geometrie vorbereitet?" *Bibliotheca Mathematica* (3), v. XIV (1914), p. 354.

lems involving loci, on the other hand, imply or invite the use of axes of reference. When expressed in algebraic symbolism, however, problems of the latter type generally involve an equation in two unknowns, one of which is thought of as a function of the other. Such functional relationships were not adequately considered at that time, for the medieval graphical representation of "forms" or functions had not been associated with algebraic geometry.

There is in the work of Viète a further contribution to the history of analytic geometry which hinges on his use of the word "analysis." In a broad sense he defined analysis as *doctrina bene inveniendi in mathematica*. More specifically, Viète regarded the "analytic art" as consisting of three parts: *zetetic*, or the determination of the properties of things required from things which are given; *poristic*, or verification; and *exegetic*, or demonstration of the proposition. Here one sees a new application of the term "analysis." As used by Plato and Pappus, the word had reference primarily to the order of ideas in a demonstration. Analysis was the path of investigation, synthesis was that of exposition. Viète, on the other hand, applied the word especially to his algebraic geometry, which he regarded as a new form of mathematical analysis.[18] His use of the term to some extent overlaps with the old, for Viète remarked that zetetic, or the algebraic attack, generally proceeds indirectly from the assumption of what is to be proved or constructed, unknown quantities being operated upon as though they were known. That is, algebra seemed to be the instrument appropriate to the analytic path in geometry. But his emphasis is more upon the use of *logistica speciosa* than upon the order of demonstration. Following this lead, his successors lost sight more and more of the Platonic meaning and came to look upon analysis as synonymous with the use of symbolic techniques, or even with algebra itself.[19]

Considerable emphasis has been placed above upon the contribution of Viète because he served as the chief connecting link between the earlier periods and the invention of coordinate geometry. In Chasles' division of the history of geometry into five epochs, Viète is the figure marking the transition from the first epoch to the second. Nevertheless, it must be borne in mind that he was by no means an isolated figure. Viète probably realized more fully than his contemporaries the fundamental character of his analytic art; but he had numerous rivals and successors, both in the application of algebra to geometry

[18] See *Introduction en l'art analytic, ou nouvelle algebre de François Viète* (transl. by Vaulezard, Paris, 1630). Cf. also *Algebre de Viète* by James Hume (Paris, 1636). See also French translations of parts of Viète's work in *Bullettino di Bibliografia e di Storia delle Scienze Matematiche e Fisice*, v. I (1868), p. 223–276.

[19] For example, Raimarus Ursus (or Reymers) in 1601 used it in this sense in his *Arithmetica analytica vulgo cosa*.

and in the geometric interpretation of the solution of algebraic equations. Moreover, only a little of his work was published during his lifetime, and then generally in small editions for private distribution, circumstances which served to limit his direct influence. Fortunately, he had a distinguished pupil, Marino Ghetaldi (1566–1627), who fell heir to some of his master's manuscript material and who continued Viète's interest in algebraic geometry.

Ghetaldi studied with Viète at Paris, and there he took an active part in the movement of the seventeenth century to restore the lost works of Apollonius. During the medieval period classical geometry had been largely forgotten, but during the sixteenth century there appeared numerous editions of the extant works of The Great Geometer. Following these, there was a flurry of attempts at the restitution of parts of the lost treatises. This trend was signalized at Paris by the *Apollonius gallus* of Viète in 1600, at Leyden by the *Apollonius batavus* of Snell (1581–1626) in 1607 and 1608, and at Venice by the *Apollonius redivivus* of Ghetaldi in 1607 and 1613. The frequency of appearance[20] of such works shows the remarkable resurgence of interest in geometry during the early seventeenth century. Throughout the preceding century, work in elementary geometry had been divided largely into two branches, one centering about Euclid's *Elements* and the other emphasizing practical geometry.[21] The treatise on conics by Werner (1468–1528) in 1522 had no direct line of descendants and interest in the curves was found mainly on the higher level, especially in connection with the treatises of Archimedes. In general such works did not make use of the newer algebraic or analytic point of view.

Toward the close of the sixteenth century the gap between practical and theoretical geometry narrowed, as did that between the elementary and higher branches. Among the distinctive features at the time was a greatly expanded use of symbols, paralleling the coordination of algebra and geometry. The Euclidean application of areas continued to be emphasized, for it served, even in the early seventeenth century, as the basis for the geometric construction of the roots of quadratic equations. The construction or geometrical solution of problems of arithmetic and algebra—such as is found, for example, in the *Di-*

[20] There was also an edition at Paris in 1612 of Ghetaldi's work, as well as a *Supplementum Apollonii redivivi* by Alexander Anderson, also at Paris in 1612. The trend declined somewhat after this, but restitutions continued to appear for almost two centuries—one by van Schooten at Leyden in 1656–1657, another by Halley at Oxford in 1706, and one by Simson at Glasgow in 1749, as well as further editions by Viète at London in 1771 and Gotha in 1795 and by Snell at London in 1772.

[21] See F. W. Kokomoor, "The Teaching of Elementary Geometry in the Seventeenth Century," *Isis*, v. X (1928), p. 21–32.

*versarum speculationum* of Benedetti (1530–1590) in 1585, or the *Algebra discorsiva numerale et bineare* of Cataldi (1548–1626) in 1618— has been hailed by Libri[22] as an anticipation of analytic geometry. Such claims are completely unwarranted.[23] The "construction of equations" had formed a traditional part of the ancient geometrical algebra, so that the innovation in this connection by Benedetti, Cataldi, and other contemporaries of Viète, was largely a carrying over of the Euclidean tradition from geometric rhetorical equations to algebraic symbolic equations. In a somewhat different sense, Cardan and Tartaglia also had applied geometric ideas in the solution of equations. However, there was in all such work no reference to the essential element of analytic geometry: the association of loci, given algebraically, with a coordinate system. This was missing in Benedetti and Cataldi, and it was overlooked also by Viète and by his disciple, Ghetaldi.[24]

Following the lead of Viète, and in line with the tendencies of the time, Ghetaldi made the reduction of determinate geometric problems to algebra a systematic device, devoting to this theme a book of his *Apollonius redivivus*. Conversely, he gave geometric proofs of such algebraic rules as $a^2 - b^2 = (a+b)(a - b)$, and he constructed geometrically the roots of determinate algebraic equations. In 1630 there was published posthumously his *De resolutione et compositione mathematica*, a work in which this topic is so extensively treated that it has been referred to as the first textbook on algebraic geometry. However, the material treated and the point of view do not differ greatly from what had appeared before. Quite incidentally, nevertheless, Ghetaldi came somewhat closer to hitting upon analytic geometry, for he considered algebraically several *indeterminate* geometrical problems. One of these, for example, called for the construction of a triangle with a given base and such that the difference of the other two sides is half the base. Calling the base $2B$ and letting $A$ be the unknown difference of the segments into which the altitude divides the base, Ghetaldi arrived at an "aequatio inutilis," $A^2 - B^2 = A^2 - B^2$. From this identity he correctly concluded that the number of triangles satisfying

[22] G. Libri, *Histoire des sciences mathématiques en Italie depuis la renaissance des lettres jusqu'a la fin du dix-septième siècle* (4 vols., Paris, 1838–1841), v. III, p. 124, and note XXVII, v. IV, p. 95.

[23] See, for example, Gino Loria's magnificent work, "Da Descartes e Fermat a Monge e Lagrange. Contributo alla storia della geometria analitica," *Atti della Reale Accademia Nazionale dei Lincei, Classe di scienze fisiche, matematiche e naturali, Memorie*, series 5, v. XIV (1923), p. 777–845.

[24] For a brief biography see M. Saltykow, "Souvenirs concernant le géomètre Yougoslave Marinus Ghetaldi," *Isis*, v. XXIX (1938), p. 20–23. For full analysis of his contribution to algebraic geometry, see Gelcich, *op. cit.*, and Wieleitner, "Marino Ghetaldi und die Anfänge der Koordinatengeometrie," *Bibliotheca Mathematica* (3), v. XIII (1912–1913), p. 242–247.

68        HISTORY OF ANALYTIC GEOMETRY

the given conditions is infinite; but unfortunately he did not notice
that the third vertex in this case traces out an hyperbola.   He included the problem among those which are "vana seu nugatoria."[25]
Like his illustrious predecessor, he shied away from the algebraic
treatment of problems on curves and loci.   The invention of analytic
geometry was made a generation later by two men who discovered that
indeterminate equations are far from "fruitless and futile," for, when
applied to curves such as the conics, they served as a more effective
bridge between algebra and geometry than the algebraic geometry
of Viète, Cataldi, Benedetti, and Ghetaldi.

Original contributions to the theory of conics in the early modern
period were primarily synthetic or kinematic in nature.   Werner, in
spite of his general return to the stereometric emphasis in connection
with the conics, nevertheless gave a variation on the planimetric
construction of the parabola by means of circles and lines,[26] and Guidubaldo del Monte (*Planisphaeriorum universalis theorica*, 1579)
supplemented Anthemius' string (or gardner's) construction for the
ellipse with the corresponding construction of the hyperbola.   Kepler
(1571–1630) in 1604 gave the usual string constructions for all
three conic sections (that for the parabola probably having been
known at least 1000 years before to Isidore of Miletus).   He also
envisioned the curves as constituting a single family: from a pair of
intersecting lines one passes through an infinite number of hyperbolas
to the parabola and thence through an infinity of ellipses to the circle.
Of all hyperbolas the most obtuse is the line-pair, the most acute the
parabola; of all ellipses the most acute is the parabola, the most
obtuse the circle.   It is surprising to note that the unification of the
conics, through Kepler's "law of continuity," should have occurred in
synthetic geometry, rather than in analytic geometry in which general
cases are the rule rather than the exception.   The parabola, Kepler
said, has a "blind" focus at infinity.   Such ideas, however, led toward
the projective geometry of Desargues and Pascal, rather than toward
the study of loci by Fermat and Descartes.[27]   Kepler (to whom we owe
the term "eccentricity" of a conic) applied the ellipse to celestial
motions (1609) and Galileo (1638) gave the application of the parabola
to terrestrial trajectories, but here too the association of conics with
practical problems probably had little influence on the origins of

[25] See Moritz Cantor, *Vorlesungen über Geschichte der Mathematik* (4 vols., Leipzig, 1880–1908), v. II, p. 737–740; or A. G. Kaestner, *Geschichte der Mathematik* (4 vols., Gottingen, 1796–1800), v. III, p. 188–195.
[26] See Coolidge, *History of the Conic Sections*, p. 26–27.
[27] *Ad Vitellionem paralipomena;* or see *Opera* (ed. by Frisch), v. II, p. 187–188.   For an excellent account of this work on conics see C. Taylor, "The Geometry of Kepler and Newton," *Cambridge Philosophical Society Transactions*, v. XVIII (1900), p. 197–219.

coordinate geometry. Kepler's use of coordinate devices was similar to that of the Greeks, based upon special lines in a given construction rather than upon general auxiliary lines. In fact, the discoveries of new properties of the conic sections during the early seventeenth century were themselves largely unrelated to algebraic geometry, for the synthetic and analytic streams of thought flowed on more or less independently of each other. Mydorge (1585–1647) in 1631 published his *Prodromus catoptricorum et dioptricorum sive conicorum,* a work which combined the ancient classical tradition of antiquity with the more recent interest in mechanical constructions and uses. (He had planned to add several books on the conics as applied to physics, especially in the reflection and refraction of light.) Yet in spite of the fact that he was a friend of Descartes and that his work appeared subsequently in several editions (1639, 1641, 1660), there is little in common between coordinate geometry and the *Prodromus* of Mydorge.[28]

The immediate path to the Cartesian method seems to have been prepared more by developments in algebra than by those in geometry. In the years 1629 and 1631 (i. e., virtually simultaneously with Ghetaldi's *De resolutione* and Mydorge's *Prodromus*) there appeared several significant works[29] in this direction—the *Invention nouvelle en l'algebre* of Girard (1595–1632); the *Artis analyticae praxis* of Harriot (1560–1621); and the *Clavis mathematicae* of Oughtred (1574–1660). All three of these books placed great emphasis upon abbreviations and symbols in algebra. The importance of the vowel-consonant convention of Viète should not obscure the weakness in his notations for operations and relationships. It was just here that the triad of works mentioned above made significant advances. The *Invention nouvelle* popularized the index notation for powers, which had come down from Chuquet and Bombelli through Stevin. Thus he wrote as ③ esgale á − 6① + 20 what would now appear as $x^3 = -6x+20$. The Viète geometric terminology has here completely disappeared, but one sees the continued lack of a symbol for equality. One striking aspect of Girard's work is the free use of negative quantities in equations and in their solution. For example, he seems to have been the first one to solve a quadratic equation with two negative roots. This was important in his anticipation of the fundamental theorem of algebra, but

[28] It may be noted incidently that Mydorge was among the first to use the word "parameter" for the quantity $2b^2/a$ which designates the length of the latus rectum of the ellipse and hyperbola.

[29] One might add also the *Nova geometriae clavis algebra* of Jacques de Billy a few years later (Parisiis, 1643). This work, non-analytic in character (even though published half a dozen years after Descartes' *Géométrie*), shows clearly that the application of algebra to geometry does not in itself constitute analytic geometry.

it was also of significance in analytic geometry, for Girard was possibly the first person to point out the geometric, as well as the algebraic, usefulness of negatives.  He said, "The negative solution is explained in geometry by a retrogression to the less advanced, while the positive advances,"[30] an idea which seems to have been adumbrated by the ancient Babylonians.  The importance of signed segments in analytic geometry was somewhat dimly recognized by Fermat and Descartes, but the idea was developed more particularly during the later seventeenth century.

Harriot's work, published posthumously in 1631, was thought out largely before 1604, not long after that of Viète, and so it did not include a recognition of negative roots.  The *Artis analyticae praxis* is significant mainly as a continuation in modified form of the Viète notations, of his theory of equations, and of the emphasis on the analytic or algebraic attack on geometrical problems.[31]  The change from the capital letters of Viète to the small vowels and consonants of Harriot was of minor importance, but the substitution of *aaaa*, for example, for the *A quad. quad.* of Viète was an advance along the lines of Girard.  Such a form as $aaa - 3bba = 2ccc$ popularized the symbol of Recorde for equality and also brought the literal calculus close to the Cartesian notation.  In fact, it was but a short step from the *aaa* of Harriot through the *a3* of Hérigone (in the *Cursus mathematicus* of 1634) to the $a^3$ of Descartes.

Girard and Harriot form but two of the links from Viète to Descartes. A third, and possibly the most influential, link is found in the *Clavis mathematicae* of Oughtred.[32]  The inspiration for the *Clavis* clearly came from Viète, as one sees from the full title: *Arithmeticae in numeris et speciebus institutio: quae tum logisticae, tum analyticae, atque adeo totius mathematicae, quasi clavis est.*  In it one finds the same tendency toward symbolism which was evident in Girard and Harriot; and as in Girard's work, minus signs are used both as symbols

[30] Albert Girard, *Invention nouvelle en l'algebre* (Amsterdam, 1629), 4th page from the end of the section on algebra.

[31] F. V. Morley, in an article on "Thomas Hariot," *Scientific Monthly*, v. XIV (1922), p. 60–66, made the unfortunate statement that in this work "There is a well-formed analytical geometry, with rectangular coordinates and a recognition of the equivalence of equations and curves."  Florian Cajori, in "A Revaluation of Harriot's Artis Analyticae Praxis," *Isis*, v. XI (1928), p. 316–324, showed that this is not the case.  Cajori's conclusion has been confirmed by D. E. Smith, *History of Mathematics*, v. II, p. 322.  See also J. L. Coolidge, *A History of Geometrical Methods* (Oxford, 1940), p. 118–119.

[32] See Henri Bosmans, "La première edition de la *Clavis mathematicae* d'Oughtred, son influence sur la 'Géométrie' de Descartes," *Annales de la Société Scientifique* de Bruxelles, v. XXXV (1910–1911), p. 24–78.  See also Florian Cajori, *William Oughtred. A great Seventeenth-Century Teacher of Mathematics* (Chicago and London, 1916).  An English translation by Oughtred of the first two books of the *Conics* of Mydorge is found in Jonas Moore, *Arithmetick in Two Books* (London, 1660).

of operation and as qualities of numbers. Few of the many new signs and abbreviations which Oughtred used have survived, the symbol x for multiplication (which he may have borrowed in modified form from Recorde, along with the symbol for equality) being an important exception. Even his abbreviations $Aq$ and $Ac$ for the second and third powers of the unknown (where Viète would have written $A$ *quadr.* and $A$ *cubus*) were replaced a few years later by the exponential notation; but Oughtred's contribution to the symbolic movement was important for its emphasis. He put great stress on the "analytical art." By this he meant essentially the same thing that Viète had had in mind. The arithmetic of numbers is contrasted with the "much more convenient" *arithmetica speciosa*, "in which by taking the thing sought as knowne, we find out that we seeke." That is, the analytic art is both a notation and an order of presentation. On the one hand, the "specious and symbolicall manner" of *analitice* is contrasted with the "verbous expressions" of the "usual synthetical manner." On the other hand, it is an "inventive way" in which "by framing like questions problematically, and in a way of Analysis, as if they were already done, resolving them into their principles, I sought out reasons and means whereby they might be effected."

Oughtred's key to mathematics involves three parts: arithmetic calculation, symbolic algebraic calculation, and applications of algebra to geometry. This is essentially the subject-matter of Viète and Ghetaldi, from whom Oughtred undoubtedly borrowed. His algebra is more formal and further removed from dependence upon geometry than that of his predecessors, but it contained the usual construction of algebraic formulas by ruler and compasses. This continued to be the chief link between algebra and geometry, and it was, in fact, destined to be the aim of the opening book of the geometry of Descartes.

The *Clavis mathematicae* went through five Latin and two English editions in the seventeenth century. It was the most influential mathematical work in Great Britain between the time of Napier and that of Wallis. Yet analytic geometry originated in France rather than England. In fact, the new geometry had been twice invented— although not published—before the appearance of the *Clavis*. The extent of the influence of Girard, Harriot, and Oughtred upon French mathematics is not definitely known, but it is clear that there was something missing in the works of these precursors of analytic geometry. As in the case of Viète and Ghetaldi, the lacuna was a consideration of locus problems. Girard tried his hand at a reconstruction of the *Porisms* of Euclid, yet he overlooked the opportunity this might have afforded of applying algebra to geometry. The ancient and medieval periods had failed to invent coordinate geometry because they lacked an algebra in which to express locus problems and the graphical representations of the latitude of forms; the early modern applications of algebra to geometry fell short of the invention because they failed to include an algebraic study of loci and of functional variability. It was probably no accident that the men who first framed analytic geometry were also the men who, independently and within a period of a dozen years, invented more new curves (or loci) than had been discovered in the whole history of mathematics up to that time.

During the early modern centuries mathematical activity had been devoted in large measure to improvement in arithmetic and algebraic technique and to the recovery of the geometry of the ancients. Here and there were to be found some new developments in the theory of curves, but the straight line and the circle continued to play the fundamental role in science as well as in geometry. Copernicus (1473–1543), for example, seems to have felt that the Ptolemaic astronomy was physically impossible because it could not be reconciled with the principle of uniform *circular* motion. However, it is reported that the imaginative Nicholas of Cusa (1401–1464) had noted the curve traced out by a point on the rim of a cart wheel as the wheel rolled along the road. Although he seems to have been unable to determine its nature or properties, this observation constituted a significant step in the study of curves, for it seems to represent the first modern instance in which a new curve was suggested by natural phenomena. The ancients had invented new curves *ad hoc* to solve specific geometrical problems; they had not discovered these, except for the line and the circle, in the world of nature. The new curve of Cusanus was followed two centuries later by other curves which were disclosed by, and useful in, the study of physical science.

During the early sixteenth century there were a number of contributions to curve theory. The study of conic sections, revived especially by the work of Werner, has ever since occupied a prominent place in mathematics and science. At about the same time, Dürer (1471–1528) made significant original additions to the theory of higher curves. He introduced the idea of an asymptotic point and illustrated it by a curve strongly resembling the logarithmic spiral. This curve, later made famous by Jacques Bernoulli, may have been suggested by the revived interest at the time in map construction; it is the plane stereographic projection of the loxodrome on the sphere, and the latter was studied in 1530 by Núñez (1502–1578). Dürer revived also the ancient kinematic definition of curves, and gave as examples an epicycloid and a new conchoid. Copernicus and Cardan similarly noticed the locus (a straight line) generated by a circle rolling on the inside of another circle with a radius twice as great—a result which had been known earlier to the Muslim Nasir Eddin. Copernicus knew also that an ellipse is generated by a point rotating in an epicycle the center of which moves along a deferent with equal angular speed in the opposite sense.[33] However, such work on curves is typical of the time in that it is casual and not systematically developed. Bovelles, for example, noted the cycloid near the beginning of the century, and Galileo referred to it again toward the end, but neither of these men made any headway in determining its equation or properties.

During the first third of the seventeenth century the study of geometry beyond the elements centered about the conics, for the number of known curves was little greater than it had been two thousand years before. In the decade from 1634 to 1644, however, the situation changed completely. This was the result of the development both of the latent possibilities in methods of curve definition previously adopted and also of new principles which were devised. The cycloid had been noted several times before, but when Mersenne (1588–1648) in 1634 and Galileo (1564–1642) in 1639 again suggested it as a curve worthy of study, its shape and properties were promptly determined through the composition of motions. This ancient method was supplemented, however, by a powerful new approach—the use of analytic geometry.

[33] For bibliographical references and further details, see my note in *Isis*, v. XXXVIII (1947), p. 54–56.

# CHAPTER V

# Fermat and Descartes

*Mathematics is the alphabet in which God wrote the world.*
—BOYLE

ANALYTIC geometry was the independent invention of two men, neither one of whom was a professional mathematician. Pierre de Fermat (*ca.* 1608–1665) was a lawyer with a deep interest in the geometrical works of classical antiquity. René Descartes (1596–1650) was a philosopher who found in mathematics a basis for rational thought. Both men began where Viète had left off, but they continued in somewhat different directions. Fermat retained the notation of Viète, but applied it in a new connection, the study of loci. Descartes adopted the aim of Viète—the geometrical construction of the roots of algebraic equations—but continued it in conjunction with modern algebraic symbolism. The two paths led to the same fundamental principle, but there continued to be a divergence in emphasis, especially during the period in which efforts were made to recover as much of Greek geometry as possible. Most of the work of Apollonius, excepting the first seven books of the *Conics*, had perished; but Viète, Snell, and Ghetaldi had joined in the effort to reconstruct some of the lost treatises on the basis of information supplied by Pappus and other commentators. Fermat, too, was fascinated by such attempts and composed a restitution of the two books of Apollonius on *Plane Loci*. This led him to the Apollonian problem of circles tangent to three circles, and he generalized this in a work on spheres tangent to four spheres. These early works he wrote in the classical style, with no reference to Viète's analytic art. He was, nevertheless, well acquainted with the content and method of Viète, Ghetaldi, and other early modern writers. By 1629 he seems to have hit upon an analytic treatment of maxima and minima, and at somewhat the same time he applied the analysis of Viète to locus problems, thus inventing the new geometry. One would like to know how the transition from the analytic art of Viète to the fundamental principles of analytic geometry took place, but Fermat gave only some incidental hints of this.

74

Fermat composed only a very short treatise on analytic geometry, and this he called *Ad Locos Planos et Solidos Isagoge.* It is a work of about a score of pages devoted to the line, circle, and conic sections. It opens with the statement that although the ancients studied loci, they must have found these difficult, to judge from the fact that in some cases they failed to state the problem in general form. Fermat proposed to submit the theory of loci to an analysis which is appropriate to such problems and which would, he asserted, open the way for a *general* study of locus problems. Without further introduction, he then states in clear and precise language the fundamental principle of analytic geometry:

> Whenever in a final equation two unknown quantities are found, we have a locus, the extremity of one of these describing a line, straight or curved.[1]

This brief sentence represents one of the most significant statements in the history of mathematics. It introduces not only analytic geometry, but also the immensely useful idea of an algebraic variable. The vowels in Viète's terminology previously had represented unknown, but nevertheless fixed or determinate, magnitudes. Fermat's point of view gave meaning to indeterminate equations in two unknowns—which previously had been rejected in geometry—by permitting one of the vowels to take on successive line-values, measured along a given axis from an initial point, the corresponding lines representing the other vowel, as determined by the given equation, being erected as ordinates at a given angle to the axis. In ancient Greek works, certain lines associated with a given curve had played a role equivalent to that of a coordinate system, and the properties of the curve had been expressed in terms of these lines by means of rhetorical algebra. The curve came first, the lines were then superimposed upon it, and finally the verbal description (or algebraic equation) was derived from the geometrical properties of the curve. Fermat's genius made it possible to reverse this situation. *Beginning* with an algebraic equation, he showed how this equation could be regarded as defining a locus of points— a curve— with respect to a given coordinate system. Fermat did not invent coordinates and he was not the first one to use graphical representation. Analytic reasoning had long been used in mathematics, and the application of algebra to geometry had become a commonplace. However, there appears to have been no appreciation before the times of Fermat and Descartes of the fact that, in general, a given algebraic

---

[1] See *Oeuvres de Fermat* (4 vols. and supp., Paris, 1891–1922), I, p. 91; III, p. 85. The Latin of the *Isagoge* is found in v. I, p, 91–110; a French translation is given in v. III, p. 85–101. The Latin is found also in the *Varia Opera Mathematica* of Fermat (Tolosae, 1679).

equation in two unknown quantities determines, *per se*, a unique geometric curve.  The recognition of this principle, together with its use as
a formalized algorithmic procedure, constituted the decisive contribution of these two men.

It will be noted that neither Fermat nor Descartes used the term
"coordinate system" or the idea of two axes.  Fermat chose a convenient line to play the role of the modern $x$-axis, and a point on it—or
"extremity," as Fermat called it—was taken as equivalent to what
later became the origin.  For a given equation in $A$ and $E$, values of $A$
were measured along this line from the fixed point.  Corresponding
values of $E$ were then erected as line-segments (later known as ordinates) making a given fixed angle with the base line.  Fermat indicated
that ordinarily this angle was taken as a right angle.  Although in some
cases a line equivalent to a $y$-axis appears, the abscissa, or quantity $A$,
is not interpreted as a line drawn from the point in question to such an
axis of ordinates.  Fermat's scheme, like that of Descartes, may be
characterized as an ordinate—rather than coordinate—geometry.
Moreover, Fermat restricted his operations to what would now be called
the first quadrant.  It will be seen that in this respect Descartes went
somewhat further than Fermat.

The analytic geometry of Fermat is surprisingly systematic for a
newly discovered subject.  It begins with the classic division of loci
into three types[2]—plane, solid, and linear—and then there follows the
important statement that if the powers of the terms in a given equation
do not exceed the square, then the locus is plane or solid.  This statement, constituting the central theme of the work, is justified by the detailed consideration of cases of equations, taken in order.  Fermat begins with a linear equation, in the terminology of Viète:

"$D$ in $A$ aequetur $B$ in $E$."  This is equivalent to $dx = by$, where $d$ and
$b$ are given constants.  From the proportion $B$ is to $D$ as $A$ is to $E$, one
sees that the locus of the point in question (point $I$ in Fig. 7) is the line

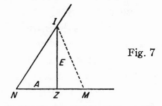

Fig. 7

(or, strictly speaking, the ray or half-line) $NI$.  The more general linear
equation, equivalent to $dx + by = c^2$, is shown similarly to correspond

---

[2] It is surprising that Fermat should have retained this ancient classification in the face
of the ordering by degrees which his own work so clearly suggested.  Descartes was less
hesitant.

to the line $MI$, where $MZ = c^2/d - A$. It is to be noted that the literal coefficients, as well as the coordinates, are taken as positive, a point of view generally persisting throughout the century. Fermat stated that all equations of first degree can, without difficulty, be shown to represent straight lines; but presumably he had in mind only such forms as are satisfied by *positive* values of $A$ and $E$.

To show the power of his new method in handling original locus problems, Fermat next announced a "très belle proposition" which he had discovered by its means:

> Given any number of fixed lines, the locus of a point, from which the sum of any multiples of the segments drawn at given angles from the point to the given lines is constant, is a straight line.[3]

No proof is given, but the proposition would follow as a simple corollary from the fact that the segments are linear functions of the coordinates of the point and from the proposition that every equation of first degree represents a straight line.

Going on to equations of second degree, Fermat showed that "$A$ in $E$ aeq. $Z$ pl." [i e., $xy = k^2$] is an hyperbola. This is obvious from the asymptotic property of the curve which was known, probably, from the time of the discovery of the conics by Menaechmus. Fermat simply indicated that the equation corresponded to the Greek verbal statement of the *symptoma;* but it should be noted again that Fermat here went from the equation—or the "specific property," as he sometimes calls it—to the curve, whereas his predecessors had proceeded inversely from the curve to a basic property or *symptoma.* Fermat added that any equation of the form $d^2 + xy = rx + sy$ can easily be reduced to the previous case of the hyperbola. The reduction is effected by substitutions which are equivalent to translations of axes, but the process is not formalized.

Fermat next considered equations involving squares of the unknown quantities, beginning with $x^2 = y^2$. This, and other quadratic equations homogeneous in $x$ and $y$—such as those in which either $A^2$ or the quantity $A^2 + AE$ is to $E^2$ in a given ratio—he interprets as a single straight line (or rather ray), for he did not consider negative coordinates.

Fermat then demonstrated that $x^2 = dy$ and $y^2 = dx$ (and the more general forms $b^2 \pm x^2 = dy$) are parabolas. After showing that $x^2 + y^2 + 2dx + 2ry = b^2$ is a circle, Fermat added that on the basis of this fact he had reconstructed all the propositions of the second book of Apollonius on *Plane Loci.* This observation would tend to confirm the

[3] The language of the original, here and in some other connections, has been considerably modified to correspond to current usage.

presumption that Fermat was led to analytic geometry through the study of loci rather than through the geometric solution of equations which occupied so much of the attention, not only of his predecessors, but also of his contemporaries, including Descartes.

After demonstrating that $b^2 - x^2 = ky^2$ is an ellipse and $b^2 + x^2 = ky^2$ an hyperbola (for which he gave both branches), Fermat considered "the most difficult of all equations [of second degree]"—one involving $x^2$, $y^2$, $xy$, and other terms. By the equivalent of a rotation of axes he converted $b^2 - 2x^2 = 2xy + y^2$ to the form of the ellipse previously given. Beyond this one reads only that any equation in $x^2$, $y^2$, $xy$, and other terms can be treated in an analogous manner "by means of a triangle known as to type"—that is, by transformations involving trigonometry.

As the "crowning point" of his treatise, Fermat then considered the following proposition:

> Given any number of fixed lines, the locus of a point, from which the sum of the squares of the segments drawn at given angles from the point to the lines is constant, is a solid locus.

The truth of this is made evident from the fact that, "according to the rules of the art," one is led in every instance to an equation of second degree. Such a problem illustrates well the power of analytic geometry as a systematic approach to loci. Had he discovered this method prior to his restitution of the *Plane Loci* of Apollonius, Fermat declared, the constructions of the locus theorems would have been rendered much more elegant.

Fermat's *Introduction to Loci* was followed by an appendix on *The Solution of Solid Problems by Means of Loci*. This represents a continuation of the work of Menaechmus, Archimedes, Omar Khayyam, and Viète on the geometric solution of cubic and biquadratic equations. The advance made by his treatise is in the fact that he was able to interpret questions of algebraic elimination directly in terms of intersecting loci, making use of his new principle that *any* second degree equation in two unknowns is a plane or solid locus. Systematic algebraic operations now replaced ingenious geometrical constructions. As Fermat pointed out, the involved methods of Viète are not needed. He boasted that "biquadratics are resolved with the same elegance, the same ease, and the same rapidity as cubics, and it is not possible, I believe, to imagine a more elegant solution." To make good his claim, he applied the method to show that all cubic and quartic problems can be constructed by means of a parabola and a circle. As an example, the equation $x^4 - z^3 x = d^4$ [or Aqq.-Zs. in $A$ aequetur Dpp., as Fermat wrote it in the terminology of Viète] is solved through the intersection of the parabola $\sqrt{2}\,by = x^2 - b^2$ and the circle $2b^2x^2 + 2b^2y^2 = z^3x + b^4 + d^4$. The method is easily extended to other cases. Fermat adds the significant remark at the close of the appendix, "He who will pay attention to the preceding will not attempt to reduce to plane problems ... the trisection of the angle and others like it." The impossibility of this ancient problem was becoming apparent not through synthetic geometry, but from the analytic study of loci.

There are three broad general approaches to the conics: as sections of a cone, as planimetric loci, and as graphs of equations of second degree. The *Isagoge* of Fermat constituted an excellent introduction to the analytic geometry of the conic sections in the third sense. Why did he not carry the work further to include curves of higher degree? Fermat was fully aware of the unlimited possibilities of the subject for the invention of new curves, for in the opening paragraphs of the work he specifically says that "the species of curves are indefinite in number: circle, parabola, hyperbola, ellipse, etc." Later he added that he was omitting the consideration of linear loci because the knowledge of these "is very easily deduced, by means of reductions, from the study of plane and solid loci." Does this mean that Fermat believed that equations of higher degree always could be reduced to those of lower degree, and hence solved by means of conics? If so, his point of view was diametrically opposed to that of Descartes, and also to his own later reflections. One respect in which the *Géométrie* of Descartes stands in sharp contrast to the *Isagoge* of Fermat is in the attention paid to the hierarchy of higher plane curves. The great contribution of Fermat to the disclosure of new curves is not found in his work on loci, but in the applications of analytic methods to infinitesimal geometry, in which he very nearly anticipated the invention of the calculus by Newton and Leibniz.

The infinitesimal methods of Archimedes had survived the medieval interlude better, perhaps, than had the geometry of Apollonius, and translations of Archimedean mensurational treatises flourished during the early modern period. The impact of algebra here was somewhat similar to that on elementary geometry, and attempts were made to arithmetize the method of exhaustion. This movement was ably represented by Stevin, among others, but it faced several serious difficulties. One of these was the lack of the limit concept in arithmetic and algebra; another was the want of an analytic theory of curves. The curves known to Kepler and Cavalieri, two of the most ardent admirers and continuators of the geometry of Archimedes, were virtually the same as those known to the great Syracusan. The number of curves was too small to encourage the search for algorithmic rules which could be applied in all cases; and, moreover, the known curves were not such as lent themselves readily to the devices which later constituted the calculus. Further developments in infinitesimal geometry clearly waited upon the rise of analytic geometry, and in this connection Fermat supplied the key. Not only was he one of the inventors of co-ordinate geometry, but he led in the application of the new methods to problems of slope and curvilinear mensuration. He introduced the curves upon which the methods of the calculus were built, and he supplied the method of tangents which served as the forerunner of differentiation. This work is found in his *Methodus Disquirendam Maximam et Minimam*,[4] a treatise composed a few years after the *Isagoge*.

The *Method of Maxima and Minima* is significant in the development of analytic geometry for the introduction of the curves $y = x^n$ and $y = x^{-n}$, the so-called higher parabolas and hyperbolas. Here, too, Fermat adopted the cumbersome terminology of Viète and used proportions rather than the newer forms of equation; but ideas are more important than notations, as one sees from his work. He retained the vowel $A$ to represent abscissae, but omitted the use of a letter representing ordinates. This necessitated an awkward semianalytic form, for the proportions defining a curve, which is in contrast to that used in the *Introduction to Loci*. The reason is perhaps to be found in the fact that his method of tangents and maxima and minima called for a variable increment equivalent to the modern $\Delta x$. Following his previous interpretation of $A$ and $E$ as algebraic variables, Fermat retained $A$ for the independent variable and used $E$ for the increment in $A$. His procedure for maxima and minima consisted essentially of forming, for a given curve or function, $f(A)$, the difference quotient $[f(A + E) - f(A)]/E$ and then finding the limit of this as $E$ tends toward zero. Neither the functional notation and idea nor the limit concept was specifically used, but the technique was virtually the same. This method was of signal importance, for it represented the earliest of the analytically formulated algorithmic rules which in the end converted infinitesimal *geometry* into infinitesimal *analysis*. Fermat found also, by analytic means, the quadratures of the parabolas and hyperbolas of higher order; but he failed to notice the inverse character of area and tangent problems, and so he missed, by a very narrow margin, the invention of the calculus. He is remembered, nevertheless, for the introduction into analytic and infinitesimal geometry of the family of curves $y = x^{\pm n}$, since known by his name—the "parabolas and hyperbolas of Fermat."

[4] See *Oeuvres de Fermat*, v. I, p. 133–179: v. III, p. 121–156.

Fermat's works contain references to numerous other curves, for he realized that in general every new equation in two unknowns represented a new curve. Nevertheless, one does not find complete graphs of these, for the equations are understood to indicate points with positive coordinates. Moreover, curves frequently were proposed solely to illustrate methods of the calculus, the interest being more in tangents and quadratures than in the shape of the curve itself. For many of these not even a partial sketch is given. For example, in a work on quadratures Fermat proposed,[5] along with many others for which diagrams are not given, the curve determined by the following equation:

$$Bc. \text{ aequalis } Aq. \text{ in } E + Bq. \text{ in } E.$$

This is the curve $b^3 = x^2y + b^2y$ which later became known as the "witch of Agnesi," although it reappeared several times in the interval between Fermat and Agnesi. Unfortunately, Fermat here was concerned solely with the question of area and so he showed no interest in the shape of the curve or in its properties as a locus.

There are other aspects of the work of Fermat which bear on analytic geometry, but they had less significance in its historical development than did those previously mentioned. A short treatise, *Isagoge ad Locos ad Superficiem*, carried the problem of loci to three dimensions, but it did not make use of the analytic method. The surfaces in question were those known in antiquity—the plane, sphere, spheroid, paraboloid, two-sheeted hyperboloid of revolution, cone, and (circular) cylinder. The single-sheeted hyperboloid of revolution, known probably to Cavalieri (or possibly earlier), is not included. The loci are not given by equations, and a coordinate system was not used. In fact, the results are not in all cases correctly stated.[6] The rise of solid analytic geometry did not take place until about a century later, even though Fermat himself was aware of the fundamental principle. In a half-page work entitled *Novus Secundarum et Ulterioris Ordinis Radicum in Analyticis Usus*, he repeated and extended his discovery of 1629:

> There are certain problems which involve only one unknown, and which can be called *determinate*, to distinguish them from the problems of loci. There are certain others which involve two unknowns and which can never be reduced to a single one; these are the problems of loci. In the first problems we seek a unique point, in the latter a curve. But if the proposed problem involves three unknowns, one has to find, to satisfy the question, not only a point or a curve, but an entire surface. In this way surface loci arise, etc.[7]

[5] *Oeuvres*, v. I, p. 279; v. III, p. 233.
[6] See Coolidge, *History of Geometric Methods*, p. 125.
[7] *Oeuvres*, v. I, p. 186–187; v. III, p. 161–162.

What a pity it is that Fermat did not extend his analytic study to problems in the latter categories! The whole history of analytic geometry might have been different, for developments in three-space later played an important role in revising the Cartesian geometry of two dimensions; and the analytic geometry of more than three dimensions, at which Fermat seems to hint, was not developed until two centuries later.

The contributions of Fermat to analytic geometry were not published during his lifetime, and hence it is difficult to determine the extent of their influence.[8] His work on loci, as well as that on maxima and minima, was known to the Parisian circle of mathematicians even before the appearance of the *Géométrie* of Descartes. The *Maxima and Minima* created quite a deep impression, especially as applied to the determination of tangents; but the *Introduction to Loci* seems to have been overshadowed by the work of Descartes. Portions of the *Maxima and Minima* promptly were included in books published by other mathematicians, but the *Isagoge* appeared in print for the first time in the *Opera Varia* of Fermat in 1679, fourteen years after the death of its author, forty-two years after the publication of the geometry of Descartes, and a half century after the treatise had been composed. By this time developments in the field had far outstripped the simple steps taken by Fermat, and the publication was largely of historical interest. Even the notation of the original was abandoned for that of Descartes, so thoroughly had Cartesian influence dominated the age. Unaware of the early date of composition, readers of the *Isagoge* missed its significance as evidence of the independent invention by Fermat of analytic geometry. The subject continued to be ascribed solely to Descartes, and it remained for later historical research to set forth the rightful claims of his rival. That the subject still is known as Cartesian geometry is unfortunate in the implications of the uniqueness of its invention, but the title does justice to the fact that it was predominantly under the influence of Descartes that the new branch of mathematics took root.

The origin of analytic geometry in the mind of Descartes is indicated by a letter which he wrote to Isaac Beekmann in October, 1628. Here he boasted that in the previous nine years he had made such strides in arithmetic and geometry that he had no more to wish for; and he substantiated the claim by giving the rule for constructing all cubics and quartics by means of a parabola. This effort to give geometric meaning to the solution of algebraic equations would make his development of

---

[8] The account by Abbé Louis Genty, *L'Influence de Fermat sur son Siècle* (Orleans and Paris, 1784), seems to overweigh this.

the subject a direct continuation of the work of Viète, and the further one traces his steps the more this view is confirmed. The geometric construction of the roots of determinate algebraic equations had been one of the chief concerns of Viète and his immediate successors, and Descartes made it the cornerstone of his own work. Like Fermat, he apparently had discovered that the interrelationship of algebra and geometry takes on added meaning through the use of coordinates in the study of equations in two unknown quantities, but the emphasis of the two men differed. Fermat had placed first the algebraic study of loci, but Descartes was concerned primarily with the construction of problems in geometry through the geometric solution of equations. The procedure was predominantly algebraic, but the significance was purely geometric.[9] The *aim* of Descartes was that of Viète and the geometers of classical antiquity; the *method* was essentially novel in that it made use of the graphical representation of indeterminate equations.

The power of the new technique was made apparent to Descartes when in 1631–1632 his attention was called by a classicist to the problem of Pappus on three and four lines.[10] Descartes indicated that he found the solution to this through calculation, and his success here, where he mistakenly thought the ancients had failed, made him aware of the importance—the universality—of his work. He became the prophet of the new geometry partly because in his own self-esteem he held a low opinion of the ancients which contrasted with the admiration which Fermat modestly felt toward the classical Greek geometers. Descartes took no part in the contemporary movement to restore the works of Apollonius. Instead he wrote and published a work which led to the virtual abandonment of synthetic geometry for almost two centuries. This important treatise, *La Géométrie*, appeared in 1637 as an appendix to the longer and better-known philosophical work, *Discours de la Méthode pour Bien Conduire Sa Raison, et Chercher la Verité dans les Sciences*. The whole was published without the author's name, although the authorship was generally known.

Cartesian geometry now is synonymous with analytic geometry, but the fundamental purpose of Descartes is far removed from that of modern textbooks. The theme of *La Géométrie* is set by the opening sentence: "Any problem in geometry can easily be reduced to such terms that a knowledge of the lengths of certain lines is sufficient for its

---

[9] *Cf.* Boyce Gibson, "La Géométrie de Descartes au Point de Vue de Sa Méthode," *Revue de Métaphysique et de Morale*, v. IV, (1896), p. 386–398.

[10] Gaston Milhaud, in *Descartes Savant* (Paris, 1921), would place the invention of analytic geometry in 1631, the date on which he attacked this problem. *Cf.* also, J. J. Milne, "Note on Cartesian Geometry," *Mathematical Gazette*, v. XIV (1928–1929), p. 413–414.

construction."[11]   Descartes in spite of his iconoclastic attitude, had not transcended the classical emphasis upon constructibility.   He first proceeds to do what mathematicians from Viète to Oughtred had done —furnish a geometrical basis for algebra.   The five arithmetic operations are shown to correspond to simple constructions with straightedge and compasses, thus justifying the introduction of arithmetical terms into geometry.   In this connection Descartes introduced his exponential notation for powers.   This had been adumbrated in various forms, but it was through the *Géométrie* that it first secured a firm foothold.   In fact, one may say that this book is the earliest mathematical text that a present-day student of algebra can follow without encountering difficulties in notation.   It represents the culmination of a century of development in symbolic algebra, and virtually the only one of the symbols used by Descartes which has since become archaic is that for equality.   The substitution of = for $\infty$ is a matter of convention only and has no significance in the development of ideas.   On the other hand, the contribution of Descartes in associating with geometry a purely symbolic algebra marks a decisive advance over earlier work, for it encouraged the development of algebraic techniques independently of geometric visualization.   The unknown quantities in the algebra of Descartes, like those used by Fermat, were variables.   They continued to represent lines, rather than numbers, but the author discouraged the interpretation of powers of these in terms of geometric dimensionality. He emphasized that by powers, such as $a^2$ or $b^3$, he means "only simple lines"—not areas or volumes, as the notation and names might imply. This is a very convenient, but by no means essential, point of view for analytic geometry.   It obviates the necessity of maintaining, through the introduction of suitable powers for the parameters or coefficients, an apparent homogeneity in a given equation or expression.   It permits one to write with impunity such an expression as $a^2b^2 - b$.   Descartes cautiously adds, nevertheless, that if one wishes to extract the cube root of this, one "must consider the quantity $a^2b^2$ divided once by unity, and the quantity $b$ multiplied twice by unity."   If unity does not enter in an equation, the lines should be of the same dimension.   That is, Descartes merely substituted homogeneity in thought for homogeneity in form.[12]   This afforded greater operational freedom to algebraic technique, and it facilitated the implicit association of the real number

[11] See *The Geometry of René Descartes* (transl. by D. E. Smith and M. L. Latham, Chicago and London, 1925), p. 2.

[12] Coolidge, "The origin of Analytic Geometry," *Osiris*, v. I (1936), p. 242, evidently overlooked this passage when he wrote "He [Descartes] made an enormous step in advance by arithmetizing his geometry.   The real objects with which he dealt were numbers.   He freed himself completely from the superstition of homogeneity."   *Cf.* Coolidge, *History of Geometric Methods*, p. 126.

system with the points on a line; but it did not seriously modify the early development of analytic geometry. Fermat saw that questions of notation were relatively unimportant as compared with ideas, and for this reason he opposed the changes by Descartes:

> I designate the unknown quantities by vowels, as did Viète, because I do not see why Descartes made a change in something which is without importance and which is purely a matter of convention.[13]

But Fermat underestimated the practical advantage to be gained in mechanical facility by abandoning the homogeneous manner of expression and by making algebra thoroughly symbolic. Successors of the two men generally retained the formal homogeneity of Fermat, for another century, but otherwise they followed the notations of Descartes.

Returning to the theme of the opening paragraph, Descartes gives directions for solving a problem in geometry. One first supposes the solution effected, giving names to the lines, both known and unknown. Then, making no distinction between known and unknown lines, one proceeds until one finds it possible to express a single quantity in two ways—that is, until one obtains a single determinate equation. This statement differs only in unessentials from the definitions of the analytic art given by Viète and Oughtred. It characterizes an analytic approach to geometry, but it does not represent coordinate geometry in the usual sense. Continuing further, Descartes states (without proof) that if the problem can be solved by ordinary geometry—that is, by straight-edge and compasses—the final equation will be a quadratic in one unknown, and "this root or unknown line can easily be found." If the equation is $z^2 = az + b^2$, for example, Descartes constructs the required line $z$ as follows: Draw a line segment $LM$ of length $b$

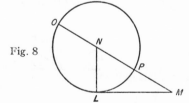

Fig. 8

(Fig. 8); at $L$ erect segment $NL = a/2$ perpendicular to $LM;$ with center $N$ construct a circle of radius $1/2a;$ and draw the line through $M$ and $N$ intersecting the circle at $O$ and $P$. Then $z = OM$ is the line desired. Descartes ignored the root $PM$ of the equation, because this is "false" [i e., negative]. Such constructions, the goal of Descartes' geometry, now are a standard part, not of analytic geometry, but of the theory of equations. They illustrate the fact that his *aim* was two

[13] *Oeuvres de Fermat*, v. I, p. 120; v. III, p. 111.

thousand years old—the search for geometric constructions of classical problems.

Book I of *La Géométrie* is on "Problems, the construction of which requires only straight lines and circles"—the old Platonic limitation. It is in Book II that one finds the more modern aspect of the work, "The nature of curved lines"; but Descartes expressly indicates that this book was written as a necessary preliminary to the third. And Book III, the last, is on "The construction of solid and supersolid problems." It is paradoxical to observe that it was largely through Descartes that the world learned that equations in two unknown quantities represent plane curves, and yet neither he nor his immediate successors showed much interest in this basic principle. Coordinates were not used by Descartes as they were by Oresme—to represent the properties of figures or functions. They were only an aid in the solution of problems of geometry.[14] His concern was not with the locus of points satisfying a given equation, but in the constructibility of these points. There is in the whole of *La Géométrie* not a single new curve plotted directly from its equation. So little interest did Descartes take in this problem that he never fully understood the significance of negative coordinates. He knew in a general sort of way that negative lines are directed in a sense opposite to that taken as positive; but he did not realize the applicability of the idea as a general principle in connection with coordinate systems. He occasionally made use of negative ordinates but not of negative abscissae. Montucla has exaggerated the part Descartes played in the geometrical interpretation of negative quantities. The folium, proposed in 1638, was truly a leaf, for Descartes regarded it as defined for the first quadrant only. Moreover, he proposed it not as an illustration of his own geometry but as a challenge to

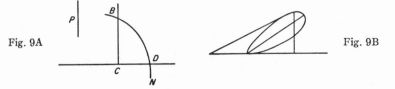

Fig. 9A                                         Fig. 9B

the method of maxima and minima and tangents of Fermat. At the time of his first reference to the curve, he was so little interested in it from the analytic point of view that he defined it synthetically as the curve *BDN* (Fig. 9A) such that the sum of the cubes of *BC* and *CD* is equal to the parallelepiped on *BC*, *CD*, and *P;* and he gave an incorrect rough sketch. Half a year later he graphed it more carefully as a leaf;[15]

[14] Loria, "Descartes Géométre," *Etudes sur Descartes* (Paris, 1937), p. 119–220.

[15] See *Oeuvres* (ed. by Adam and Tannery, 12 vols. and supp., Paris, 1897–1913), v. I, p. 490; v. II, p. 274. *Cf.* Loria, "Da Descartes e Fermat a Monge e Lagrange," p. 790–791.

(Fig. 9B) and half a dozen years after his death Fermat continued to draw the folium as Descartes had visualized it.

Descartes had been much impressed by the power of his method in dealing with the three- and four-line locus of Pappus, and this problem, running like a thread of Ariadne through the three books, well illustrates where he placed the emphasis. It is in connection with this problem that, about midway in Book I, coordinates enter in *La Géométrie*, and hence it is here that Cartesian geometry, in the strict sense of the word, appears. However, the essential principle that indeterminate equations in two unknowns correspond to loci is first clearly enunciated later in Book II.

> For the solution of any one of these problems of loci is nothing more than the finding of a point for whose complete determination one condition is wanting. ... In every such case an equation can be obtained containing two unknown quantities.[16]

Note here (as also in Fermat) the emphasis upon *two* unknowns for a plane locus, in contrast to the Apollonian use of *several* unknowns (all but one of which, however, were in reality dependent variables).

The problem of Pappus, in somewhat simplified form, is essentially this: Given $2n$ (or $2n + 1$) lines, find the locus of a point which moves so that the product of its distances to $n$ of the lines is equal (or proportional) to the product of the distances from the other $n$ (or $n + 1$) lines. For three or four lines Descartes knew (as did also the ancients) that the locus is a conic section. For five lines the locus is a cubic curve, and one should have expected Descartes to consider the variety of shapes these curves afford. However, the question which was of immediate concern to him was not the shape of a given locus but its constructibility. For five lines, not all parallel, he remarked triumphantly that the locus is elementary in the sense that, given a value for one of the co-

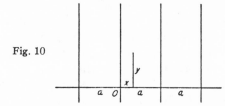

Fig. 10

ordinates of a point on the curve, the line representing the other coordinate is constructible by ruler and compasses alone. If, for example, four of the lines are parallel and equal distances $a$ apart and the fifth is perpendicular to the others (Fig. 10), and the constant of proportionality is taken as $a$, then the locus is a cubic which Newton called

[16] Book II, p. 334–335.

the Cartesian parabola (or trident)—$x^3 - 2ax^2 - a^2x + 2a^3 = axy$. This curve reappears frequently in *La Géométrie*, but Descartes at no point gives a complete sketch of it. His triple interest in the curve was limited to the following three aspects: (1) deriving its equation as that of a Pappus locus; (2) showing its constructibility by kinematic means; (3) using it in turn to construct the roots of equations of higher degrees. Descartes considers the Pappus problem or "question" constructible in the above case because one can assign successive values for the line $x$ and in each case one is able to construct the corresponding value for the line $y$. Whereas Viète had been interested in the constructibility of determinate problems, Descartes went further and applied the criteria to loci as well. It was here that he found it necessary to use a coordinate system. One may say that, in a general sense, the invention of analytic geometry by Descartes consisted in the extension of the analytic art of Viète to the construction of *indeterminate* equations, just as in the case of Fermat it was the study of loci, by the analytic art, which led to the same result. But Descartes continued to regard the construction of *determinate* equations as his ultimate purpose.

The plotting of curves in the now customary manner was not a part of Cartesian analytic geometry. Even the Pappus loci are not sketched. Descartes knew that an equation in two unknowns determines a curve, but oddly enough, he seems not to have regarded such an equation as an adequate definition of the curve, and felt constrained to exhibit an actual mechanical construction in each case. It has been conjectured that the ancient Greeks stressed constructions because these served as existence theorems. One is tempted to apply this idea to Descartes and say that he doubted the existence of a curve corresponding to an equation unless he could supply a kinematic construction for it. Like the ancient Greeks, he felt that a locus had to be legitimized by associating it geometrically or kinematically with another known curve. Perhaps it was the traditionally axiomatic form of geometry that led him in this direction. Viète had suggested adding new postulates which would make possible the systematic construction of the roots of cubic and biquadratic equations. Descartes wished to systematize geometry on a higher level so that there should be no limitation on the degree or dimensionality of a problem. He might have done this simply by admitting into geometry all curves given by algebraic equations, but he preferred a kinematic basis. Descartes would therefore add to Euclid's postulates one further assumption: "Two or more curves can be moved one upon the other, determining by their intersection other curves."[17]

[17] Book II, p. 316.

This represents, of course, a clear-cut break with the Platonic limitation of instruments to compasses and straight-edge, and Descartes makes free use of various linkages and mechanical devices. The concept of movement plays a far more prominent role in his work than in that of Fermat. In a sense Descartes had not freed himself from the ancient kinematic definition of curves, and he would therefore admit to geometry only such curves as "can be conceived of as described by a continuous motion or by several successive motions, each motion being completely determined by those which precede; for in this way an exact knowledge of the magnitude of each is always obtainable."[18]    To make clear the distinction between Descartes' idea of *constructing* a curve and the modern attitude toward the *plotting* of a curve, the following passage is significant:

> It is worthy of note that there is a great difference between this method, in which the curve is traced by finding several points upon it, and that used for the spiral and similar [i. e., mechanical or transcendental] curves. In the latter case one does not find indifferently all of the points of the curve sought, but only those which can be determined by means of a more elementary construction. ... [The former] method of tracing a curve by determining a number of its points taken at random applies only to curves that can be generated by a regular and continuous motion.[19]

Among the "regular and continuous motions" which are admissible in geometry, Descartes would include the "gardner's construction" of the ellipse and other similar loci determined kinematically by the lengths of strings or by moving straight lines. The ovals of Descartes, for example, are handled exclusively as loci, and the equations of these curves are not given in analytic form. However, he would not include loci based upon the lengths of curved lines for the reason that he believed rectifications "cannot be discovered by human minds." Had he lived a decade longer than he did, he would have had to revise this opinion!

The designation "Cartesian curve" still is applied to the members of a family used by Descartes as an illustration of the way in which one

Fig. 11

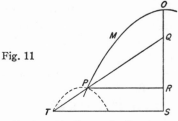

builds up the "family tree" of an algebraic curve. Let *OM* be a curve previously constructed, let *O* be a point on it and *Q* a point not on it,

[18] *Ibid.*
[19] Book II, p. 339–340.

both fixed with respect to the curve (Fig. 11). Let $S$ be a fixed point on the line $OQ$ and let $T$ be a fixed point on the line perpendicular to $OQ$ at $S$. Let $P$ be a point of intersection of the curve with the line $TQ$. Then as the curve (and hence also $Q$) moves with a rigid motion of translation in a direction parallel to $OS$, the point $P$ will describe a new curve $PT$ which may be regarded as a successor of the original curve. If the given curve is a straight line, the new curve will be an hyperbola; if it is a parabola, the derived curve will be the Cartesian parabola [or trident] referred to above. This kinematic hierarchy of curves Descartes transformed into an algebraic classification through his principle that "all points of a geometric curve [as defined by motions] must have a definite relation expressed by an equation."[20] That is, he found the equation of the locus of the point $P$ above as follows: Let $OQ = a$ and $ST = b$, and let the curve be given by $z = f(x)$, where $z = OR$ and $x = PR$. Then if $RS = y$, we have

$$\frac{z-a}{x} = \frac{z-a+y}{b}.$$

The equation of the locus of $P$ is therefore

$$f(x) = \frac{xy + ab - ax}{b - x}.$$

If $z = f(x)$ is linear, the locus of $P$ is of second degree. If the curve $z = f(x)$ is of second degree, the locus of $P$ is of third or fourth degree. If $z = f(x)$ is a cubic or quartic curve, Descartes said that the locus of $P$ should be of fifth or sixth degree; "and so on to infinity."[21] In this manner a hierarchy by pairs of degrees was established for the new loci. Fermat, however, pointed out[22] that there were inconsistencies here, for if the sliding curve is $y^3 = b^2x$, then the curve generated is of fourth degree—i. e., it belongs not to the next, but to the same, pair of degrees.

Descartes' classification into orders of two degrees each nevertheless was generally adopted throughout the century. It was based on the fact that the algebraic solution of the quartic leads to a resolvent cubic from which Descartes rashly concluded—incorrectly, as Hudde later showed[23]—that an equation of degree $2n$ would in all cases lead to a resolvent of degree $2n - 1$. From Descartes' statements in *La Géométrie*, one sees that his classification was suggested also by other considerations. It arose naturally in the Cartesian curves and in the

[20] Book II, p. 319.
[21] Book II, p. 319–323.
[22] *Oeuvres*, v. I, p. 121–123; v. III, p. 112–113.
[23] *De reductione aequationum*, Book I, p. 488–489.

problem of Pappus, and it was confirmed by the goal of his work—the
geometrical construction of the roots of polynomial equations through
the use of intersecting curves.  Cubics and quartics are both solvable by
conics; and quintics and sextics are solvable by cubics.  Descartes adds
the phrase, "and similarly for others,"[24] the definite implication being
that cubics do not suffice for equations beyond sixth degree, whereas in
reality they can be used for degrees up to nine.  To conform to his
ideas of constructibility Descartes might better have grouped curves by
orders with degrees corresponding to the perfect squares rather than to
the even numbers.

Having indicated his classification of curves, Descartes returned to
the problem of Pappus.  For three or four lines the equation which
serves to determine [or construct] the points of the locus is of second
degree; for not more than eight lines it is at most a biquadratic; for
not more than twelve lines it is of sixth degree or lower; "and so on for
other cases."  Descartes now examined in some detail the locus for
three or four lines, and his treatment is equivalent to a discussion of the
general equation of second degree.  Descartes took four lines in general
position, $EABG$, $TG$, $ES$, and $AR$ (Fig. 12).  From a variable point $C$

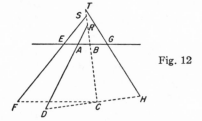

Fig. 12

on the locus required, he drew lines $CB$, $CH$, $CF$, and $CD$ to the given
lines at the appropriate given angle.  Taking $AB$ as $x$ and $BC$ as $y$, he
then expressed the distances $CD$, $CF$, and $CH$ as linear expressions in
$x$ and $y$ with coefficients determined by the fixed distances and by the
fixed angles between the lines, both variable and fixed.  In arriving at
these expressions, Descartes used ratios equivalent to the trigonometric
law of sines.  Setting $BC \cdot CF = CD \cdot CH$ and introducing some abbreviations, Descartes arrived[25] at an equation of the form $y^2 = ay -
bxy + cx - dx^2$.  This is the general equation of a conic passing through
the origin of coordinates, but under the view of Descartes, the literal
coefficients presumably were to be taken as positive.

Solving for $y$, one obtains the form $2y = a - bx + \sqrt{Kx^2 + lx + a^2}$,
where $K = b^2 - 4d$ and $l = 4c - 2ab$.  Descartes uses only a single

[24] Book III, p. 389.
[25] See Book II, pp. 325f.  The notation of Descartes has been somewhat modified for purposes of exposition.

sign before the radical, but he mentions that for various positions of the given lines, some of the terms may vanish or be reversed in sign.  He now shows how, for arbitrarily selected points on the line of abscissae, the corresponding ordinates of the locus are to be constructed with straight-edge and compasses.  Incidentally, however, he indicated the nature of the locus for the various cases.  If, for example, the expression under the radical sign vanishes, or is a perfect square, the locus will be a line.  This is the only reference in *La Géométrie* to the fact that the equation of a straight line is of first degree.  If the coefficient of $x^2$ is zero, the locus is a parabola; if it is "preceded by a plus sign," it is a hyperbola; if it is "preceded by a minus sign," it is an ellipse—except for the special case $b = 0$, $d = 1$, when, for rectangular axes, the curve is a circle.  These conditions are equivalent to a recognition of what is now known as the "characteristic" of the equation of a conic.  In this respect the work of Descartes is more general than that of Fermat, but it is less well adapted as an introduction to analytic geometry because of the omission of separate treatment of the simpler special cases of the straight line and conic sections.  The equations $x^2 = y^2$ and $xy = k^2$, for example, had been given by Fermat, but they are missing in the geometry of Descartes for the reason that they did not arise specifically in his study of the problem of Pappus.  Possibly it never occurred to him to represent a line by an equation for the reason that equations served him as a means of referring a *curve* to a straight line— "all the points of a geometric curve have the same exact measure or ratio to all the points of a straight line."  Coordinates themselves were straight-line segments.  Perhaps the omission of the equation of the straight line was the result of his emphasis upon generality, for he justified the omission of plane loci by the fact that "they are included under solid loci."[26]  Descartes likewise was familiar with the general idea of the transformation of coordinates,[27] which Fermat had used; but in this connection the *Geometry* includes only the remark that by a proper choice of origin and axis a simplest form of equation is obtained, and that the *type*—that is, the degree—of the equation will be the same for any other choice.  Because of such elliptic remarks, the successors of

[26] *Cf.* Book II, p. 319.  It should be remarked here that Descartes tacitly makes use of the Cantor-Dedekind axiom, i. e., he assumes that a one-to-one correspondence can be established between the points on a line and the real numbers, and likewise that a perfect correspondence can be set up between the points of a plane and pairs of line-segments (or real numbers.)  This tacit assumption was not a new idea at the time, for it had been broached over 2000 years earlier by the Pythagoreans who sought to associate number with all geometric magnitudes.

[27] In 1638 Descartes proposed to Roberval the curve equivalent to $\dfrac{y^2}{x^2} = \dfrac{l - x}{l + 3x}$, hoping to be able to laugh at him for not recognizing it as a folium rotated through an angle of 45°.  See Loria, *Kurven*, v. 1, p. 52–59.

Descartes found his work difficult to understand, and this fact later invited men to compose texts and commentaries on a somewhat lower level.

The discussion by Descartes of the equation of second degree includes also the determination of properties of the curves for various cases—centers, foci, vertices, and latera recta. The methods are given generally, and then are applied to the special case $y^2 = 2y - xy + 5x - x^2$. In contrast to textbooks of the present day, the consideration of such a specific numerical instance was quite unusual at that time. Descartes, however, seems to have had little interest in the analytic theory of conics. Apparently he felt that the future of geometrical study lay in higher plane curves rather than in the plane and solid problems of the ancients.

Descartes concluded the discussion of the locus to three or four lines with the statement that all such loci lead to equations of second degree and are therefore plane or solid loci. If the equation is of higher degree, the curve may be called "a supersolid locus"—what would now be called a higher plane (algebraic) curve. Descartes then adds the cryptic remark that "If two conditions for the determination of a point are lacking, the locus of the point is a surface which may be plane or spherical or more complicated."[28] This hint of an analytic geometry of three dimensions[29] reappears at the close of Book II where Descartes indicates that his remarks on plane curves "can easily be made to apply to all those curves which can be conceived of as generated by the regular movement of the points of a body in three-dimensional space." Here also, as in two-space, the emphasis is on the kinematic point of view rather than on arbitrarily given equations. The method Descartes proposed for the study of the properties of a space curve is to project it upon two mutually perpendicular planes and to consider the two curves of projection. Unfortunately, the only illustrative property given here is erroneous, for one reads that *the* normal to a curve in three-space at a point $P$ on the curve is the line of intersection of the two planes through $P$, determined by the normal lines to the curves of projection at the points corresponding to $P$. This would be true of the tangent line, but does not in general hold for a normal; but even the captious Roberval did not notice this error, and it was repeated almost a century later by the commentator Rabuel. Descartes, in these casual remarks,

[28] Book II, p. 335.
[29] Bell is not correct in the statement in *Men of Mathematics* (New York, 1937), p. 63, that "Fermat was the first to apply it [analytic geometry] to space of three dimensions. Descartes contented himself with two dimensions." Francisco Vera, *Breve Historia de la Matemática* (Buenos Aires, 1946), p. 82, makes the same mistake. If anything, the *analytic* remarks of Descartes with respect to three-space go further than those of Fermat, even though the latter did contribute a brief synthetic work on surface loci.

seems not to have been aware of the fact that for space of more than two dimensions a normal is not uniquely determined for a point on a curve. Apparently he did not envision the difficulties which mount up on increasing the number of degrees of freedom.

Fermat's *Isagoge* was a brief but systematic exposition of the analytic geometry of second-degree equations, but the *Géométrie* of Descartes is more especially concerned with higher plane curves. Having disposed easily, through the quadratic equation, of the problem of Pappus for three or four lines, the author proudly passed on to the case of five lines. In Book I he already had remarked that if the lines are all parallel, the locus is not in general constructible by ruler and compasses alone, even though the points lie on one or three straight lines. If four of the lines are parallel and the fifth is perpendicular to these, the Pappus-locus (for variable lines drawn at right angles to the given lines) is a cubic curve generated by the motion of a parabola—the Cartesian parabola or trident previously referred to. If the fifth given line is oblique to the other four, or if non-parallel lines are substituted for the other four given lines, the nature of the curve changes, but Descartes assures the reader that it can be handled by the methods he has presented.

Descartes next digresses from the problem of Pappus to explain that "all other properties of curves depend only on the angles which these curves make with other lines"—that is, the properties are determined by the equation of the curve. This is equivalent to the implication in Fermat's use of the phrase "specific property" for the equation of a curve. To illustrate this point Descartes chose what he regarded as "not only the most useful and most general problem in geometry that I know, but even that I have ever desired to know."[30] This is the determination of the normal to a given curve. In somewhat simplified form, the method of Descartes is as follows:

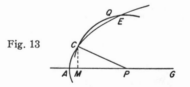

Fig. 13

Let the equation of the curve $ACQ$ (Fig. 13) be given with respect to $A$ as origin and $AG$ as axis. Let the rectangular coordinates of $C$ be $AM = y$ and $CM = x$, and let $CP$ be the desired line normal to the curve at $C$ and intersecting the axis in the point $P$, where $AP = v$ and $CP = s$. Then (from the Pythagorean theorem) $\overline{PC}^2 = s^2 = x^2 + v^2 - 2vy + y^2$, and the ·equation of

[30] Book II, p. 342.

the circle through $C$ with $P$ as center is $y = v + \sqrt{s^2 - x^2}$. Eliminating either $x$ or $y$ in the equations of the curve and the circle, one obtains an equation in one "unknown quantity," $x$ or $y$, and the quantity $v$. If the circle cuts the curve in two points $C$ and $E$, the final equation above will have two unequal roots. But "the nearer together the points $C$ and $E$ are taken, the less difference there is between the roots; and when the points coincide, the roots are exactly equal, that is to say, the circle through $C$ will touch the curve $CE$ at the point $C$ without cutting it." That is, in modern terminology, one finds the value of $v$ by setting the discriminant of the equation equal to zero, and $v$ then determines the normal line $PC$, and hence also the tangent line.

Descartes very laboriously applies his awkward method to the ellipse $x^2 = ry - (r/q)y^2$, finally obtaining an elaborate equation for $v$ in terms of known quantities. In view of the algebraic complications involved, his concluding remark, "I see no reason why this solution should not apply to every curve to which the methods of geometry are applicable," is more a theoretical boast than a practical reality. However, it should be borne in mind that the method of tangents of Descartes was the first such general method —i. e., the first anticipation of the idea of a tangent as the limiting position of a secant—to appear in print. Fermat was at the time in possession of his unpublished and simpler linear method and hence was in a position to criticize Descartes' circular device. The result was an unnecessarily acrimonious interchange of challenges and criticisms which served no very useful end but which did incidentally bring forth the folium of Descartes and which may have popularized the use of analytic methods.

The study of tangency led Descartes to include a long section devoted to the ovals which bear his name and to their use in optics. These served to relate *La Géométrie* to *La Dioptrique*, and *Les Météores*, the other appendices of the *Discours de la Méthode*. The ovals again show the emphasis of Descartes on loci, for he takes pains to describe the ways in which they can be generated and used, but he nowhere gives their analytic form in terms of equations.

Book III of the *Geometry* is the *raison d'être* of the work, to which the other books served as an introduction. The purpose of the work is the graphical solution of equations beyond the second degree, with particular emphasis on the cubic and quartic. The heading of the third book is "On the construction of solid and supersolid problems," and this title was used for somewhat over a century to designate what most writers regarded as the primary aim of Cartesian geometry. Descartes had abandoned the Platonic apotheosis of line and circle, but he substituted for it a fetishism which enslaved his successors for several generations. Book III opens with the pronouncement, "While it is true that every curve which can be described by a continuous motion should be recognized in geometry, this does not mean that we should use at random the first one we meet in the construction of a given problem. We should always choose with care the simplest curve that can be used in the solution of a problem, but it should be noted that the simplest means not merely the one most easily described, nor the one that leads to the easiest demonstration or construction of the problem, but rather the one of the simplest class that can be used to determine the required quantity."

The Cartesian principle of simplicity of class is a natural consequence of the hierarchy of curves, which in turn is an extension of the ancient classification of loci. Pappus[31] had objected to the "inappropriate" solution of plane problems through the use of solid loci, or of solid problems through linear loci. Descartes continued this important idea of the order of complication appropriate to a problem, but he did not state it clearly and did not investigate it carefully. He spoke of the use of curves of an unnecessarily high class as "a geometric error," supplementing this with the complementary warning that "It would be a blunder to try vainly to construct a problem by means of a class of lines simpler than its nature allows."[32] Much of the book consequently is devoted to what now is contained in works on algebra, for, as Descartes observed, "the rules for the avoidance of both these errors" calls for a study of "the nature of equations."

Book III is the most systematic of those in *La Géométrie*, but it is not analytic geometry in the strict sense of the word. It is an elementary course in the theory of equations, written in a language and notation almost identical with that in modern textbooks. Beginning with a pseudodefinition of equation, rules are given for combining, factoring, transforming, and solving equations, illustrated by examples with specific numerical coefficients. "Descartes' rule of signs" is here published for the first time in general form for positive and negative roots.[33] Increasing and decreasing the roots, changing their sign, multiplying or dividing them by a constant, removing the second term of an equation, testing for rational roots by an abbreviated method of division, the algebraic solution of cubics and quartics, the notion of irreducible equation—all these are found in Book III of *La Géométrie*. Because much of this material had been given earlier, Descartes was accused of

[31] *La Collection mathématique* (Book IV, prop. 30), v. II, p. 208–209.
[32] Book III, p. 371.
[33] Quite probably his discovery of the general rule was a consequence of the fact that he was among the first to make a systematic practice of bringing all terms of an equation to one side, equating these to zero.

plagiarism, especially from Viète and Harriot; but Descartes here made no special claim to originality. Following this algebraic introduction, Descartes proceeds to complete the problem begun in Book I—to construct geometrically the roots of algebraic equations. Having shown that linear and quadratic equations are constructible by ruler and straight-edge, he now demonstrates at length that the solution of cubics and quartics (i. e., of "solid problems") can always be found "by any one of the three conic sections, or even by some part of one of them however small, together with only circles and straight lines."

In this respect, Descartes did much to nourish the fetish of conics for another generation. He showed that the equations $x^3 = \pm pz \pm q$ and $z^4 = \pm pz^2 + qz \pm r$ can be solved, for real roots, through the intersections of a parabola with various lines and circles. For example, he solved $z^3 = pz + q$ graphically as follows: Draw the parabola $FAG$ with axis $ADKL$ and semiparameter $AC = {}^1\!/_2$ and take $CD = p/2$ (Fig. 14.)

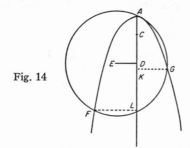

Fig. 14

Draw $DE = q/2$ perpendicular to $AD$. With $E$ as center and with radius $AE$ draw circle $FG$. Then the intersection point $F$ on the left of the axis gives the "true" [i. e., positive] root; any on the other side correspond to "false" [i. e., negative] roots. In modern symbolism, this method consists of finding the intersections of the parabola $x^2 = y$ and the circle $x^2 + y^2 = qx + (p + 1)y$. With slight modifications in procedure, Descartes applied the method to other cases of cubics and quartics with real roots. So pleased was he with these solutions by conics that he felt there was nothing more to be desired in this connection. The nature of the roots does not permit expression in simpler terms nor their determination by any construction which is both easier and more general.[34]

In going on to equations of degree more than four, it was clear to Descartes that, in general, the geometrical solution called for curves beyond the plane and solid loci. Even here, however, the conic-complex was continued. Instead of solving the quintic and sextic through

[34] *La Géométrie*, p. 334, 402.

a simple application of the parabola of second order defined by the equation $y = x^3$, Descartes, true to the postulate of hierarchical construction, made use of cubic curves defined through the intersections of moving conics and lines. These complicated geometric constructions of the roots of algebraic equations are on a level far above the simple problems of Fermat's elementary *Isagoge*.

The contemporaries and successors of Descartes spent much time and energy discussing the curves of lowest degree needed for the geometrical solution of a given polynomial equation. Fermat, in articles subsequent to the *Isagoge*, devoted much space to the same graphical schemes for solving polynomial equations as had Descartes. He proposed solving $x^3 + bx^2 = c^2b$ through finding the intersections of the parabola $x^2 + bx = by$ with the hyperbola $c^2 = xy$, adding[35] that "the method will be the same for all cubic equations." The geometrical solution suggested for the biquadratic $x^4 + c^2x^2 + b^3x = d^4$ is to determine the intersections of the parabola $x^2 = cy$ and the circle $d^4 - b^3x - c^2x^2 = c^2y^2$. Fermat saw that "the same procedure can serve to solve all biquadratic equations"; but he added also solutions through the intersections of two parabolas, or of a parabola with a hyperbola. Echoing the judgment of Descartes on such procedures, Fermat boasted that "it is not possible, I believe, to imagine a solution more elegant."

Fermat in 1660 composed a dissertation in three parts—*De Solutione Problematum Geometricorum per Curvas Simplicissimas*—pointing out that the Cartesian classification by pairs of degrees is not transferable from determinate to indeterminate equations. The equation $x^{11} = b^{10}d$, for example, is solved by the intersection of the curves $x^3a = y^4$ and $x^2y = b^{10}da$, one a quartic, the other a cubic, whereas the rule of Descartes would call for a curve of sixth degree.[36] Similarly, the ninth degree equation, $x^9 = b^8d$, is solvable by means of two cubics; and $x^{257} = b^{256}d$ is solved by the intersections of $x^{17} = dy^{16}$ and $b^{16} = x^{15}y$. Fermat adds the general rule covering such problems: If the degree of the given equation is greater than $n^2$, then a curve of degree greater than $n$ is needed in the geometrical construction of the roots. In commenting on such graphical solutions, Fermat held with Descartes that "it is a fault in true geometry to take for the solution of any problem curves too complex or of too high a degree, leaving out the simpler which are suitable." At that time any curve whatsoever of degree $n$ was regarded as geometrically "simpler" than one of higher degree. The folium of Des-

---

[35] *Oeuvres de Fermat*, v. I, p. 103–110; v. III, p. 96–101.

[36] Fermat misunderstood Descartes here to require an equation of degree 9 or 10. See *Oeuvres de Fermat*, v. I, p. 118–131; v. III, p. 109–120. This treatise appears to be almost a polemic against the work of Descartes.

cartes, for instance, was considered superior in simplicity to the curve $y = x^4$. Thus the conic-section rule was generalized into an understanding that curves used in the graphical solution of equations are to be of lowest possible degree, a view which Fermat seems to have shared.[37]

Book III of *La Géométrie* is of less significance in the development of analytic geometry than it is in the history of the classical problems of antiquity. In the first place, it put undue emphasis upon the geometric construction of roots of algebraic equations at the expense of the analytic study of curves. On the other hand, it marked a milestone in the attempts to duplicate the cube and to trisect the angle, for it boldly stated the impossibility of these problems. "Solid problems in particular cannot, as I have already said, be constructed without the use of a curve more complex than the circle."[38] Viéte had paved the way for this conclusion by showing that a solid problem is reducible either to the Delian problem or to an angle trisection. Fermat, too, had made an unpublished statement[39] similar to that of Descartes, but the influence of *La Géométrie* was most effective. Unfortunately Descartes was unable to give a satisfactory proof of his assertion, limiting himself to the weak inductive argument that if geometers will list all the ways of finding the roots, it will be easy to prove his method "the simplest and most general."

Estimates of the work of Descartes and Fermat differ widely. Bell has said[40] that "Descartes saw that an infinity of distinct curves can be referred to one system of coordinates. In this particular he was far ahead of Fermat, who, apparently, overlooked this crucial fact. Fermat may have taken it for granted, but nothing in his work shows unequivocally that he did;" yet Bell adds later that, "With the exception already noticed..., Fermat's analytic geometry appears to be as general as that of Descartes. It is also more complete and systematic." Loria points out that the fundamental idea of the equation of a curve is more clearly set forth by Fermat, and that his treatment of analytic geometry is more systematic than that of Descartes, and nearer to ours.[41] Coolidge, on the other hand, holds[42] that the work of Descartes "affords a far broader base for future development than either the writings of the Greeks or those of Fermat. He had a far more workable algebra

---

[37] See Fermat, *Varia Opera Mathematica* (Tolosae, 1679), p. 110–115.

[38] Book III, p. 401.

[39] See Fermat, *Oeuvres*, v. III, p. 101.

[40] E. T. Bell, *The Development of Mathematics* (New York, 1940), p. 125–127.

[41] "Sketch of the Origin and Development of Geometry Prior to 1850" (transl. by G. B. Halsted), *Monist*, v. XIII (1902–1903), p. 80–102, 218–234; also "Pour une Historie de la Géométrie Analytique," *Verhandlungen des III. Internationalen Kongresses in Heidelberg*, 1904, p. 562–574.

[42] J. L. Coolidge, *A History of Geometrical Methods* (Oxford, 1940), p. 127–128.

and a far wider vision of the importance of what he had done and of himself the doer.... Fermat recognized the relation between all sorts of equations and curves, but lacked the curiosity to go beyond the quadratic case and the study of conics. [Coolidge here overlooks Fermat's analytic works beyond the *Isagoge*.] Descartes showed that if any curve were mechanically constructible, we could translate the mechanical process into algebraic language, and so find the equation of the curve." Wieleitner's judgment is expressed as follows:

> Descartes' work is so differently presented that there is little likelihood of dependence on Fermat. In one respect Descartes gave less than Fermat, in other respects much, much more. One does not find the collection of simple equations and their geometrical representations; but he gave much more of algebraic form and the relation between it and geometry.[43]

Even in the seventeenth century Cartesian geometry had been received with varying reactions. Fermat seems not to have realized fully the significance of his own invention, and so he undervalued also the analytic geometry of his rival. At one point Fermat implies that the methods of Descartes are very nearly the same as those of Viète except for an unimportant change in notation.[44] Such an estimate may have been due in part to the lack at the time of a clear distinction between algebraic geometry and the calculus. Fermat's own contributions to the former were largely in the nature of application to the latter; but Descartes did not play an active role in the analytic transformation of infinitesimal geometry. The Cartesian method for tangents, for example, was definitely inferior to that of Fermat. It may have been such considerations that later led Leibniz also to view the work of Descartes somewhat coolly as but an application of equations to curves of higher degree which Viète and the ancients had neglected.[45] Leibniz spoke of Descartes' work as going back to the ancients, and this was the opinion of many contemporaries. The revolution which Comte and historians of the nineteenth century—especially Chasles—saw in *La Géométrie* was largely an illusion, due possibly to the striking change in the subject during the closing years of the eighteenth century.[46]

Some of the more conservative mathematicians of the century rejected analytic geometry completely and continued to use the synthetic method and representation. Others who accepted the work of Descartes emphasized the material in the third book, and so they looked

---

[43] *Geschichte der Mathematik*, v. II (2), p. 5.
[44] See *Oeuvres de Fermat*, v. I, p. 118–131; v. III, p. 109–120.
[45] *Philosophische Schriften* (ed. Gerhardt), v. IV, p. 347.
[46] It is to Michel Chasles, *Aperçu Historique sur L'origine et le Développement des Méthodes en Géométrie* (new ed., Paris, 1875), p. 94, that we owe the unfortunate characterization of analytic geometry as "proles sine matre creata."

upon *La Géométrie* as primarily a contribution to algebra, overlooking the essential similarity between the second book of Descartes' *Geometry* and Fermat's *Introduction to Loci*. Even now surprise is often expressed that the third book of *La Géométrie* comes so close to the traditional course, not in analytic geometry, but in advanced or college algebra. The answer to this paradoxical situation is easy to find. Descartes was not interested in curves as such. He derived equations of curves with one purpose in mind—to use them in the construction of determinate geometrical problems which had been expressed by polynomial equations in a single variable.[47] For this reason he had to consider in detail the transformation of equations and their reducibility. The *method* of Descartes is that of coordinate geometry, but his *aim* is now found in the theory of equations rather than in analytic geometry.

For about two centuries following 1637 analytic geometry was generally regarded as the invention of one man, but it is now quite clear that, years before the appearance of *La Géométrie*, Fermat used essentially the same methods. However, his work circulated largely through correspondence in manuscript form until its publication in 1679. By this time the geometry of Descartes had been popularized through the Latin editions of van Schooten. Had the Cartesian influence not predominated, certain aspects of analytic geometry might have developed more rapidly, for while Fermat's *method* was similar, his *object* was nearer to the modern one than was that of Descartes. It may well be that the absence of any priority controversy here—in contrast to the unpleasant episode in the calculus—was due to the difference in aim. Fermat proposed more clearly than Descartes the basic principle that an equation in two unknowns is an algebraic expression of the properties of a curve; and his work is devoted to the elaboration of this idea. Where Descartes had suggested classes of new curves generated by simple motions, Fermat introduced groups of curves given by algebraic equations. Unlike *La Géométrie*, the *Isagoge* of Fermat had as its purpose to show that linear equations represent straight lines and quadratic equations correspond to conics. To a large extent one may say that where Descartes had begun with a locus problem and from this *derived an equation* of the locus, Fermat conversely was inclined to *begin with an equation* from which he derived the properties of the curve. —Descartes repeatedly refers to the generation of curves "by a continuous and regular motion"; in Fermat one finds more frequently the

---

[47] As Eneström has well said, an appreciation of the objective of *La Géométrie* will make one less likely to see in it the influence of Oresme. See "Kleine Mitteilungen," *Bibliotheca Mathematica* (3), v. XI (1911), p. 241–243.

phrase, "Let a curve be given having the equation...."[48]   These are the
two inverse aspects of the fundamental principle of analytic geometry,
in much the same sense that differentiation and integration are inverse
aspects of the calculus.   It is now customary to speak of an "integral
in the sense of Leibniz" and of an "integral in the sense of Newton,"
according as the emphasis is on the summation concept (associated with
quadratures) or on the idea of rate of change (associated with tangent
problems).   Similarly it would be appropriate to introduce phrases to
mark the difference in emphasis indicated by the two inventors of coordinate geometry: "analytic geometry in the sense of Descartes" and
"analytic geometry in the sense of Fermat."   The one admitted curves
into geometry *if* it was possible to find their equations, the other
studied curves *defined* by equations.   As an indication of this distinction, successors of the two men looked back to Descartes as the one who
derived the equations of loci and to Fermat as the one who introduced
the equations of the generalized hyperbolas, parabolas, and spirals.
The distinction cannot, of course, be carried too far.   It is essentially a
matter of relative emphasis, for both men were aware of the dual aspect.   Elementary analytic geometry as now taught usually covers four
main topics in Cartesian plane coordinates: the derivation of formulae
on points, lines, angles, and areas, together with the application of these
to problems and theorems; the sketching of curves; the derivation of
equations of loci; and the study of the properties of curves, especially
of linear and quadratic equations.   Of these topics Descartes emphasized the third and considered briefly some aspects of the last; Fermat
emphasized the last and solved a few problems connected with the
third.   The second topic did not come into its own until the early years
of the eighteenth century, and the first topic not until the very close of
that century.   Descartes and Fermat discovered the two aspects of the
fundamental principle of analytic geometry, but they did not make the
subject what it is today.

[48] *Cf. Oeuvres*, v. I, p. 255f; v. III, p. 216ff.

# CHAPTER VI

# The Age of Commentaries

*Mathematical proofs, like diamonds,*
*are hard as well as clear.*
—JOHN LOCKE

THERE is a widely held opinion, in large part due to Montucla, Chasles, and Comte,[1] that the analytic geometry of Descartes and Fermat brought about a rapid transformation of mathematics; but a survey of the period immediately following their work does not substantiate such a view.[2] For one thing, the new geometry was not everywhere welcomed. The literary controversy of that time on "ancients vs. moderns" had its mathematical counterpart in misdirected attacks on algebraic geometry (and later also on the calculus) by men who overvalued the classical methods of antiquity. Then, too, the inventors of analytic geometry were themselves largely responsible for the failure of the new subject to make rapid gains. Mathematics was for Fermat only a hobby, the satisfaction of which was not increased through publication; and hence his *Isagoge* appeared posthumously, in 1679, after the stream of commentaries on analytic geometry was well under way. At that late date it was received with some indifference. Moreover, there is a strong presumption that Fermat did not fully realize, or was too modest to stress, the value of his method as a tool for professional mathematicians. Descartes realized all too well the significance of his contribution, but he was a poor expositor. He did not arrange his work in the orderly and systematic manner to be expected in an introduction of novel methods; nor did he go into detail to make the thread of his argument clear. One gets the impression that Descartes wrote *La géométrie* not to explain, but to boast about the power of his method. He built it about a difficult problem, and the most important part of his method is presented, all too

---

[1] See N. Saltykow, "La géométrie de Descartes. 300e anniversaire de géométrie analytique," *Bulletin des sciences mathématiques* (2), LXII (1938), 83–96, 110–123.
[2] Cf. Gaston Milhaud, "Descartes et la géométrie analytique," *Nouvelles études sur l'historie de la pensée scientifique* (Paris, 1911), p. 155–176.

concisely, in the middle of the treatise for the reason that it was necessary for the solution of this problem.  In concluding the work, Descartes justifies the inadequacy of exposition by the incongruous remark that he has left much unsaid in order not to rob the reader of the joy of discovery.  Either this was sarcasm or else the author grossly misjudged the abilities of his readers to profit by what he had written.  It is no wonder that the number of editions of his *Géométrie* was relatively small during the seventeenth century and has been still smaller since then; and it is not surprising that his early successors were in most cases professional geometers of marked ability.[3]

One contemporary who did seize upon the meaning of the work of Descartes was Gilles Persone de Roberval (1602–1675), a man who held his position at the Collège Royal through superiority demonstrated in competitive examinations every few years.  He composed two memoirs on algebra and geometry which might have served as introductions to the work of Descartes.  The *De recognitione aequationum* of Roberval is a theory of equations along the lines of Viète and Descartes in which the vowel-and-consonant convention of the former is combined with the small letters and operational signs of the latter.  The *De geometrica planarum et cubicarum aequationum resolutione* is an excellent example of analytic geometry in the sense of Descartes.  It is concerned with two problems: the representation of loci by means of equations and the use of intersecting loci to solve equations.  It omits the Fermatian study of the graphical representation of equations. Roberval well expressed the crux of Cartesian geometry when he wrote: "It is said that any geometrical locus can be reduced to an analytic equation, since from one or more of the specific properties can be derived an analytic equation in which one or two or three at most of the quantities are unknown."  Instead of applying this idea to difficult cases of the Pappus problem, as Descartes had done, he derived the equations of simple familiar curves.  He began with the equation of a circle, with respect to a diameter as axis and with an end-point of this as origin.  Let $A$ be the center of a circle with diameter $BC = 2b$, and let $EG$ be a line perpendicular to $BC$ cutting the circle in $D$. (Fig. 15). Then if $DE$ is $a$ and $BE$ is $e$, "the rectangle $BEC$" [i. e., $BE \cdot EC$] will be $2be - e^2$.  Hence the equation of the circle is $2bc - e^2 = a^2$ [since $BE \cdot EC = DE^2$].  Roberval, like Descartes, made use of a single

<hr/>

[3] One gets a very unfavorable impression of Descartes from a letter which he wrote to Mersenne in 1648.   Here he says that his geometry is what it should be to keep people like Roberval from slandering it without ending in confusion because they can not understand it. He says he intentionally omitted what was easiest because of malicious spirits, and were it not for them, he would have written quite otherwise.   Descartes added that he might clarify it further himself some day, but he died two years later without having done this.   See Léon Brunschwicg, *Les étapes de la philosophie mathématique* (Paris, 1912), p. 125 f.

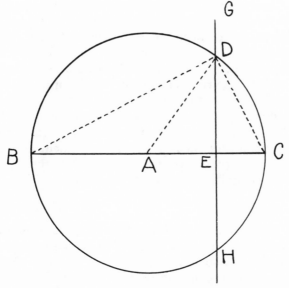

Fig. 15

axis. He also adopted the symbol of Descartes for equality, but this was not widely used by other mathematicians.

Roberval next derived the equations of the parabola, ellipse, and hyperbola with respect to an axis and a vertex, and also equations (one for each branch) of the conchoid. The form $ae = b^2$ he recognized (as had Fermat) as that of an hyperbola with respect to its asymptotes; but there is, strangely, no reference to the straight line. The equation of the circle is equivalent to a distance formula, but such preliminary formula-work as is given in modern textbooks did not appear in any of the early commentaries on Cartesian geometry. As an appendix Roberval added the familiar Cartesian solution of solid problems (cubic equations) through intersecting loci. In other works, notably his *Traité des indivisibles*, he contributed to analytic geometry indirectly through the development of the calculus. In this connection he seems to have had an analytic method for finding tangents, but he abandoned it for one based upon the composition of motions, possibly because of his interest in the cycloid and other transcendental curves. The period of Roberval's work, however, is uncertain and its influence is doubtful inasmuch as it was published posthumously in volume VI of the *Mémoires de l'Académie des Sciences* for 1666–1699, long after Cartesian geometry had been popularized by others. The substance may, however, have been given at an early period in his lectures at the Collège de France.

Descartes must have realized soon after the appearance of his *Geometry* that readers were experiencing difficulties with it, for he took the trouble to announce in his correspondence that a friend of his, a "Dutch gentleman," had written an "Introduction" to it. This *Introduction*, composed in 1638 but not published at the time,[4] opens with a simple exposition of the rules of elementary algebra, making use of the Cartesian notations. At that time the work of Descartes was looked upon as contributing as much to algebra as to geometry, a fact which shows how closely related to each other developments in the two fields were. The commentator then repeats the argument of Descartes that non-homogeneous expressions are justified by the assumption of suitable powers of unity. The expression $a^2b^2 - b$, for example, is really $\dfrac{a^2b^2 - bc^3}{c}$, where $c$ is the unit. It is interesting to note here the avoidance of more than three dimensions. The author then explains that in applying the Cartesian method to geometrical problems, one must find as many equations as unknowns. If this cannot be done, then there are numerous points satisfying the conditions and these make up a plane, solid, or linear locus if only one equation is lacking; and surface loci if two are lacking, "and similarly for others." The last phrase seems to envision the possibility (which Roberval specifically had rejected) of an analytic geometry of more than three dimensions; but, as in the case of Oresme, this important suggestion is not carried further. In fact, it is significant that, of the four examples given by this anonymous commentator as illustrations of the Cartesian method, three are determinate geometrical problems and only one is a locus. This confirms the impression one gets that Cartesian geometry at that time did not mean so much the study of loci as the solution of geometrical problems by algebraic means. The one locus problem (Problem 3) is that of finding a point such that the sum of the squares of its distances from four given points is equal to a given square—a problem solved by Fermat in his restoration of the plane loci of Apollonius. The distance formula was not then known—or, at least, was not specifically expressed—so the problem was handled geometrically in the following manner:

Let the four given points be $A$, $D$, $E$, and $F$, and let $C$ be the point to be determined (Fig. 16). Draw the line $AD$, then draw $EK$, $CB$, and $FG$ perpendicular to $AD$, and draw $EH$ parallel to $AD$. Let $AB = x$

[4] See René Descartes, *Oeuvres* (ed. by Charles Adam and Paul Tannery, 12 vols. and supplement, Paris, 1897–1913), X, 659–680. Presumably this is described in Wieleitner, "Über zwei algebraische Einleitungen zu Descartes' Géométrie," *Bl. f. d. Gymn.-Schulw.* hrsg. v. bayr. *Gymn.-Lehrverein*, XLIX (1913), 299–313; but I have not had access to Wieleitner's article.

and $BC = y$.  Then $BD = c - x$, $GF = b$, $GB = x - a$, $CH = y - g$, $BK = f - x = HE$, where $a$, $b$, $c$, $f$, and $g$ are constants.  From these values and the Pythagorean theorem the locus is easily seen to be a circle.  This illustration by the anonymous commentator[5] is a fair ex-

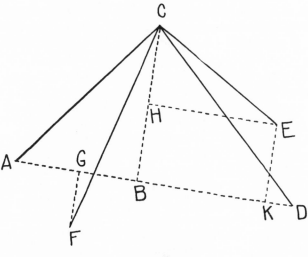

Fig. 16

ample of Cartesian geometry at that time and shows how greatly it differed from modern analytic geometry.

About a year later Descartes received and approved a more extensive commentary on *La géométrie* composed by Florimond Debeaune (1601–1652) under the title *Notae breves*.  This work opens with the statement that *algebra speciosa* includes not only the algebra of numbers and the geometric analysis of the ancients, but also the study of all quantities which have interrelationships or proportions, as Descartes asserted.  It therefore includes the consideration of the ratios of lines, i. e., analytic geometry.[6]  The early sections of the *Notae breves* do not depart from the Cartesian rationale.  They represent a paraphrase of Book I of *La géométrie* (the construction of algebraic quantities, including the roots of quadratic equations).  The commentary on Book II, however, is especially noteworthy for its emphasis upon the Fermatian aspect of analytic geometry, the systematic consideration of cases

[5] Loria has suggested that he may have been Godefroy de Haestrecht.
[6] For the *Notae breves* see the 1659–1661 edition of the *Geometria* of Descartes, v. I, p. 107 ff.

of second degree equations in two variables in which various terms are missing. With respect to an arbitrary coordinate angle, he showed that the cases $y^2 = xy + bx$, $y^2 = -2dy + bx$, and $y^2 = bx - x^2$ represent hyperbolas, parabolas, and ellipses respectively. In connection with the hyperbolas $xy + bx + cy - df = 0$, he considered separately seventeen cases according as the various coefficients are positive, negative, or zero. This tedious multiplication of cases shows how far his century was from a realization that one of the great values of analysis lies in the possibility of sweeping generalities. The only consideration given by Debeaune to equations of first degree is the comment that if for the general equation of second degree the terms in $x^2$, $y^2$, and $xy$ vanish, then the figure is a straight line. This fact, here explicitly stated for the first time, had been known to Descartes, but neither he nor Debeaune regarded the analytic study of the straight line as of any importance. The *Notae breves* in general follows Descartes with respect to constructibility, loci, tangents, and the interpretation of coordinates. Like Descartes, he did not adopt any conventional direction for the axis or axes, and he used $x$ and $y$ interchangeably as dependent or independent variable. The interchangeability of the axes or variables is not, in the strict sense, to be ascribed to any one individual. Often it was tacitly assumed, even by those who made use of a single axis, but it did not become a clearly recognized principle until almost a century after the time of Debeaune.

The historian Montucla referred to Debeaune as "the first to penetrate the mystery of analytic geometry."[7] This reference undoubtedly is an oversimplification of the situation, but it does indicate that Debeaune was one of the leading Cartesian commentators. Nevertheless, the *Notae breves* is typical of the seventeenth century in its failure to demonstrate the power of algebraic geometry as an instrument of discovery. Men of the time were too easily satisfied to work over again much of the same old material found in Apollonius, Descartes, and Fermat; and Descartes is said[8] to have felt, not without some justification, that there had been no great progress in mathematics.

The work of Debeaune was given wide publicity through its inclusion in the Latin translation of the Cartesian *Geometry* issued by Frans van Schooten (1615–1660) in 1649, 1659–1661, 1683, and 1695. It is probably not too much to say that it was these Latin editions which established the place of Cartesian geometry in the seventeenth century,

    [7] Étienne Montucla, *Historie des mathématiques* (new ed., 4 vols., Paris, 1799–1802), II, 103, 147.
    [8] See A. E. Bell, *Christian Huygens and the Development of Science in the Seventeenth Century* (New York and London, 1947), p. 18. Lagrange, a century and a half later, expressed pessimism about the future, rather than the past, of mathematics.

for they not only made the vernacular work of Descartes available in the universal language of the time, but included also a tremendous amount of additional material clarifying the original cryptic account and building up a spirit of enthusiasm for the new subject. The 1649 edition included, besides Debeaune's *Notae*, the still more extensive *Commentarii* by the editor. Van Schooten's *Commentaries* is purely Cartesian in inspiration and served to popularize *La géométrie* through the inclusion of additional proofs, algebraic-geometrical problems and constructions, and new loci. The straight line enters only in connection with the construction of the Pappus "locus to two lines," and the only equation given for the circle is that with respect to the center as origin. However, where Descartes had given his ovals only as kinematic loci, van Schooten studied them analytically in terms of their equations. As in Debeaune, there is much more consideration given to various cases of quadratics in two variables. An important part of van Schooten's commentary is on the Cartesian graphical solution of cubic and quartic equations.

In 1651 Erasmus Bartholinus (1625–1698), literary heir of Debeaune and student of van Schooten, arranged and published—under the title *Principia matheseos universalis seu introductio ad geometriae methodum Renati Des Cartes*—some lectures van Schooten had delivered. These deal with the "logistic" of quantities—that is, with algebra in the notation of Descartes. They do not include any analytic geometry, but the work closes with the statement that it suffices as an introduction to the geometry of Descartes. For this reason the *Principia matheseos* was included by van Schooten in his subsequent Latin editions of Descartes. It shows again that commentators of the time were more attracted by the algebra of *La géométrie* than by the geometry.

What van Schooten and his associates did for Cartesian geometry on the Continent was effectively accomplished in England by John Wallis[9] (1616–1703), one of the most important figures in the early history of analytic geometry. Wallis did not publish either new editions of *La géométrie* or commentaries on it, but in his own works he seized upon the methods and aims of Cartesian geometry with as great an understanding and forceful originality as any other figure of his century. It is sometimes claimed that Descartes arithmetized geometry; but it would be nearer to the truth to say that whereas Descartes made such an arithmetization a possibility, Wallis made it a fact. Descartes had justified his disregard of homogeneity geometrically rather than

---

[9] A summary of his work by J. F. Scott, *The Mathematical Work of John Wallis* (London, 1938), is available, but this does not give an adequate account of his analytic geometry. Historians seem to be attracted more by anticipations of the calculus in the work of Wallis than his analytic geometry.

arithmetically, pointing out that powers of the unit line segment could always be introduced where desired; his coordinates were not numbers but lines. The aim of his geometry was not to express properties of curves by algebra, but to use algebra to facilitate geometrical constructions; and so he had used conics to solve algebraic equations, but not algebraic equations to study conics. Wallis, on the other hand, boldly replaced geometrical concepts by numerical wherever possible,[10] maintaining that proofs by algebraic calculation were as valid as deductions using geometric lines. Proportions, he held, are not to be interpreted geometrically but as purely arithmetic concepts. In line with such views, his *Tractatus de sectionibus conicis* of 1655 presented the earliest systematic algebraic treatment of the conic sections to appear in print. Descartes and Debeaune had shown that certain quadratics are conic sections, inasmuch as they possess the properties given by Apollonius; but Wallis first did in algebraic symbols what Apollonius had done in words. Beginning with a brief stereometric consideration of the sections of a cone and substituting letters for geometric lines, he deduced the well-known properties (*symptomae*) expressed by $e^2 = ld - \dfrac{ld^2}{t}, p^2 = ld$, and $h^2 = ld + \dfrac{ld^2}{t}$, where $e$, $p$ and $h$ are ordinates of the ellipse, parabola, and hyperbola, respectively, corresponding to abscissas $d$ measured from a vertex, and where $l$ is the latus rectum and $t$ is the "diameter" or axis. This is the first time that these important equations of conics (known in essence to Apollonius and possibly to Menaechmus) appeared in algebraic form; but it is interesting to note that Wallis did not adopt a standard convention in designating the ordinates. Wallis then considered the conics "absolutely"—i. e., "as though having nothing whatsoever to do with the cone."[11] For example, he defined the ellipse purely analytically as follows: "I therefore call the ellipse the plane figure characterized by the property $e^2 = ld - \dfrac{l}{t} d^2$." This appears to be the first time that the conic sections were defined, neither stereometrically nor kinematically, but simply as instances of equations of second degree. Wallis then took these equations, as given, and proved conversely that the curves defined by them were the conics of the ancients. From these equations and "without the embranglings of the cone" Wallis then deduced other properties, such as tangents and conjugate diameters. If analytic geometry were simply the translation of Apollonius into the language

---

[10] For example, he showed how all the theorems of Euclid II and V could without difficulty be derived arithmetically. See A. Prag, "John Wallis," *Quellen und Studien zur Geschichte der Mathematik, Astronomie und Physik*, Part B, *Studien*, I (1931), 381–412.

[11] *Operum mathematicorum* (2 vols., Oxonii, 1656–1657), II, 28.

of algebra, Wallis would have a strong claim as the inventor. Descartes and Fermat had invented the methods, but Wallis gave the first systematic application of them to the study of the conic sections. This represents a striking exception to the English predilection at that time for synthetic methods. Wallis had a fine appreciation of the generality afforded by algebraic forms, and so he realized well the power of the new analysis. He said that from the coefficients of an equation of second order one could calculate all the magnitudes, such as center and axes, connected with the conic thus determined. Had he pursued this idea further, he would have been led to the "characteristic" of the second degree equation.

Wallis contributed further to the advance of Cartesian geometry through its association with infinitesimal analysis in his *Arithmetica infinitorum* of the same year. This work expressed the method of indivisibles of Cavalieri in terms of arithmetic and algebra, and so popular did it become that the arithmetization of the calculus overshadowed that of geometry. It may be for this reason that an important advance made by Wallis in analytic geometry went largely unnoticed. In an appendix to his *Conics* Wallis had considered the cubic parabola $p^3 = l^2d$, a curve which he mistakenly took to have line symmetry with respect to an axis, as does the ordinary or Apollonian parabola. By the end of 1656, however, he had discovered the correct form of the curve through an algebraic study of the intersections with it of a family of parallel lines.[12] It is interesting to note that in the early days the correct interpretation of negative coordinates was derived from the known shape or algebraic properties of a curve or equation, a situation which is now reversed. Wallis then generalized his discovery to parabolas of higher orders and showed that for even orders the two halves lie on the same side of the tangent at the origin and for odd orders on opposite sides. Fermat's earlier consideration of the parabolas (and hyperbolas) of higher orders had been concerned primarily with tangents and quadratures and so was limited to the first quadrant. It had been known in a general way by Descartes and his early successors that negative ordinate lines were plotted in a direction opposite to that taken as positive, but Wallis seems to have been the first one consciously to introduce negative abscissas and to associate them correctly with positive and negative ordinates. The significance of this step was not appreciated by his contemporaries, many of whom continued throughout the century to make mistakes through overlooking or misinterpreting negative coordinates. This work of Wallis

[12] Wallis, *Opera* (3 vols., Oxonii, 1693–1699), I, 229–290, especially p. 249–250. Loria (1923–1924) refers to the error of Wallis but not to its later correction. Progress in curve tracing was slow in the seventeenth century.

may well have been influential in the later use of negative coordinates
by Newton, just as Wallis' method of interpolation inspired Newton's
discovery of the binomial theorem.   In England, however, Wallis'
arithmetization of geometry did not achieve great popularity.   Hobbes
and Barrow disapproved of it very strongly, the former calling the
*Conics* a scab of symbols;  and British mathematicians continued for a
century and a half to prefer synthetic methods.   On the Continent
Wallis became persona non grata through his chauvinism in attributing
to English mathematicians most of what Descartes had written.   *La
géométrie* he characterized, most unfairly, as practically a transcription
from the work of Harriot.   Moreover, Wallis and Fermat engaged in a
sharp controversy over the former's method of induction or interpolation.   As a consequence, Continental developments in analytic geometry were bound up with the more conservative views of van Schooten
than with the arithmetization of Wallis.

In 1656–1657 van Schooten published his *Exercitationes mathematicae*,
in which algebraic calculations are applied to geometric problems and
the method of Descartes is used in an attempted reconstruction of the
lost *Loci* of Apollonius.   One of the loci leads to a linear equation, and
van Schooten indicates that this represents a straight line.   There are
other applications of analytic methods, but no distinctly new point of
view.   The folium of Descartes appears as a leaf only,[13] indicating that
the author did not know of the contemporary work of Wallis on negative coordinates.   The *Exercitationes* contains an interesting book devoted to various organic descriptions of the conic sections, a topic which
played a large part in the geometry of that day.   Although "organic
descriptions" of curves go back to the ancients and were found also in
synthetic treatises of Ubaldo, Stevin, Mydorge, and others, such descriptions of conics were, in a sense, a natural addition to the work of
Descartes who had emphasized the kinematic construction of higher
plane curves.

One of the outstanding events in the development of analytic geometry was the appearance in 1659–1661 of van Schooten's second Latin
edition of the *Geometria* of Descartes.   The work of Descartes occupies
but the first hundred or so pages in volume I;  the remainder of the two-
volume work, including almost a thousand pages in all, is made up of
supplementary treatises.   The *Commentarii* of van Schooten is well
over twice as long as the *Geometria* itself.   It contains additional locus
problems—one of which led to the linear equation $y = a - x$—derivations
of the equations of conics, and the study of quadratic equations.   Van
Schooten's adherence to geometrical tradition is shown in his directions

[13] *Exercitationum mathematicarum* (Lugduni Batavorum, *1557*), p. 498.

for carrying out the converse problem of finding a cone containing a given parabola, ellipse, or hyperbola. However, a definite advance is indicated by his use of a transformation of coordinates as follows: If $(x, y)$ are the rectangular coordinates of a point $C$ with respect to $A$ as origin and $DAB$ as axis of abscissas, then the rectangular coordinates of $C$ with respect to $D$ as origin and $DFGH$ as axis are $DG = \dfrac{a^2 + ax - by}{\sqrt{a^2 + b^2}}$ and $CG = \dfrac{ab + bx + ay}{\sqrt{a^2 + b^2}}$, where $AD = a$ and $AF = b$. (Fig. 17). This is equivalent to the translation $\begin{Bmatrix} x' = x + a \\ y' = y \end{Bmatrix}$ followed by the rotation $\begin{Bmatrix} x'' = x' \cos\theta + y' \sin\theta, \\ y'' = -x' \sin\theta + y' \cos\theta \end{Bmatrix}$ where $\theta$ is measured positively in a counterclockwise direction. He used such transformations to remove linear terms and also to associate the asymptotic equation of an hyperbola with that with respect to the axes. It is important to note, however, that such transformations were not formalized and generalized, but were, in each instance, carried out *de novo* in connection with appropriate geometrical diagrams.

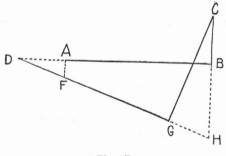

Fig. 17

Van Schooten's 1659–1661 *Geometria* of Descartes contained also a work of Johann Hudde (ca. 1633–1704), *De reductione aequationum*, which contributed more to Cartesian algebra than to analytic geometry; but it was of some importance as the first instance of the systematic recognition of literal coefficients as either positive or negative, regardless of the sign attached. Newton seems to have been the first one to extend Hudde's view to letters used as exponents. This advance made possible the elimination of numerous special cases through the consideration of general forms, and encouraged the use of universally applicable formulas, a tendency which developed gradually throughout the following century. Hudde's work on maxima and minima and Heuraet's on rectification of curves also form part of this *Geometrica*, but they are of significance only for the application of analytic methods to the early stages of the calculus. A further algebraic treatise by Debeaune, *De aequationum natura, constitutione, et limitibus,* and an old-fashioned quasi-analytic work by van Schooten, *Tractatus de concinnandis demonstrationibus geometricis ex calculo algebraico,* undoubtedly served to help bridge the gap between the older and newer points of view; but undoubtedly the most significant addition to Descartes made by the second van Schooten edition was the *Elementa curvarum linearum* of de Witt.

Jan de Witt (1623–1672), the well-known Burgomaster of Amsterdam, composed the *Elementa curvarum* when he was but twenty-three years of age; but the delay in its publication led Wallis to claim that it was an imitation of his own *Conics.* The two works are, indeed, comparable in some respects; but they open quite differently. Where Wallis had begun with stereometric and analytic considerations of the conic sections, de Witt began kinematically with the Keplerian constructions of the curves. The whole first book is devoted to these and other organic descriptions *in plano,* and it is written in the language of synthetic geometry. Among the alternative loci defining the ellipse, he gave the construction (known in essence to Mydorge and probably to Stevin and Archimedes) in terms of two concentric circles with the eccentric angle as parameter. He also extended the theorem of Proclus (known as well to Ubaldo and Stevin) on the description of an ellipse by the points on a line segment, the ends of which move along two intersecting lines, to the case where the two lines are not necessarily perpendicular. De Witt was familiar also with the ratio definitions of the conics, and to him is due the name "directrix."

The second book of de Witt's *Elementa,* by contrast, is so systematic a treatment of analytic geometry that it has been described as the first textbook on the subject.[14] His application of analysis resembled Fermat's more than it did that of Wallis and Descartes, for he began with equations, rather than with curves or loci. Going beyond Fermat, he explicitly stated that an equation of first degree represents a straight line,[15] and as examples he gave the equations $y = c$, $y = \dfrac{bx}{a}$, $y = \dfrac{bx}{a} + c$, $y = \dfrac{bx}{a} - c$, and $y = -\dfrac{bx}{a} + c$. The omission of $y =$

[14] See Wieleitner, *Geschichte der Mathematik,* II (2), 26.
[15] See *Geometria* of Descartes (1659–1661 ed.), II, 243.

$-\dfrac{bx}{a} - c$ is significant as showing that de Witt, like many of his contemporaries, had not mastered the meaning of negative coordinates. His
diagrams are line segments or rays limited to the first quadrant. De
Witt does little with the linear equation from the analytic point of view,
and the persistence of the geometric tradition is apparent in the fact
that he gave instructions for the construction of lines corresponding to
given equations. For example, he constructed the equation $y =$
$\dfrac{bx}{a} + c$ by laying off the segment $AB = a$ along the axis (with $A$ as the
"fixed initial point"), constructing parallel line segments $CB = b$ and
$FA = c$ at any desired angle (the angle of obliquity for the coordinate
system), connecting points $A$ and $C$ (Fig. 18), and finally drawing from
$F$ a half-line $FG$ parallel to the directed segment $AC$ and in the same
sense. (The phrases half-line and directed segment are not used by de
Witt, but these ideas are implied by his constructions.)

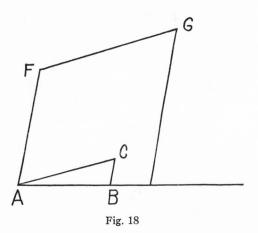

Fig. 18

De Witt's treatment of the second-degree equation is similar in that
he did not begin with the most general form, but with numerous
special cases. He reconciled the equations $y^2 = ax$, $y^2 = ax \pm b^2$, and
$y^2 = -ax + b^2$ with the properties of parabolas already established
synthetically in book I. He regarded these equations as corresponding
only to segments of the parabola defined for positive values of $x$ and
$y$, and he failed to consider the forms $y^2 = -ax$ and $y^2 = -ax - b$
inasmuch as these are not real for positive abscissas. Because he continued the Cartesian-Fermatian use of a single axis, he felt it necessary

to give a separate treatment of those forms obtained by interchanging the variables $x$ and $y$. Equations of the form $\dfrac{ly^2}{g} = f^2 - x^2$ or $\dfrac{lx^2}{g} = f^2 - y^2$ he showed to be ellipses. For hyperbolas he gave the asymptotic equation $xy = f^2$, and again not one, but two axial forms, $\dfrac{ly^2}{g} = x^2 - f^2$ and $\dfrac{lx^2}{g} = y^2 - f^2$. It should be borne in mind that these equations refer to general Cartesian coordinates, oblique as well as rectangular, so that $y^2 = f^2 - x^2$ was not necessarily a circle. By means of transformations of the form $\left\{\begin{array}{l} v = x - h \\ z = y + \dfrac{bx}{a} + c \end{array}\right\}$ de Witt was able to reduce other equations of second degree to his canonical forms. Such transformations, however, were particularized with respect to specific cases and diagrams.

De Witt seems to have come very close to the idea of the "characteristic" of a general quadratic equation, for he gives rules for reducing parabolas given by $yy + \dfrac{2bxy}{a} + 2cy = bx - \dfrac{bbxx}{aa} - cc$ to his standard forms by a rotation of axis. It is apparent from his general forms that, for a parabola, the square of the $xy$ coefficient must equal four times the product of the coefficients of the other terms of second degree (if all terms are brought to one side of the equation). For the corresponding general forms of the ellipse and hyperbola, he gave a rule on inequalities equivalent to the modern $B^2 - 4AC \lesseqgtr 0$, but his form of statement is considerably more awkward.[16] The *Elementa curvarum* closes with the claim that all loci of less than "three dimensions" [third degree] have been covered—an aspect of algebraic geometry which had been wanting in the work of Descartes.

The *Elements* of de Witt is in a sense complementary to the *Conics* of Wallis. Wallis first expressed the conic sections in analytic form and from these equations derived the properties of the curves; but de Witt first derived the properties of the conics geometrically and then showed analytically that second-degree equations represent curves with these properties. Were one to combine the analytic portions of these two works, the result would be a fair approximation to the material in modern textbooks. One would even find the familiar locus of points for which the sum (or difference) of the distances to two fixed

[16] *Geometria* (1659–1661), II, 283.

points is constant, for this proposition was given by de Witt in analytic, as well as kinematic, form. However, the political fame of de Witt overshadowed that which he might have gained in mathematics, and his *Elementa curvarum* was not much better known than Wallis' *Conic Sections*.

There is in the early works of Wallis and de Witt a significant omission, that of the Cartesian emphasis on geometrical constructibility by the use of conics. This situation, however, was not representative of the period. The year 1659 saw the publication not only of de Witt's treatise, but also of the *Mesolabum* of René de Sluse (1622–1685). The title[17] and theme of the latter book was suggested by the ancient solution of the Delian problem by the construction of mean proportionals in the manner of Eratosthenes, or through the use of intersecting curves, as adopted by Menaechmus. Sluse's "book of means" furnished a new stimulus to the Cartesian theory of geometrical constructibility. Adopting the *e* and *a* of Fermat as the unknowns and using the composition of proportions, he reduced "solid" problems to the determination of intersections of circles and conic sections. He proved that for any (determinate) cubic or quartic equation and any given conic, he could determine a circle which would, through its intersections with the conic, solve the equation. This is a direct continuation of the Cartesian tradition; but whereas Descartes had not given the key to the method by which he determined the equation used in his constructions, Sluse gave a systematic procedure. Beginning with the equation to be solved and with the equation of the given conic, he simply manipulated these—by means of substitution and the rational algebraic operations— until he arrived in the end at the equation of a circle.

The *Mesolabum* of Sluse achieved wide popularity, in spite of its unusual notations, and it appeared in a second enlarged edition in 1668. A review of this in the *Philosophical Transactions* for the following year praises the book as "the most excellent Advancement made in this kind of Geometry, since the famous Mathematician and Philosopher Des Cartes."[18] It had been praised highly by Blaise Pascal (1623–1662) and it influenced James Gregory (1638–1675) and Christiaan Huygens (1629–1695) in their search for intersecting conics which might solve the problem of Alhazen.[19] But if on the one hand analytic geometry in

[17] The full title is *Mesolabum seu duae mediae proportionales inter extremas datas per circulum et per infinitas hyperbolas vel ellipses et per quamlibet exhibitae, ac problematorum omnium solidorum effectio per easdem curvas.* I have used the edition of 1668, Leodii Eburonum.

[18] *Philosophical Transactions* (1669), p. 903–909, esp. p. 909.

[19] See H. W. Turnbull, *James Gregory Tercentenary Memorial Volume* (London, 1939), p. 435–440.

the sense of Descartes received a new stimulus through the *Mesolabum*, on the other hand analytic geometry in the sense of Fermat was advanced also to some extent by the correspondence of Sluse with Huygens on a new class of curves. The so-called "pearls of Sluse," defined as curves given by the equivalent of the equations $y^m = kx^n(a - x)^p$, are interesting as illustrations of the mistakes of the time with respect to negative coordinates. The first pearl-curve suggested by Sluse in 1657 was the cubic $b^2y = x^2(a - x)$. This is a simple cubic polynomial curve but Sluse's knowledge of its shape was limited to the portion lying in the first quadrant, and he mistakenly assumed the existence of a branch symmetric to it with respect to the axis of abscissas. Similarly Sluse did not recognize $ay + y^2 = ax - x^2$ as a circle, for he drew instead the arc in the first quadrant and then sketched its image in the axis of abscissas, giving the curve a special name. In 1658 he suggested such further cases as $ay - y^2 = x^2$, $ay - y^3 = a^2x$, $ay^2 - y^3 = ax^2$, and $ay^3 - y^4 = a^2x^2$, again under the impression that they were pearl-shaped.[20] It may be that the widespread use of oblique coordinates at that time obscured the now customary tests for symmetry with respect to the axes. Upon finding the point of inflection and the critical points in the case of the cubic polynomial curve $ax^2 - x^3 = a^2y$, however, Huygens (who had acquired the reputation of being van Schooten's best pupil[21]) saw the error into which they were falling and was able to draw the correct form. Like de Witt, Sluse was not aware of the fact that interchanging the variables $x$ and $y$ results, for rectangular coordinates, in the mirror image of a curve with respect to the line $y = x$, and so he gave independent treatments of the two cases. That the two coordinates are essentially on the same footing had been implied by Descartes and Debeaune. This principle, overlooked by Sluse, was more specifically recognized by Philippe de Lahire (1640–1718), one of the important contributors of the century to synthetic and analytic geometry.

De Lahire's father was a close friend of Gérard Desargues (1593–1662), and so the son undoubtedly came in contact with the new work in synthetic geometry. It is true that in general during the seventeenth century the projective and analytic schools, although they sprang up almost simultaneously, had little in common. Gregory of St. Vincent (1584–1667), for example, in 1647 published a voluminous work, *Opus*

---

[20] For Sluse's correspondence on these curves see Christiaan Huygens, *Oeuvres complètes* (22 vols., La Haye, 1888–1950), II, 47, 76, 88 ff., 93, 106, 121. For a general account of the pearls see Gino Loria, *Spezielle algebraische und transcendente ebene Kurven. Theorie und Geschichte* (Leipzig, 1902). The early days of analytic geometry and curve sketching were marked by surprisingly wide-spread error and misconception.

[21] See A. E. Bell, *op. cit.*, p. 19.

*geometricum*, in which a multitude of properties of the conics are derived by a method of perspective.[22] This work anticipated relationships between the circle and the hyperbola which later were developed analytically by Riccati, and it gave the quadrature of the hyperbola, an important contribution to the calculus. But the *Opus geometricum*, composed in the older language of proportions, had little influence on coordinate geometry. Similarly the *Brouillon projet*[23] of Desargues in 1639 blazed a new trail in the study of conics; but his work, too, written in a bizarre language, seemed to have little to offer to contemporary analytic geometry and it was soon virtually lost and forgotten.

Lahire was one of the very exceptional geometers who were able to appreciate both the analytic and synthetic developments in the theory of conic sections. Evidence of the inspiration of Desargues appears in two works by Lahire: *Nouvelle méthode en géométrie pour les sections des superficies coniques et cylindriques* of 1673 and *Sectiones conicae* of 1685.[24] These highly valuable and original treatises, both synthetic in method, have led to the impression that Lahire opposed the new geometry of Descartes; but a trilogy of 1679 shows him as an important link in the continuation of Cartesian traditions. The first of the three books bound together in this latter work carries the title *Nouveaux elemens des sections coniques*. It is a planimetric, but nonanalytic, treatment such as was found in the first book of de Witt's *Elements*. Beginning with the plane definitions in terms of the sum and difference of focal radii, Lahire deduced the properties of the ellipse and hyperbola. The theorems on the parabola are derived from the equality of distances to focus and directrix. The second of the three books is called *Les lieux géométriques*, a work which corresponds in a sense to the second book of de Witt's *Elements*, with less emphasis on the graphical interpretation of equations and more upon the representation of indeterminate problems by means of coordinates. In the latter connection he hinted (as had Oresme and the anonymous author of the Cartesian *Introduction* of 1638) at a generalization of analytic geometry for spaces of more than three dimensions, for he defined a geometrical

---

[22] See Karl Bopp, "Die Kegelschnitte des Gregorius a St. Vincentio in vergleichender Bearbeitung," *Abhandlungen zur Geschichte der mathematischen Wissenschaften*, XX (1907), 87–314. A long general account of his work is given in the "Discours preliminaire" in the French translation by Rondet of Edmund Stone's *Integral calculus—Analise des infiniment petits, comprenant le calcul integral* (Paris, 1735). A shorter biography is given in the paper by H. Bosmans, "Grégoire de Saint-Vincent," *Mathesis*, XXXVIII (1924), 250–256.

[23] See Desargues, *Oeuvres*, ed. by Poudra, 2 vols., Paris, 1864.

[24] For a summary of this work see Ernst Lehmann, "De La Hire und seine Sectiones Conicae," *Jahresbericht des Königlichen Gymnasiums zu Leipzig*, 1887–1888, pp. 1–28; or refer to Coolidge's *History of the Conic Sections and Quadric Surfaces*, p. 40–44.

locus as a straight line, or a curve, or a surface, etc., the points of which bear the same relationship to the points of a given line. For two dimensions this is a clear expression of the Cartesian view of coordinates with respect to a single axis, although Lahire's terminology stems from Desargues and the architectural tradition to which he at first belonged. The given line he calls the "trunk" and on it he takes a fixed point $O$ known as the "origin." He calls the points on the trunk "knots" and the distances of these from $O$ the "parts of the trunk." The lines drawn at a given fixed angle to the trunk he designates as "branches," and the ends of these are the loci. This was the first systematic application to a coordinate system of a conventional terminology. Previous writers had largely extemporized, and although at various times since Apollonius certain phrases—including the words abscissa and ordinate—had been used, especially in Latin translations of the *Conics*,[25] in connection with conic sections and coordinates, such terms were not recognized as names necessarily corresponding to their present application. Frequently terms like "segment" or "portion" or "intercepted diameter" were used instead of "abscissa"; and throughout the eighteenth century the phrase *ordinatis applicata*, used by Descartes and Fermat, appeared and was abbreviated in French either as *appliquée* or as *ordinée*, the latter finally winning out at the end of the century. It is not the purpose here to indicate the variety and sources of terms used with reference to coordinate systems,[26] but it is well to

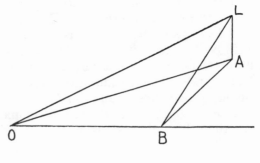

Fig. 19

[25] Borelli's translation of 1661, for example, regularly used the word "abscissa."

[26] The reader who is curious about these details should consult Johannes Tropfke, *Geschichte der Elementar-Mathematik*, v. VI (2nd ed., Berlin and Leipzig, 1924); or cf. F. Dingeldey, "Coniques" (translated by E. Fabry), *Encyclopédie des sciences mathématiques*, III (3), 1–256, especially p. 1–15.

remark that by 1679 analytic geometry had reached the point where an appropriate technical language was felt necessary. Of Lahire's notations, only the word "origin" was retained by his successors; but by the end of the century the terms axis, abscissa, coordinates, and ordinate (or appliquée) had come to be used generally in the present sense. Lahire himself, in a memoir composed thirty years later, adopted the words ordinate and abscissa.[27]

Descartes and Fermat both had indicated that an equation in three unknowns is represented by a surface, without giving further details. Lahire went further in this connection and showed that an indeterminate problem lacking two conditions could be represented as follows:

Let the problem be to find the point $L$ in space such that if a perpendicular $LB$ is dropped from $L$ to a fixed line $OB$ in a given plane, the line $LB$ shall exceed the line $OB$ by a fixed distance $a$. To illustrate this geometrically, Lahire dropped a perpendicular $LA$ from the point $L$ to the given plane, and then drew $AB$ perpendicular to $OB$ (Fig. 19). Then with $LA = v$, $OB = x$, and $AB = y$, he was led to the equation $a^2 + 2ax + x^2 = y^2 + v^2$ as that satisfied by the coordinates of the point $L$. This is important as the first example of a surface given analytically by means of an equation; but Lahire unfortunately did not continue the problem further. He was chiefly concerned with pointing out the degree of indeterminacy of the problem, and so did not bother to describe or sketch the locus of $L$. It is to be presumed that he recognized this as a surface, for later he expressly said that the points $L$ of a surface are related to the points of a given line $OB$ by passing a plane through $OB$, drawing lines $LA$, parallel to each other, from the points of the surface to the points of the plane, and then from the points $A$ in the plane drawing lines $AB$ parallel to each other. This reference to a single axis and to oblique coordinates in space resembles the corresponding treatment at the time in the plane. When, in the following century, interest in three-dimensional analytic geometry was resumed, the convention of a single axis was retained, but the developments were then limited largely to the case of rectangular systems.

The twenty-year interval between de Witt's *Elementa curvarum* and Lahire's *Lieux géométriques* seems to have brought little improvement in the matter of negative coordinates. Lahire referred to the fact that in determining the intersection of a line with a parabola a root of the resulting equation may be false (negative), but this had been described by Descartes long before. Wallis claimed, without adequate justification, that the *Lieux géométriques* was an imitation of his *Conics;* but

---

[27] "Remarques sur la construction des lieux géométriques & des equations," *Mémoires de l'Académie des Sciences*, 1710, p. 7–45.

one wishes indeed that Lahire had followed Wallis' later work in the correct use of signed coordinates. It is often asserted that Lahire was the first one to recognize that the two axes in a plane Cartesian system are on the same footing and that hence the variables are interchangeable, but such a claim is too strong. Lahire made use of only a single axis, as shown above; and his reference to the interchanging of $x$ and $y$ is nothing more than what was implied by Descartes[28] and appeared more clearly in Debeaune long before—that either unknown can be measured along the reference line or axis. Lahire repeated the question raised by Wallis of determining the form of a curve from an inspection of the coefficients of the equation, but he did not attempt a general solution.

The third book of Lahire's work of 1679 is *La construction des équations analytiques*. This covers the graphical solution of equations by means of intersecting curves in the manner of Descartes. Lahire retained the Cartesian dyadic classification of equations, but he pointed out, as had Fermat and Huygens, the error Descartes had made in assuming that the construction of equations of degree $2n$ and $2n - 1$ requires curves of degree $n$. He laboriously listed correctly the curves of minimum order which sufficed to construct equations the degrees of which are not greater than 64. Lahire's *Construction* breaks with the Cartesian tradition in another respect, for in solving quintic and sextic polynomial equations the construction is not traced back to a moving conic but is effected by a direct application of "the parabola of the second kind," $x^3 = aay$. It is possible that this change was inspired by Wallis who had given the correct form of the cubic parabola more than a score of years before.

The contributions of Wallis to analytic geometry were continued in 1685 through the publication of his well-known *Treatise on Algebra*. This is interesting for a twenty-chapter account of the algebra of Harriot; but a more relevant portion of the book is the section on the customary geometrical construction of polynomial equations beyond the second degree.[29] For the solution of the cubic and quartic by conics, one is referred to a work published the preceding year by Thomas Baker, *The Geometrical Key: or the Gate of Equations Unlock'd*, in which only parabolas and circles are used. Wallis, however, suggested an alternative non-conical construction as follows: Let the equa-

---

[28] See *La géométrie*, p. 385, for the statement that, given an equation in the two unknowns $x$ and $y$, one can take either of these as the independent variable and determine the other in terms of it.

[29] For the continuing importance of this topic in the seventeenth and eighteenth centuries see the works of Favaro and Matthiessen cited in chapter I. In this sense algebra and geometry were more closely related then than they are now.

tion be a cubic; or, if it be a quartic, let it be reduced to the resolvent cubic. Then transform to remove the second term. Let the final equation be $R^3 - mR = \pm n$ or $R^3 + mR = \pm n$. Now draw the "Two Cubick Paraboloeids" $y = x^3$ and $y = -x^3$ (Fig. 20). Let $P$ be any point on the $y$-axis and let the horizontal line through $P$ cut one of the curves in the point $\alpha$. Let $F$ be the point in which the tangent at $\alpha$ intersects the $y$-axis. Now take points $H$ so that $\sqrt{\dfrac{m}{3}} : \dfrac{n}{m} = \alpha F : FH$,

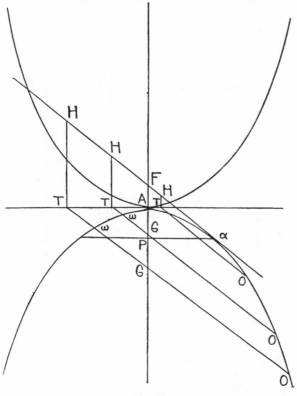

Fig. 20

where $H$ is to the right of $F$ for $+n$ and to the left for $-n$. Draw the ordinates for the points $H$ and let the feet of these be $T$. Then through the points $T$ draw lines parallel to $F\alpha$, cutting the cubic parabolas in points $O$ and $\omega$ and the axis in $G$. Then the segments $GO$ of these parallels are roots of $R^3 - mR = \pm n$, and the segments $G\omega$

are roots of $R^3 + mR = \pm n$.   Moreover, the root is positive when $O$ or $\omega$ falls below the $x$-axis and negative when it falls above this axis.

The present-day objection to this graphical method would be that it is unnecessarily intricate.   One might more easily have solved these cubics by finding the intersections of $y = x^3 \pm mx \pm n$ with the $x$-axis, or the intersections of $y = x^3$ with the lines $y = \pm mx \pm n$.   But Wallis in his day anticipated criticisms of a different nature:

> It may be objected against this Construction, that I here make use of a Line more compounded for a Problem which may be constructed by a Conick Section.
>
> But this Objection, I take to be (in this case) of no great weight; because it is compensated by cutting this with a Straight line, instead of a Circle. Which makes the Construction no more compounded than when a Circle cuts a Parabola.[30]

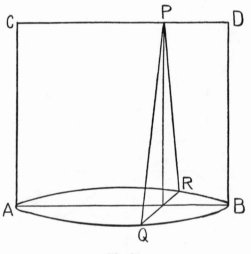

Fig. 21

To compute the extent of this "compounding," Wallis takes a straight line to be of weight one, a circle of weight two, a conic section of weight three, a cubic "paraboloeid" of weight four, and so on further according to the degree of the equation of the curve used.   Hence the weight of the compounding in the construction of Wallis is four plus one, or five, whereas that of canonical solutions making use of a conic and a circle is three plus two, or five also.   Hence the methods are on the same footing in this respect.

[30] *Treatise on algebra* (London, 1685), p. 275.

The *Algebra* of Wallis included in an appendix a short treatise he had published the year before on a solid figure which he called the cono-cuneus. It is described somewhat as follows: Given a rectangle *ABCD* (Fig. 21) and a circle with diameter *AB* drawn in the plane perpendicular to ABCD. Let a third plane, perpendicular to the other two, move so as to intersect the line segment *CD* in point *P* and the circle in points *Q* and *R*. The cono-cuneus is then the figure bounded by the totality of lines *PQ* and *PR*, continued indefinitely in both directions. This is significant as probably the first curvilinear figure other than solids of revolution, to be studied in the geometry of three dimensions. It would now be called a conoid, but at the time of Wallis this name still was applied in the Archimedean manner to the figures obtained by revolving a segment of a parabola or hyperbola about its axis. One must be cautious about the anachronistic use of terminology. Wallis in his *Conics* had used the words elliptoid, paraboloid, and hyperboloid to designate segments of the three types of *conic sections*, but some historians[31] have misread these as referring to general *quadric surfaces*. Wallis may indeed have recognized quadric surfaces, but he did not refer to them by the modern names. The lines on the hyperboloid of one sheet had been described[32] by Sir Christopher Wren (1632–1723) in 1669, and Wallis again pointed them out in his *Mechanica* of 1670.[33] Wallis noted the parabolic sections also, but the name given to the solid figure was not "hyperboloid" but "hyperbolic cylindroid." Wallis suggested that conic sections be substituted for the circular base of his cono-cuneus; and he proposed also[34] various "pyramidoids" or "conoids" in which the sections are similar conics with the ordinates being altered in a given ratio. (Here one finds a departure from the Archimedean use of the word conoid.) In this work Wallis was led to general quadrics, an important step beyond the surfaces of revolution with which geometers had been largely preoccupied;[35] but it should be noted that neither he nor his predecessors studied three-dimensional

---

[31] See Heinrich Wieleitner, *Geschichte der Mathematik* (new ed., Berlin, 1939), I, 121; E. Kötter, "Entwickelung der synthetischen Geometry von Monge bis auf Staudt (1847)," *Jahresbericht der Deutsche Mathematiker-Vereinigung*, V (1896), part 2, Leipzig, 1901, 65 f; Coolidge, "The Beginnings of Analytic Geometry in Three Dimensions," *The American Mathematical Monthly*, LV (1948), 76–86. Kötter points out that in Kepler's *Stereometria* there is a figure which seems to indicate an hyperboloid of one sheet. Cavalieri also may have known of this surface. Kötter's article includes considerable material on the early history of surfaces.

[32] *Philosophical Transactions* (1669), p. 961–962.

[33] See vol. I of his *Opera mathematica* (3 vols., Oxonii, 1693–1695).

[34] *Opera mathematica*, II, 23–42, 101–112.

[35] Roberval, for example, had classified loci as either plane or surface, saying that the latter are obtained from the former by revolution. Developments in the recognition and sketching of surfaces were at first extraordinarily halting.

figures analytically in terms of the equations of surfaces. In fact, mathematicians of the time were more interested in surfaces as the boundaries of solids, the volumes of which are to be determined, than as two-dimensional entities with analytic properties of their own.

Two years after the publication of Wallis' *Algebra* there appeared at Paris an unoriginal work by Jacques Ozanam (1640–1717) which continued the Continental tendencies in analytic geometry. This work of 1687 appeared in the same three parts as Lahire's, with titles as follows: *Traité des lignes du premier genre; Traité des lieux géométriques; Traité de la construction des équations.* This triple division of the new geometry—made up of first a general theory of conics; then a study of equations, especially of second degree; and finally the application of intersecting curves to the solution of equations—had become a tradition which persisted well into the following century. Ozanam's work helped to establish this tradition, but it added little to the material of analytic geometry. It served rather to strengthen the unfortunate impression left by Descartes that the study of curves had no intrinsic value but served only to facilitate the geometrical construction of the roots of algebraic equations. In fact, the author states in the introductory "Au Lecteur" of the *Traité des lignes* that this treatise "was composed chiefly in favor of those who wish to know how to solve equations of more than two dimensions by means of conic sections." However, Ozanam followed Wallis and Lahire in supplying also non-Cartesian solutions of cubics and quartics through the use of the familiar cubic parabola.[36]

Ozanam's treatment of linear and quadratic equations is unexceptional in its lack of generality. He did not show that an equation of first degree in all cases represents a straight line, nor did he consider the general equation of second degree. There is, however, an original touch where the author derives the standard equation of the parabola $y^2 = px$ from the corresponding forms of the ellipse and hyperbola $y^2 = px \pm \dfrac{px^2}{d}$ by allowing the axis $d$ to become infinite. In such suggestive but uncritical language one may see the influence of contemporary work in the calculus. In general, however, it is surprising that the two fields of analytic geometry and infinitesimal geometry remained so distinctly separate at that time. Even the Newtonian and Leibnizian infinitesimal methods of determining tangent lines seem frequently to have been neglected by Cartesian geometers who continued to use the awkward circular construction given in *La géométrie.* Perhaps one

---

[36] Loria, "Da Descartes e Fermat a Monge e Lagrange," p. 806–807, incorrectly implies that Ozanam was first to give this.

reason for this is to be found in the surprising fact that neither field had yet developed a general theory of curves beyond the conics.

The uninspired work of Ozanam was one of the last systematic commentaries of the century,[37] and the next dozen years or so leave the definite impression that enthusiasm for the calculus was responsible for a neglect of other branches of mathematics. The scientific journals of the time carried numerous articles on infinitesimal methods and problems, but few on either analytic or synthetic geometry. The formula for radius of curvature, for example, appeared frequently in L'Hospital's calculus of 1696,[38] but the distance formula did not appear in his analytic geometry of 1707. The indefatigable Bernoullis shared this tendency, but they found time nevertheless to add in a significant way to the history of analytic geometry. Polar coordinates had been implicit in a number of earlier works, such as that of Archimedes on the spiral. The comparison of spiral curves and parabolas was a favorite topic of the seventeenth century—especially for Cavalieri, Torricelli, Gregory of St. Vincent, Fermat, Roberval, Pascal, and Sluze—and this work bears definite resemblance to the use of polar methods. Jacques Bernoulli (1654–1705) in 1694 seems to have glimpsed the possibility of vectors in a general coordinate system, for he derived a formula for radius of curvature in polar coordinates.[39] He took the "applicata," $y$, as a radius measured from a fixed pole or "umbilic," and as abscissa, $x$, the arc of a circle of radius $a$, and center at the pole, intercepted by the radius and a given fixed line [polar axis]. To put his coordinates in modern form, one needs simply to replace the symbol $y$ by $r$ and $x$ by $a\theta$. He applied his new theorem (which is easily converted to the modern formula of the calculus) only to the spiral of Archimedes, $y = ax:c$. In the *Acta Eruditorum* a few years earlier[40] (in 1691) Bernoulli had suggested a somewhat different scheme related to polar coordinates. He took the equation of the parabola $yy = lx$ and inquired what the curve would be like if one were to measure the abscissas along the circumference of a fixed circle, the ordinates being taken along the corresponding normals to the circle. The resulting curve, obtained by bending the axis of the parabola around the circle,

[37] However, Paul Tannery, "Notes sur les manuscrits français de Munich 247 a 252," *Annales internationales d'histoire* (Congrès de Paris, 1900), 5th section, *Histoire des sciences*, pp. 297–310, has described a large unpublished work of 1982 pages, composed about 1700, with the title "Application de l'algebre et des lieux geometriques pour la solution des problémes de geometrie." This manuscript is attributed to Ozanam, and the theme of it is similar to that of his published work.

[38] It was, of course, known earlier to Huygens, Leibniz, Newton, and the Bernoullis. See J. L. Coolidge, "The unsatisfactory story of curvature," *Am. Math. Monthly*, LIX (1952), 375–379.

[39] See Jacques Bernoulli, *Opera* (2 vols., Genevae, 1744), p. 578–580; or *Acta Eruditorum*, 1694, p. 264–265; or *Bibliotheca Mathematica* (3), XIII (1912–1913), 76–77.

[40] See *Opera*, I, 431 f.

he called a parabolic spiral or helicoidal parabola. Its equation in modern polar coordinates would be $(a - r)^2 = la\theta$, where $a$ is the radius of the circle. The work of Bernoulli seems to be the earliest published application of the idea of polar coordinates in analytic geometry,[41] but the new system went largely unnoticed.

In 1695 Jacques Bernoulli saw to the publication of the fourth Latin edition of van Schooten's translation of Descartes, and with this he included extensive comments on a favorite Cartesian topic—on the graphical solution of polynomial equations.[42] In this connection he added his name to the long list of those who, like Fermat, had corrected Descartes' inference that a curve of order $n$ is necessary for the construction of an equation of degree $2n$. In his treatment of the topic Bernoulli took exception also, as had Wallis before him, to the Cartesian insistence on solutions by curves of lowest possible degree, as well as to the system of classification of Descartes. He preferred to arrange curves by degree, and he said that geometers had given no reasons to support the authority of Descartes on the matter of simplicity. Bernoulli pointed out that if one were to attempt the construction of a ninth-degree equation such as $x^9 + mx^7 + nx^6 + px^5 + qx^4 + rx^3 + sx^2 + tx + v = 0$ in terms of cubics, the usual procedure of substituting $a^2y = x^3$ in the first five or six terms results in a cubic non-polynomial curve of the form $a^6y^3 + a^4mxy^2 + a^4ny^2 + a^2px^2y + a^2qxy + a^2ry + sx^2 + tx + v = 0$ which would be exceedingly difficult to draw. He therefore proposed an alternative method of solution as follows: Let the equation to be solved be $x^5 = ax^4 + b^2x^3 - c^3x^2 - d^4x + e^5$. Divide this by $x^4$, obtaining $x = a + \dfrac{b^2}{x} - \dfrac{c^3}{x^2} - \dfrac{d^4}{x^3} + \dfrac{e^5}{x^4}$. Now construct [for positive abscissas] the points of the curve $y = a + \dfrac{b^2}{x} - \dfrac{c^3}{x^2} - \dfrac{d^4}{x^3} + \dfrac{e^5}{x^4}$, a simple quartic polynomial in the reciprocal of $x$ (Fig. 22). This construction can easily be made by means of proportions of various orders, inasmuch as only rational operations are involved. Then the ordinates of the points of intersection of this curve with the line $y = x$ are lines representing the positive roots of the original quintic equation. Bernoulli indicated that this procedure can be applied to polynomial

[41] An earlier hint of somewhat the same idea is found in the work of James Gregory who suggested that a curve be bent in such a way that all the ordinates become concurrent radii through a point, while the length remains unchanged. See James Gregory Tercentenary memorial volume (ed. by H. W. Turnbull, London, 1939), p. 493 f. (Cf. J. L. Coolidge, "The Origin of Polar Coordinates," *Proceedings of the International Congress of Mathematicians*, 1950, V. I, p. 749.

[42] *Notae et animadversiones tumultuariae in universum opus geometriam Cartesii*, p. 423–468. Cf. also *Acta Eruditorum* for 1688.

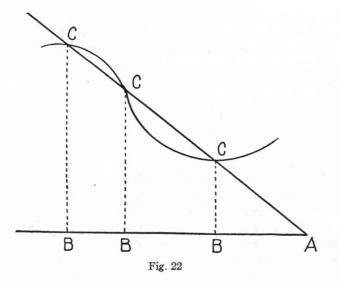

Fig. 22

equations of any degree.[43]    The attitude of the time on the shibboleth of constructibility is seen in Bernoulli's assertion that he "is not ashamed to use this method."

Two years later, in a letter to Leibniz of April 3, 1697, Jean Bernoulli (1667–1748) likewise broke with the Cartesian tradition by suggesting an alternative construction of equations.[44]    In this letter he drew a curve, presumably of a quartic polynomial, and indicated that the intersections of this with the axis (of abscissas) gave the roots of the equation.    Here one sees the modern graphical solution of polynomial equations; but the graphical representation here was incidental and not specifically for the purpose of solving the equation.    Moreover, this work was not published at the time; but it is quite possible that it was known to L'Hospital, an eager student of Jean Bernoulli, who published a somewhat similar method in the following century.

The correspondence of Jean Bernoulli with Gottfried Wilhelm von Leibniz (1646–1716) tells of a general development in mathematics which in the end was found to influence analytic geometry.    The seventeenth century saw the old language of proportions give way to the symbolism of equations in two unknowns, so that in 1693 Leibniz expressed the definitive opinion: "I have always disapproved of the fact that special signs are used in ratio and proportion, on the ground that for ratio the sign of division suffices and likewise for proportion the

[43] See Jacques Bernoulli, *Opera*, II, 689–691.  Cf. I, 343–351.
[44] See *Leibnizens mathematische Schriften* (ed. by C. I. Gerhardt), III (part I), Halle, 1855, p. 390–391 and Fig. 5.

sign of equality suffices." Equations had become the recognized form
of representation for functional relationships, and the particular symbolism of equality which was finally adopted was the familiar two
parallel lines of Recorde which Viète, Newton, and Leibniz popularized. At first Leibniz and Bernoulli used the word function with
various connotations tending toward the modern meaning. Originally
it designated certain variable geometrical quantities—such as ordinates, tangents, and radii of curvature—connected with a given
curve; sometimes the term indicated powers of algebraic variables.
By 1718 Bernoulli had come to apply the expression generally to
"quantities formed in any manner whatever of an independent variable
and constants." Leibniz already had rechristened the "geometric"
and "mechanical" curves of Descartes, using instead the names "algebraic" and "transcendental," and Bernoulli carried this terminology
over into the now familiar classification of functions. The notation
$f(x)$ was not used at the time but entered about 1734 with Clairaut and
Euler, two men who were to play decisive roles in the next period in the
history of analytic geometry.

While Leibniz and the Bernoullis were spreading the new calculus on
the Continent, there appeared in Scotland a little-noticed writer who
published works important in both analytic and infinitesimal geometry.
John Craig (†1731) is said to have been one of the first two (Jacques
Bernoulli being the other) to take up the study of the calculus, and in
1685 and 1693 he published two treatises on the subject. The first,
*Methodus figurarum lineis rectis et curvis comprehensarum quadraturas
determinandi*, is devoted especially to the calculus (only a year after the
cryptic paper of Leibniz had appeared!), but it is of some interest also
in analytic geometry. In finding the areas of parabolas of varying degree, Craig uses the figure of an Apollonian parabola in all cases, including the cubical and semicubical. The author apparently was not
familiar with the correct use of negative coordinates made by Wallis
almost thirty years before. However, the second work by Craig,
*Tractatus mathematicus de figurarum curvilinearum quadraturis et locis
geometricis*, made an important positive contribution to analytic
geometry. It contained a section, entitled "Nova methodus determinandi loca geometrica,"[45] in which a new method is proposed for
determining the nature and properties of the conic section represented
by any equation of second degree with respect to Cartesian axes, rectangular or oblique, without the reduction of the equation by means of
the geometrical transformations emphasized by de Witt and van
Schooten, to whom Craig refers in complimentary terms. The method

[45] *Tractatus mathematicus*, p. 62–76.

depends upon the derivation of four standard forms—one for the ellipse, one for the parabola, and two for the hyperbola—and inasmuch as the approach is similar in all four cases, it will suffice to illustrate it here for the parabola.   Let $A$ be a "certain and immutable fixed point" (i. e., the origin) and $AE$ a "straight line in any given position extended indefinitely" (i. e., the axis of abscissas).   Let $G$ be the vertex of a parabola $GD$ (Fig. 23) with diameter $GH$ and latus rectum $r$.   Let $AED$ be the given or assumed angle which the coordinates make with each other.   Through $A$ draw $AF$ parallel to $GH$ and $AK$ parallel to $ED$. Let $BC$ be a fixed line drawn parallel to $ED$ and let $AB = m$ and $BC = n$.   Then if $AE = x$, $ED = y$, $AC = e$, $AK = k$, $KG = l$, the equation of the parabola is found to be

$$ y^2 + \frac{2nxy}{m} - 2ky + \frac{nnxx}{mm} - \frac{2nkx}{m} + kk - \frac{rex}{m} + rl = 0. $$

This is, of course, the general equation of a parabola with vertex $(l, k)$ and axis inclined to the $x$-axis at $\arctan\left(-\dfrac{n}{m}\right)$.   If one compares it with the usual modern form, $Ax^2 + Bxy + Cy^2 + Dx + Ey + F = 0$,

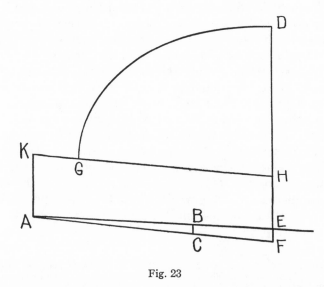

Fig. 23

it is apparent that, for any parabola, $B^2 - 4AC$ must be zero.   Craig, like de Witt, did not specifically refer to this quantity, now known as the characteristic, but it seems likely that he was aware of its significance. From Craig's general form it is also clear that the angle of inclination of the axis is given by $\tan 2\theta = \dfrac{B}{A - C}$, a fact which again he undoubtedly recognized.

After repeating the above procedure with the $x$ and $y$ coordinates interchanged, Craig then used the method of equating coefficients to compare the equation of a given parabola with one or the other of his two standard forms in order to determine analytically the properties of the curve. Given the equation $y^2 - \dfrac{bxy}{a} + \dfrac{bbx}{4aa} - bx - dd = 0$, for example, he noted that $\dfrac{2n}{m} = -\dfrac{b}{a}$, indicating that the tangent of the angle between the axis of the curve and the axis of abscissas is $\dfrac{b}{2a}$. Moreover, it is clear from the comparison of coefficients that $k = 0$, that $r \csc \theta = b$ or $r = b \sin \theta$, and that $rl = -d^2$ or $l = \dfrac{-d^2 \csc \theta}{b}$, so that the orientation, vertex, and latus rectum of the parabola are known without recourse to geometrical diagrams.[46]

Craig similarly derived standard forms for the ellipse and hyperbola, first with respect to the $x$-axis and then with respect to the $y$-axis. In order to avoid a division by zero, he treated separately the case of the hyperbola in which the coefficients of $x^2$ and $y^2$ (but not of $xy$) are zero. Through a comparison of the coefficients of the equation of a given conic with those of his canonical forms, Craig was able to derive expeditiously the properties of the curve. His work thus represents the most thoroughly analytic treatment of the general equation of second degree to appear in the seventeenth century. His forms may strike a modern reader as unduly awkward, but part of this impression is due to the fact that he lacked the trigonometric symbols and formulas which today make the rotation of axes a simpler procedure. The method he proposed compared favorably with those of his contemporaries and was adopted by L'Hospital, the most successful textbook writer of the eighteenth century.[47]

---

[46] The language and notation of Craig have been slightly modified, for purposes of exposition. He did not use the modern trigonometric symbols, but made use instead of the sides of an auxiliary triangle, as was customary in his day. The more modern approach to trigonometry entered about half a century later through the work of Euler.

[47] Yet the work of Craig has been almost completely overlooked by historians, and much of the credit he deserves has gone instead to L'Hospital. Craig is not mentioned by Coolidge, Loria, or Tropfke, three of the outstanding historians of analytic geometry. Wieleitner, however, gives a fair account of his work in *Geschichte der Mathematik*, Vol. II; part II (Berlin and Leipzig, 1921).

In spite of Craig's error with respect to negative coordinates, in connection with the higher parabolas, mathematical correspondence and articles of the last part of the seventeenth century show that ideas on this point where becoming clearer.[48]   The folium of Descartes had betrayed the erroneous views of geometers with respect to negative coordinates.   In 1663 Huygens considered only the portion in the first quadrant, but his letters to L'Hospital in 1692 and 1693 show the complete curve sketched correctly.[49]   The correct graph of the lemniscate, given by Jacques Bernoulli in *Acta Eruditorum* of 1694, shows skill in the plotting of curves.   Both of the Bernoulli brothers referred to the radius vector in the sense of a coordinate, anticipating the use of polar coordinates, and Jean suggested a scheme of plane coordinates making use of the radius vector and the ordinate—an interesting compromise between the rectangular and polar systems—but such ideas were developed more particularly in the following century.   Jean Bernoulli in 1692 used the name "Cartesian" for geometry based on a coordinate system, and it is interesting to note that he interprets this as the determination of the equation of any curve given by an assigned property. The converse Fermatian aspect, the graphical representation of equations, had not achieved a prominent place during the age of commentaries, but it was, at least for algebraic curves, to be a focal point in the development of analytic geometry throughout the first half of the eighteenth century.   Jean Bernoulli[50] and Leibniz encouraged the study of transcendental curves, especially of the form $x^x = y$, $x^y + y = x^x + x$, etc.   Leibniz emphasized that his calculus, unlike that of Viète and Descartes, was applicable alike to algebraic and transcendental curves, and Jacques Bernoulli said that geometry should study curves "which nature herself can produce by simple and expeditious motion";   but non-algebraic curves were not well-adapted to the methods of Cartesian geometry and so generally were not incorporated into the subject.   Leibniz in letters of 1694 first used the word "coordinates" in the strictly modern sense and recognized the two coordinates as on the same footing.   Correspondence in 1697 and 1698 of

[48] Carlo Renaldini, *Opera mathematum* (3 vols., Patavii and Venetiis 1684) seems to indicate a lack of progress along these lines in Italy as compared with that in other countries of Europe.   Numerous equations of "Medicean curves" are proposed, but they are not plotted. In Spain also there seem to have been few contributors to coordinate geometry.   P. A. Berenguer in "Un géometra espanol del siglo XVII," *El Progreso Matemático*, V (1895), 116–121, cites Antonio Hugo de Omerique as a precursor of modern analytic geometry; but the second half of his *Analysis geometrica* (Cadiz, 1698) was not published, making an estimate of his work impossible.
[49] See Huygens, *Oeuvres complètes*, X, 351 f., 378–417.   Cf. IV, 238, 246, 312, 316.
[50] See Jean Bernoulli, *Opera omnia* (4 vols., Laussanae and Genevae, 1744), I, 179.   He sometimes is referred to as the inventor of "the exponential calculus."   Transcendental curves consistently have played a larger role in the calculus than in analytic geometry.

Jean Bernoulli with Leibniz and L'Hospital on geodetic lines shows that he was undoubtedly familiar with the use of space coordinates, another aspect of analytic geometry to be developed during the following century.

It is surprising to note that the development of a general theory of curves also was a contribution of the eighteenth century. The invention of analytic geometry had opened the door to the easy definition and classification of an indefinitely great variety of curves, but the age of commentaries did not exploit this possibility. Coordinate methods at first were used largely to study *old* curves, the conics especially, rather than to invent new ones. In fact, the *new* curves which did appear in the interval between Fermat and Newton frequently were defined by other than analytic means, and these did not become a part of coordinate geometry until Euler later developed a general theory of functions. When in the crucial decade from 1634 to 1644 the simple logarithmic and sine curves had made their appearance, they were not plotted as the graphs of the equations $y = \sin x$ and $y = \log x$, for the function concept had not been introduced in these situations. The definition was in terms of superimposed motions and of geometrical transformations, ideas developed especially by Roberval and Evangelista Torricelli (1608–1647), the men to whom it appears we owe the first graphs of these curves.

The method of the composition of movements frequently is ascribed to Galileo, but in physics it goes back at least to the time of Aristotle, and in mathematics it is found still earlier in the quadratrix of Hippias. At about the time that Galileo was applying the method to projectile motion, the study of the cycloid again brought it into prominence in mathematics as a means of defining curves. It may well have been the cycloid which suggested in turn the new logarithm and sine curves.[51] Logarithms had been defined by Napier in terms of velocities of moving points—a line representing the number (or sine) decreased with a speed diminishing in geometrical progression while another line representing the logarithms of the number increased with uniform speed. Torricelli seems to have been the first one to modify this idea by imposing these two types of motion upon a single moving point: that is, he took abscissas at equal distances and drew the corresponding ordinates, beginning from a fixed initial ordinate, in continued geometric progression, with ratio less than one. This gave him a monotonically decreasing curve which, because of its form and generation,

---

[51] See Gino Loria, "Le ricerche inedite di Evangelista Torricelli sopra la curva logaritmica," *Bibliotheca Mathematica* (3), I (1900), 75–89; and Evelyn Walker, *A Study of the Traité des indivisibles of Roberval* (New York, 1932).

Torricelli called the *hemhyperbola logaritmica*. He showed that the subtangent of this curve is of constant length, and he found also the area between the curve, its asymptote, and a given ordinate.

The composition of motions had been one of the foremost means of defining *new* curves, but Roberval and others extended it to all the higher plane curves then known, as well as to the conic sections. The kinematic tangent methods of Descartes, Roberval, and Torricelli were so widely and effectively applied that they rivaled the analytic methods of Descartes and Fermat. It may well be that the two points of view in the early development of the calculus resulted directly from this rivalry between two methods of curve definition, with the composition of motions leading through Barrow to the fluxions of Newton, and with the analytic view culminating in the differentials of Leibniz.

During the seventeenth century the kinematic and analytic approaches were rivaled, in the discovery of new curves, by a method at once new and old—that of geometric transformation. The Ptolemaic and Mercator projections in the construction of maps, and the methods used by artists (especially Leonardo da Vinci and Dürer) in enlarging or

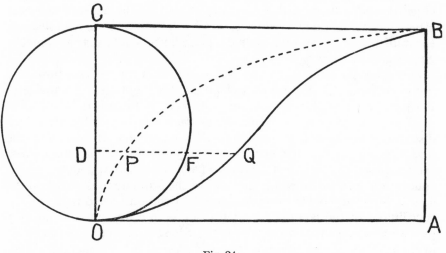

Fig. 24

reducing the scale of a design, represented simple examples of such transformations. The work of Desargues and Pascal in transforming conics projectively was but one instance among numerous types of curve transformation widely used at the time. In determining the quadrature of the cycloid, for example, Roberval had occasion to make use of a new curve defined by the following transformation: Let the ciccle $OCF$ roll along the line $OA$ and generate the cycloid $OPB$ (Fig. 24). For any point $P$ on the cycloid draw the line $DPF$ parallel to $OA$, and on this line take $PQ = DF$. The locus of $Q$ is a sine curve which divides the rectangle $OCBA$ into two parts, each equal in area to the generating circle. Since the area $OPBQ$ is equal to the semicircle $OCF$, the area of $OPBA$ is 3/2 the area of the generating circle. Roberval did not call the curve $OQB$ a sine curve, but merely the "companion of the cycloid." However, it is probable that he was aware of its connection with the trigonometric functions, for in another connection he constructed part of the same curve by plotting the sine lines of a quadrant of a circle as functions of the corresponding arc lengths. That the first consciously constructed graphs of a trigonometric function were motivated by the geometrical transformation of sine lines rather than by the analytical function concept probably explains the surprising fact that the notion of periodicity entered so slowly into the theory of goniometric functions. Wallis in the *Mechanica* of 1670 described the periodicity of the sine and cosine curves, drawing two full cycles of the sine curve;[52] and De Lagny in 1705 and Cotes in 1722 indicated the periodic nature of the tangent and secant curves; but only after 1748, when Euler's *Introductio in analysin infinitorum* emphasized the multiple angle formulas, was the periodicity of all the trigonometric curves generally recognized.

Several of the new curves discovered in the seventeenth century were found to have striking tangential properties. For the "parabolas of Fermat," $y = kx^n$, the ratio of subtangent to abscissa was found to be $1/n$; and the subtangent of the exponential curve was constant. Such discoveries suggested to geometers the possibility of reversing the question—of seeking new curves the tangents of which should possess certain properties specified *a priori*. Debeaune in 1637 sought, in this connection, a curve for which the ratio of the ordinate to the subtangent should for every point be proportional to the difference between the coordinates of the point. This deliberate search for curves exhibiting preassigned tangential properties may be looked upon as a further step in the systematic definition of curves.[53] Moreover, the merging of in-

[52] *Opera*, I, fig. 201 opposite p. 542.   Cf. p. 504–505.
[53] The following supply a wealth of information on curves in general: Gino Loria, *Spezielle algebraische und transzendente ebene Kurven* (2 ed. 2 vols., Leipzig and Berlin, 1911); F. Gomes Teixeira, *Traité des courbes spéciales remarquables planes et gauches* (transl. from the Spanish, 2 vols., Coimbre, 1908–1909); H. Brocard, *Notes de bibliographie des courbes géométriques* (Bar-le-duc, 1897); D. Joaquin de Vargas y Aguirre, "Catálogo general de curvas," *R. Acad. de ciencias exactas de Madrid, Memorias*, XXVI (1908); R. C. Yates, *A Handbook on Curves and Their Properties* (Ann Arbor, 1947).

verse tangent problems with the kinematic definition of curves did much to inspire Newton's invention of the calculus. The calculus then reciprocally served as a still further means of curve determination, for every differential (or fluxionary) equation in one function of a single independent variable (subject to given boundary conditions) implies a distinctive plane curve. During the following century algebraic geometry and the calculus grew to maturity together, but their close association was not entirely advantageous to the older of the two. That the development of analytic geometry from Newton to Monge was so surprisingly halting appears to have been due in part to the fact that this subject was overshadowed by the more novel and powerful—and hence more attractive—methods of infinitesimal analysis.

## CHAPTER VII

# From Newton to Euler

*There was far more imagination in the head of Archimedes than
in that of Homer.*                                                  —VOLTAIRE

THE eighteenth century was in some respects a prosy period in the
history of science and mathematics. The preceding age had run
off with the honors in the form of the great law of the universe
and the mathematical means of exploiting it. To the eighteenth century was bequeathed the task of sharpening the new analytical instruments; but in pursuing this assignment there appears to have been,
until the very close of the century, far more enthusiasm for the calculus
than for algebraic geometry. The work of Sir Isaac Newton (1642–
1727) in the calculus belongs properly in the seventeenth century, with
the publication in 1687 of the *Principia;* but his chief contribution to
analytic geometry appeared in the eighteenth, with the publication in
1704 of the *Opticks.* As the rules of the calculus were given unobtru-
sively as a lemma in the middle of the former book, so the work in
analytic geometry was relegated to an inconspicuous place as one of
two appendices in the latter volume. This appendix, the *Enumeratio
linearum tertii ordinis*, had been composed at least by 1676 and had
been revised in 1695, but publication was delayed by Newton's aver-
sion to putting things into print. In the *Enumeratio* analytic geometry in the sense of Fermat may be said to have come into its own.

Newton is reported to have mastered the *Géométrie* without any
preliminary study, but he also extended it in a new direction. In the
geometry of Descartes curves were defined (except in the case of those
of first and second order) as loci of moving points, and they were considered exclusively as auxiliaries in the solution of determinate algebraic equations. Newton, on the other hand, was more concerned
with the converse or Fermatian aspect of the subject. The *Enumeratio* opens with a brief description of the meaning of coordinates and of
their use in determining curves by means of equations.[1] Then, for

[1] An extensive account of this work is given by W. W. R. Ball, "On Newton's Classification of Cubic Curves," *Proceedings of the London Mathematical Society*, XXII (1890), p.
104–143. A summary of this appears in *Bibliotheca Mathematica*, new series, V (1891), p.
35–40. The *Enumeratio* was given in an English translation by Talbot, but I have not seen
this. I have used the Latin edition of 1787.

138

the first time since Fermat, a whole class of new curves defined by cubic equations in two unknowns are plotted as a matter of intrinsic interest.   It is, in fact, the first instance of a work devoted solely to the theory of curves as such.   Newton noted 72 species of cubics (half a dozen are omitted), and a curve of each species is carefully drawn. In this respect the work opened up an essentially new field, that of higher plane curves.   Moreover, one finds for the first time the systematic construction of two axes.   The second line is not called an axis, but only the principal ordinate, and it is not used in quite the same sense as the axis of abscissas.   The origin is looked upon as the initial point for abscissas only, inasmuch as ordinates are not measured along the principal ordinate line.   There is no hesitation, however, with respect to negative coordinates, and the curves are plotted completely and correctly for all four quadrants.   Newton is sometimes given sole credit for the correct use of negative coordinates, but he had been anticipated to some extent by others, notably Wallis and Lahire.   Such use seems to have developed but gradually in the period following Descartes.   In 1692 Christiaan Huygens (1629–1695) in a letter to L'Hospital correctly sketched the folium of Descartes, together with its asymptote, showing familiarity with negative coordinates.[2]   Men of the eighteenth century continued in many cases to use a single axis, but the sketching of curves and the use of negative values of the coordinates were fairly well established before the middle of the century.   Coordinate axes in the *Enumeratio*, as in many other works of the period, generally are assumed to be oblique.   Transformations of axes are not specifically given, but Newton was obviously familiar with them inasmuch as they are necessary for the reduction of equations to his canonical forms.   He noted, as had Descartes, the invariance of the degree of an equation under the usual transformations, and he interpreted the degree as the number of possible intersections of the curve with a straight line.   He abandoned the Cartesian classification for the modern designation according to degree, thus making way for the idea of order of a curve.   This work also provided for a new distinction between algebraic and transcendental curves.   Descartes had thought of curves as "geometrical" or "mechanical" according as the motions were algebraically determined or not—that is, according as $\dfrac{dy}{dx} = \dfrac{dy}{dt} \div \dfrac{dx}{dt}$ is algebraic or not.   Newton described a curve as transcendental or algebraic according as it does or does not intersect some straight line in an infinity of points, real or imaginary.

[2] *Oeuvres complètes* (22 vols., La Haye, 1888–1950), X, p. 351, 378.

Newton did not discuss the conics except as a guide in finding analogous properties for the cubics.   Thus, he defined a diameter of a cubic as the locus of points on a set of parallel chords such that from these points the sum of the two segments of the chord intercepted by the curve on one side is equal to the segment on the other side.   If the diameters go through a single point, this is known as the center.   Other properties of conics related to vertices, axes, latus rectum are modified suitably so as to apply to cubics.   As the hyperbola has two asymptotes, so Newton pointed out that a curve can have only as many asymptotes as is indicated by the order; and as all conics are projections of the circle, so all plane cubics can be obtained by projection from the five divergent parabolas given by $y^2 = ax^3 + bx^2 + cx + d$.   (Newton was one of the few early figures generally to abandon the homogeneous form of expression for equations, although Descartes had suggested this.)   In general, Newton saw that the order of a curve under projection is invariant, and he went further to define two curves to be of the same genus if one can be obtained from the other under projective transformation.   He accordingly classified curves of third order into five genera including 72 species.

Newton's interest lay especially in the Fermatian graphical study of given equations, but he contributed also to the kinematic generation of curves in a manner strongly reminiscent of Descartes' hierarchy of "Cartesian curves."   Book I of the *Principia* contains dozens of problems on the determination of conic sections satisfying given conditions —having given foci, passing through given points, and tangent to given lines.   Among them one runs again into the Pappus problem which loomed so large in *La géométrie* of Descartes; but Newton's solution of "that famous Problem of the ancients concerning four lines, begun by Euclid and carried on by Apollonius," is synthetic rather than analytic.[3]   Of greater interest is the Newtonian "organic description of curves," a topic likewise reminding one of the Cartesian construction of loci.   Newton had two angles of fixed magnitude and with fixed vertices $O$ and $O'$ rotate about $O$ and $O'$ so that the intersection $P'$ of one side $P'O$ of the one angle with one side $P'O'$ of the other should lie on a given curve $P'C'$ (Fig. 25).   Then as $P'$ moves along $P'C'$, the point of intersection $P$ of the other two sides of the angles will describe a curve $C$.   Newton pointed out that if $P'C'$ is a line, then $PC$ is a conic; if $P'C'$ is a conic passing through $O'$, then $PC$ is a cubic passing through $O$ with a double point at $O'$.   If $P'C'$ is an arbitrary conic, then $PC$ is either a cubic or a quartic.

[3] Sir Isaac Newton's *Mathematical Principles* (Cajori's revision of the Motte translation, Berkeley, California, 1946), p. 80–81.   Cf. p. 76–80.

Newton's contribution to the theory of curves was not limited to the *Enumeration of Cubics* and the *Principia*, for he had composed still earlier a work which Horsley published in the *Opera* of 1779 as *Artis analyticae specimina vel geometria analytica*. The title of this treatise includes the phrase which later was to become the standard designation

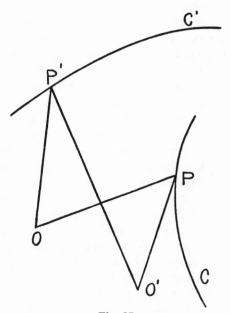

Fig. 25

of Cartesian or algebraic geometry, but here it designated a Newtonian work collated from three manuscripts essentially the same as that published by Colson in 1736 as the *Method of Fluxions*.[4] It is primarily on the calculus, but it includes material in other fields as well,[5] and considerable of it touches upon coordinate geometry. The opening sections are on the algebraic solution of equations, after which Newton describes a "diagram" which later came to be known as Newton's parallelogram. This well-known diagrammatic representation of polynomial equations of the form $f(x, y) = 0$ was used by New-

[4] *The Method of Fluxions and Infinite Series; with Its Application to the Geometry of Curve-Lines* trans. with commentary by John Colson (London, 1736).

[5] See *Opera quae exstant omnia* (5 vols., Londini, 1779–1785), I, 391 f. A brief description is found in H. W. Turnbull, *Mathematical Discoveries of Newton* (London and Glasgow, 1945). The *Methodus fluxionum* will also be found in Newton's *Opuscula* (3 vols., Lausannae and Genevae, 1744), I, p. 29–200.

ton to get successive approximations for one variable expressed as a function of the other through a power series convergent for small values of the latter; but in the early eighteenth century it became popular in the theory of curves as a quick method of obtaining graphical approximations for the curve about the origin.

The *Geometria analytica* includes maxima and minima, tangents, radii of curvature, and quadratures, all treated in an analytic manner which is in marked contrast to the synthetic form of the *Principia*. The term analytic geometry is applied, apparently, to "the application of algebra to geometry," especially "through the equations of curves." The work is full of locus problems and of curves sketched from equations. In some cases the curves are given kinematically or by differential equations, but generally they are sketched for all quadrants. Among the graphs one finds the Cartesian parabola and the parabolas and hyperbolas of Fermat. Newton derived also the equation $y = a^3/(a^2 + z^2)$ for a locus which he described as "conchoidal," but which later was known as the witch of Agnesi; but in studying and sketching this he had been anticipated by Fermat and Huygens.[6]

The contributions of Newton to coordinate geometry are not well known, and one aspect in particular has been completely overlooked by historians—his use, in the *Geometria analytica*, of polar coordinates. Having shown how to apply the method of fluxions in finding tangents to curves given analytically in Cartesian coordinates—oblique as well as rectangular—Newton added:

> However, it may not be foreign from the purpose, if I also shew how the problem may be perform'd, when the curves are refer'd to right lines, after any other manner whatever: so that having the choice of several methods, the easiest and most simple may always be used.

A little later he indicated that

> The problem is not otherwise perform'd, when the curves are refer'd, not to right lines, but to other curve-lines, as is usual in mechanick curves.[7]

To illustrate this point, he suggested eight further types of coordinate system. One of these, the "Third Manner" of determining a curve analytically, is what would now be called bipolar coordinates. In this connection Newton considered the "ellipses of the second order"—i. e., the ovals of Descartes. In *La Géométrie* Descartes had proposed these curves in problems on refraction, but he handled them, as Newton says, "in a very prolix manner," without the use of coordinates.

---

[6] Fermat gave no diagram, but Huygens (*Oeuvres complètes*, X, 370 ff) gave enough of the graph in the first quadrant to show clearly the point of inflection.

[7] *Opera*, I, p. 435, 441; *Method of Fluxions*, p. 51, 57.

Newton therefore seems to have been the originator of bipolar co-ordinates in the strict sense of the word. Representing by $x$ and $y$ the "subtenses" (or distances) of a variable point from two fixed points (or poles), Newton wrote "their Relation" for the ovals as $a + (ex/d) - y = 0$. For $a - (ex/d) - y = 0$, Newton observed that a contrary sense is indicated in the construction; and if $d = e$, he noted that the curve becomes a conic section. He closes this topic with the remark that "it would be easy . . . to give more Examples of it."

Newton suggested also other combinations of pairs of distances measured radially from given fixed points or obliquely to given fixed lines, or along arcs of circles. If, for example, $x$ is the distance to a fixed point and $y$ is the oblique distance to a given axis, then the equation $aa + bx = ay$ represents a conic section, and $xy = cy + bc$ a conchoid. If, on the other hand, $x$ is an arc measured along a unit circle and $y$ is the abscissa of the point, then $x = y$ is the equation of the quadratrix. No more general point of view with respect to co-ordinates is found in the history of analytic geometry before the nineteenth century.

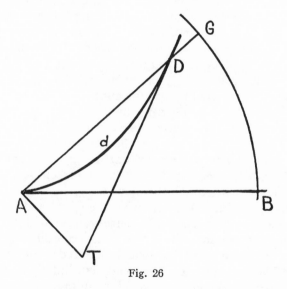

Fig. 26

Newton used polar coordinates in three distinct portions of his book on fluxions: first in connection with tangents, again for curvature, and finally in the rectification of curves. On the first two occasions the system he used was as follows: Let $A$ be the center and $AB$ a radius of the fixed circle $BG$, and let $D$ be any point on the curve $AdD$ (Fig.

26).   Then, designating $BG$ by $x$ and $AD$ by $y$, the curve $AdD$ is determined by a relationship between $x$ and $y$.   Newton suggests $x^3 - ax^2 + axy - y^3 = 0$ as an illustration, and then determines, from the proportion $\dot{y}:\dot{x}::AD:At$, the polar subtangent $AT$ for any point $D$ of this curve.   Similarly Newton found the polar subtangents of $y = (ax/b)$, "which is the equation to the spiral of Archimedes," and of $by = xx$; and, he concluded, "thus tangents may be easily drawn to any spirals whatever."[8]

Following the calculation of the radius of curvature for rectangular Cartesian coordinates $x$ and $y$—$r = \dfrac{\overline{1 + zz} \, \sqrt{1 + zz}}{\dot{z}}$, where $z = \dot{y}$ and fluxions of independent variables are taken as unity—Newton again turned to the corresponding problem in polar coordinates. Using a diagram and a notation similar to those applied in connection with tangent problems—but with the radius $AB$ of the reference circle taken as unity—he derived the result $r \sin \psi = \dfrac{y + yzz}{1 + zz - \dot{z}}$, where $z = \dot{y}/y$ and $\psi$ is the angle between the tangent and the radius vector (fluxions of independent variables again being taken as unity).   Newton applied this formula, virtually the same as the modern equivalent, to the spiral of Archimedes and to the curves $ax^2 = y^3$ and $ax^2 - bxy = y^3$.[9]   In conclusion he added, "And thus you will easily determine the curvature of any other spirals; or invent rules for any other kinds of curves."   That he realized the significance of his introduction of polar coordinates seems to be implied by his comment that he had "made use of a method which is pretty different from the common ways of operation."   In fact, Newton gave the equivalent of the transformation from rectangular to polar coordinates—$xx + yy = tt$ and $tv = y$, where $t$ is the radius vector and $v$ is a line representing the sine of the vectorial angle associated with the point $(x, y)$.

The comparison of the parabola with the spiral had been a favorite topic throughout the seventeenth century, and in his treatment of this question Newton made use of a polar coordinate system yet a third time.   Here, however, his scheme differed from that previously presented.   The notation, too, was modified, but this may have been done in order to avoid confusion in the simultaneous use of polar and Cartesian coordinates.   If $D$ is any point on a curve $ADd$, he took the coordinates of $D$ as $z$ and $v$, where $z$ is the radius vector $AD$ (Fig. 27) and $v$ is the circular arc $DB$.   That is, his coordinates were, in modern

[8] *Opera*, I, p. 440; *Method of Fluxions*, p. 56.
[9] *Opera*, I, p. 452–453; *Method of Fluxions*, p. 68–70.

notation, $(r, r\theta)$ instead of $(r, \theta)$ or $(r, a\theta)$. Then if the relation between $z$ and $v$ is given "by means of any equation"; and if a new curve $AHh$, given in rectangular coordinates $AB = z$ and $BH = y$, is so determined that, for all corresponding positions of $D$ and $H$, the arc $AD$ is equal to the arc $AH$; then Newton showed that $\dot{y} = \dot{v} - (v\dot{z}/z)$,

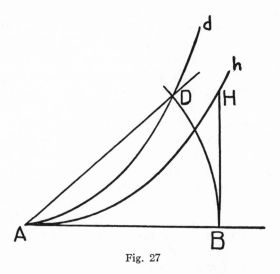

Fig. 27

or, if $z$ is taken as unity, $\dot{y} = \dot{v} - (v/z)$. In particular, "if $zz/a = v$ is given as the spiral of Archimedes," then $\dot{v} = 2z/a$ and hence $z/a = \dot{y}$ and $zz/2a = y$. The lengths of the spirals $z^3 = av^2$ and $z\sqrt{a + z} = v\sqrt{c}$ are shown similarly to correspond, respectively, to lengths measured along the semi-cubical parabola $z^{3/2} = 3a^{1/2}y$ and the curve[10] $(z - 2a)\sqrt{ac + cz} = 3cy$.

Evidence indicates[11] that the *Method of Fluxions* was composed by 1671, when Jacques Bernoulli was in his teens; and there seems to be no reason for suspecting the sections on polar coordinates of being a later interpolation. The three pertinent passages would appear to be a natural part of the whole, and Horsley, after his editorial examination of three different manuscript copies of the work, apparently saw no reason to question the date or authenticity of this material. It is therefore strange that this contribution to coordinate geometry should

[10] *Opera*, I, p. 511–512; *Method of Fluxions*, p. 132–134.
[11] See, e. g., H. G. Zeuthen, *Geschichte der Mathematik im XVI und XVII Jahrhundert* (Leipzig, 1903), p. 374. See also the opening page of the preface of the French edition of *La méthode des fluxions, et des suites infinies* (Paris, 1740).

have been so completely overlooked that the use of polar coordinates invariably is attributed to others of later periods.   Newton is not entitled to priority of publication, but he probably does deserve to be known as the first one to develop a system of polar coordinates in strictly analytic form.   Moreover, his work in this connection is superior, in generality and flexibility, to any that appeared during his lifetime.   The earliest publication of polar coordinates seems to be that of Jacques Bernoulli in 1691, mentioned above.   It is of some interest to recall that Bernoulli's second use of polar coordinates, in 1694, was identical, both in conception and notation, with the first system of Newton.   This same scheme and symbolism appeared again in 1704 in a memoire by Pierre Varignon (1654–1722) on "Nouvelle formation de spirales."[12]   Fontenelle, the secretary of the Académie, gives a glowing account[13] of this work of Varignon, apparently unaware of the anticipations by Newton.

Beginning with the equation of a curve given in Cartesian coordinates, Varignon raised the question as to what the curve would be if one took $y$ to be the radius vector and $x$ to be the arc of a fixed circle. The higher parabolas, for example, became spirals of Fermat.   Varignon gave an elaborate classification of other spirals obtained in the same way, but his treatise is tedious and unimaginative in comparison with the then unpublished work of Newton.

The *Geometria analytica* or *Method of Fluxions* remained unpublished until 1736, and hence it exerted but a limited influence on the course of the history of analytic geometry.   The *Enumeratio linearum tertii ordenis* meanwhile reappeared in 1706 in the Latin *Optice* of Newton and also in 1711 in the *Analysis per quantitatum series* of Wm. Jones.[14]   At first it did not attract much attention, but after about a dozen years it became the basis for a new trend in analytic geometry in the development of a general theory of algebraic curves by Stirling, Maclaurin, Nicole, De Gua, and Cramer.   In England this was almost the only aspect of algebraic geometry which attracted attention during the eighteenth century, for the Cartesian tradition had failed to take effective hold there.   This may be due in part to Newton's attitude toward the geometrical constructions of Descartes.   In the *Enumeratio* there is a section on the application of intersecting cubics

---

[12] *Académie des Sciences, Mémoires*, 1704, p. 69–131.
[13] *Ibid.*, p. 47–57.   It is interesting to note that Varignon, in referring to the work published by Jacques Bernoulli in 1691, ascribes the *idea* of polar coordinates to the younger brother, Jean Bernoulli.
[14] The work appeared also in the *Opuscula* of 1744, in the *Opera* of 1779, in a Latin edition of 1797, and in an English translation of 1861, as well as in the edition of 1717 with commentary by Stirling.

and quartics to the solution of polynomial equations of degree not greater than twelve, which would almost seem to indicate that Newton followed Descartes in feeling that curves had to be used for something. The equation $a + cx^2 + dx^3 + ex^4 + fx^5 + (g + m)x^6 + hx^7 + kx^8 + lx^9 = 0$, for example, is solved by the pair of cubic curves $x^2y = 1$ and $ay^3 + cy^2 + dxy^2 + ey + fxy + g + m + hx + kx^2 + lx^3 = 0$. Possibly here Newton was influenced by Fermat's later work, as well as by Descartes. But if Newton at one time embraced the Cartesian view on geometric constructions, it is clear that he renounced it in his *Arithmetica universalis*.

The *Géométrie* of Descartes contained an extensive treatment of algebra necessary in the solution of problems involving geometrical constructions, and so, conversely, the *Arithmetica universalis* of Newton devoted much space to geometrical questions and the "linear construction [graphical solution] of equations." This work was written in lecture form in the period 1673–1683, but it did not appear until 1707. In it there is a discussion of the plane, solid, and linear loci of the Greeks, in connection with which Newton gave a clear-cut statement of the aims of Cartesian geometry:

> But the Moderns advancing yet much farther, have received into Geometry all Lines that can be expressed by Equations...; and have made it a Law, that you are not to construct a Problem by a Line, i. e., a curve of a superior Kind, that may be constructed by one of an inferior one. ... In Constructions that are equally Geometrical, the most simple are always to be preferred. This Law is beyond all Exception.[15]

Newton went on to show, as had Wallis, that the contemporary notions of relative simplicity had not been clearly defined. The equation of the parabola (with reference to vertex and axis) is *algebraically* simpler than that of the circle and of the same degree; yet the circle was universally accepted as exceeding the parabola in *geometrical* simplicity and in ease of construction. In the geometric solution of equations Newton preferred—because of the ease with which they are constructed—the ellipse to the parabola, and the conchoid to the conics. Newton consequently proposed abandoning the ancient emphasis on conics and the then current insistence on curves of minimum degree. He would substitute for the rules of simplicity of Descartes and Wallis the policy that, "We are always to aim at Simplicity in the Equation, and Ease in the Construction." But where Wallis had eagerly accepted the Cartesian association of algebra and geometry, Newton expressed a point of view which would contradict the

[15] Sir Isaac Newton, *Universal arithmetick* (transl. by Raphson and revised by Cunn, London, 1769), p. 468–469 of the appendix. Cf. *Opera*, I, 200 ff.

whole purpose of *La géométrie*, and which would, moreover, exclude also the Fermatian conception of analytic geometry. In the *Arithmetica universalis* one reads:

> Equations are Expressions of Arithmetical Computation, and properly have no place in Geometry. . . . Therefore these two Sciences ought not to be confounded. The Ancients did so industriously distinguish them from one another, that they never introduced Arithmetical Terms into Geometry. And the Moderns, by confounding both, have lost the Simplicity in which all the Elegancy of Geometry consists.[16]

In this very same work Newton made expeditious use[17] of undetermined coefficients in determining the values of $e, f, g$, and $h$ so that the parabola $y = e + fx \pm \sqrt{gg + hx}$ should pass through four given points; yet he went on to express the opinion that "The modern Geometers are too fond of the Speculation of Equations." He asserted that

> Therefore the conic Sections and all other Figures must be cast out of plane Geometry, except the right Line and the Circle, and those which happen to be given in the State of the Problems. Therefore all these descriptions of the Conicks *in plano*, which the Moderns are so fond of, are foreign to Geometry.[18]

This reminds one of the attitude toward algebraic geometry of Newton's teacher Barrow; and it recalls the sharp criticism by Hobbes of "the whole herd of them who apply their algebra to geometry." The sentiments Newton expressed in the *Arithmetica* would indicate that he was here further from analytic geometry in the modern sense than either Descartes or Fermat; and yet it is clear from the *Enumeratio* and the *Geometria analytica* that he realized fully the value and power of coordinate methods. This paradox perhaps is resolved by the fact that Newton apparently would deny the validity of algebraic methods in *elementary*, but not higher, geometry. In this respect his attitude would not be greatly different from that of his contemporaries. Unfortunately, however, it was the *Arithmetica* which most influenced his countrymen, for it appeared during the century in at least three English editions (1720, 1728, 1769) and five Latin editions (1707, 1722, 1732, 1752, 1761), as well as a French edition of 1802. Perhaps it was for this reason that the next steps in the development of Cartesian geometry were taken on the Continent.

[16] *Universal arithmetick*, p. 470; *Opera*, I, p. 202.
[17] See Coolidge, *History of Conic Sections*, p. 75.
[18] *Universal arithmetik*, p. 494–496.

The state of analytic geometry in Europe at that time is well typified by two popular French works of 1705 and 1707. The first of these was the *Application de l'algèbre à la géométrie* of N. Guisnée (†1718). The title of this book was adopted throughout the eighteenth century as the customary name for what Jean Bernoulli had called "Cartesian geometry." The work is a direct continuation of the tradition set in the preceding century by Descartes, de Witt, Lahire, and Ozanam. It follows the Cartesian classification of curves as geometric or mechanical (the relations between the coordinates of which are not expressible geometrically). Guisnée, like Descartes, usually begins the discussion of a given curve, not with its Cartesian equation, but with a kinematic or geometric definition or property from which the equation is subsequently derived. The author recognizes two main problems in analytic geometry: (1) to construct the roots of determinate equations through intersecting curves; (2) to construct indeterminate equations (or loci). The word construct is here used in the narrow sense of Descartes and does not refer to plotting or sketching in the modern manner. Thus Guisnée suggests the following description of the curve $x^4 - ayxx + byyx + cy^3 = 0$: Let $az = x^2$, and substitute $az$ for $x^2$ in the first two terms of the given equation, obtaining $zz - yy + [(byyx + cy^3)/aa] = 0$. This last equation is, for a given value of $y$, a parabola in $z$ and $x$. The intersection of this parabola with the parabola $az = x^2$ will determine the value of $x$ corresponding to the given value of $y$; or, if one prefers, he can combine the equations of the parabolas to get a circle, after which the intersection of the circle with either parabola can be used. Repeating this for other values of $y$, corresponding values of $x$ can be found, and hence the indeterminate equation or locus is constructed. Determinate equations are solved in an analogous manner. The equation $a^6 = x^6 + a^4x^2$, for example, is constructed by the intersections[19] of $a^2z = x^3$ and $a^2 = z^2 + x^2$. Guisnée did not stop, as did Descartes, with the geometric construction. He then demonstrated synthetically, in the manner of the ancients, that this satisfies the given problem. The treatment of conics is extensive, but as in earlier works it is not entirely analytic. Incidentally, Guisnée was perhaps the first one to use the letters $a$ and $b$ for the semiaxes of the ellipse, the equations of which (with respect to center and vertex) he wrote as $(aa - xx) = aayy/bb$ and $(2ax - xx) = aayy/bb$. Variants of these have remained as standard forms ever since. Guisnée's work illustrates the gradual advance made in the use of coordinates in that he (like Leibniz) used two axes and that he dis-

---

[19] See *Application de l'algèbre à la géométrie*, p. 228 ff.

cussed the circle $y^2 = a^2 - x^2$ for both positive and negative values of the coordinates. Moreover, this book seems to be the first one in which *both* the $x$ and $y$ coordinates in a rectangular Cartesian system are interpreted as the segments cut off on the two axes by perpendiculars from a given point. However, in Guisnée and other books of the first half of the century, linear equations such as $ay = bx$ are regarded as determining half-lines in the first quadrant only. By the use of a variety of transformations,[20] the author reduced equations of the form $ax - by = aa$ or $ax = bc + by$ to the canonical form $ax = bz$.

The year in which Newton published the anti-Cartesian *Arithmetica* there appeared in France a conspicuously successful textbook on Cartesian geometry along the lines of that of Guisnée. This was the *Traité analytique des sections coniques* of the Marquis de l'Hospital (1661–1704), a book which contains less original material than that of Guisnée, but which is more extensive and closer to the modern manner of treatment. The work had been intended for publication at the time the author's famous calculus textbook appeared in 1696, but L'Hospital's illness apparently led to delay and it appeared posthumously in 1707. It is Cartesian in emphasis and although it consists of but one volume, it follows generally the tripartite plan of Lahire and Ozanam: first an algebraic quasi-analytic treatment of the conic sections along the lines of the Apollonian theory; then an analytical study of loci; and finally a long section on the customary construction by conics of the roots of cubic and quartic polynomial equations. The last was still the goal of the analytic geometry of that time. L'Hospital sometimes used two axes and seems to have recognized the interchangeability of these, but he betrays some hesitation. He did not measure coordinates along the axis of ordinates (appliquées), and he sought to confine himself to the first quadrant. Likewise he knew that a locus must pass through the end-points of all the lines representing both true (positive) and false (negative) values of $y$ corresponding to the true and false values of $x$; but he listed only the usual four cases of linear equations, $y = bx/a$, $y = (bx/a) \pm c$, and $y = c - (bx/a)$. The form $y = -(bx/a) - c$ is omitted, presumably because the line contains no points in the first quadrant. The line $y = bx/a$ and the circle $y^2 = a^2 - x^2$ are correctly given, but in other cases the figures are limited to portions corresponding to positive coordinates. L'Hospital expressly says that if ordinates are "supposed tending toward one side" of the axis, they are to be taken as negative on the other; and if one has supposed the abscissas "to fall on one side of the initial point,"

[20] *Ibid.*, p. 141 ff.

they become negative on the other; but he then adds an *avertissement:*

> When, in what follows, one deals with the construction of the locus of a given equation, one always supposes that the $x$'s and $y$'s are positive, that is, that all the points fall in the same quadrantal angle. And one takes as the locus of the given equation the portion of the locus which is included within this angle.[21]

He refers to the two branches of the hyperbolic curve as "opposed hyperbolas," possibly because of his timidity in the use of negative coordinates.

L'Hospital opens his treatment of the conics with a plane definition and kinematic construction for each type. Only later (in book VI) does he consider them briefly from the stereometric point of view. The ellipse is defined by the familiar string construction, and the parabola by the well-known string-and-square generation. The hyperbola is defined by a mechanical adaptation of the property that the difference of the focal radii is constant. From these defining properties, the equations of the curves are derived analytically. For the parabola the forms $yy = px$ and $yy = 4mx$ are given. For the central conics L'Hospital gives the standard equations with respect to the center. In this connection he used the semiaxes, as is now the common practice, but he wrote the equations in less symmetrical form as $y^2 = c^2 - (c^2x^2/t^2) = \frac{1}{2}pt - (px^2/2t)$ and $y^2 = (c^2x^2/t^2) \mp c^2 = (px^2/2t) \pm \frac{1}{2}pt$, where $t$ is the semimajor axis, $c$ the semiminor axis, and $p$ the parameter. The principal properties of the conic sections are derived in part from these equations and in part from numerous geometrical diagrams. Later the conics are generated in various ways as loci, the Newtonian organic description being included.

An appropriate illustration of the hesitancy of L'Hospital with respect to negative coordinates—as well as the operation of the Pythagorean theorem as a distance formula—is found in his derivation (without the present customary use of radicals) of the equation of the ellipse with respect to its axes, starting from the gardner's construction.[22] Let $M$ be a point on the ellipse (Fig. 28) with center $C$ and vertices $A$, $a$, $B$, and $b$. Let the lengths of the major axis $Aa$ and the minor axis $Bb$ be denoted by $2t$ and $2c$, respectively, and let the distance $Ff$ between the foci be $2m$. Setting $MF = t - z$ and $Mf = t + z$ (so that $2z$ is the difference between the focal radii), L'Hospital wrote $MF^2 = t^2 - 2tz + z^2 = y^2 + m^2 - 2mx + x^2$, and $Mf^2 = t^2 + 2tz + z^2 = y^2 + m^2 + 2mx + x^2$, using the Pythagorean theorem for

[21] *Traité analytique des sections coniques* (Paris, 1707), p. 208.
[22] *Traité analytique*, p. 22–25. Cf. Coolidge, *History of Conic Sections*, p. 77.

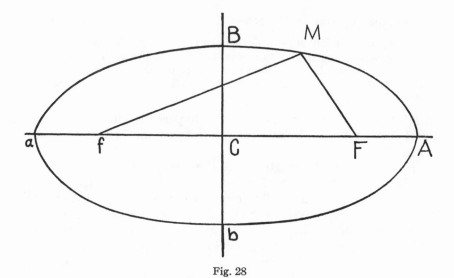

Fig. 28

the distances. Subtracting these equations, he found $z = mx/t$, and on eliminating $z$ in either of the above equations, he obtained $y^2 = c^2 - (c^2x^2/t^2)$, which "expresses perfectly the nature of the ellipse." It is of interest to note that for positions of $M$ on "the other side of the center" (L'Hospital thus avoids reference to an axis of ordinates), he wrote $MF = t + z$ and $Mf = t - z$, thus implying that values of $x$ are to be taken as positive, whether measured to the right or to the left of the center. Moreover, it is to be noted that here, and throughout the book, line segments are designated both in the synthetic manner (in terms of letters denoting the end-points) and in the notation of analysis (using variables and parameters).

The eclectic nature of L'Hospital's textbooks is apparent, but the author fails to give credit for his sources. His treatment of the general quadratic equation in two unknowns, for example, leans so heavily upon the work of Craig as to risk charges of plagiarism. There is a pronounced similarity, both in figures and in notation, between Craig's *Tractatus mathematicus* and the corresponding treatment in L'Hospital's *Conics*.[23]   Beginning with one of each of the three types of conic sections, chosen in arbitrary position with respect to a coordinate axis, a general standard equation is derived; and then this standard equation is used as a basis with which to compare the coefficients of

[23] *Traité analytique,* p. 213 ff.

given special cases of quadratic equations in order to determine the position and shape of the conic in question. As was the case with Craig, so also L'Hospital considered it necessary to give two separate treatments of each curve according as it was referred to the $x$-axis or the $y$-axis, showing how slowly the idea of interchangeability of axes developed. However, L'Hospital did state explicitly the property of the characteristic which had been implicit in the work of Craig and de Witt. That is, if, in the general quadratic equation in $x$ and $y$, the coefficient of $y^2$ is unity, it is stated as a general rule[24] that the curve is an ellipse, a parabola, or an hyperbola according as the square of half the $xy$ coefficient is less than, equal to, or greater than, the coefficient of $x^2$.

L'Hospital, like Guisnée, recognized two uses for the conics, as the full title indicates: *Traité analytique des sections coniques et de leur usage pour la resolution dans les problêmes tant déterminez qu'indéterminez.* Having first treated of loci or indeterminate problems leading to conics, he then devoted a fifty-page section to the second aspect, the then customary "construction of equalities" by means of conics. For equations of degree greater than four, L'Hospital proposed a rule for the simplicity of the construction similar to that of Fermat, Lahire, and Bernoulli, depending upon the square root of the degree of the equation.[25] However, in a couple of pages at the very end of the work, L'Hospital proposed the following significant modification: "To construct any equality of whatsoever degree by means of a straight line and a locus of the same degree." This may have been suggested by the work of his teacher, Jean Bernoulli. It is illustrated by the solution of $x^5 - bx^4 + acx^3 - aadx^2 + a^3x - a^4f = 0$ through the determination of the points of intersection of the polynomial curve $y = \dfrac{x^5}{a^4} - \dfrac{bx^4}{a^4} + \dfrac{cx^3}{a^3} - \dfrac{dx^2}{a^2} + \dfrac{x}{a}$ with the straight line $y = f$. Here one sees essentially the modern graphical representation of polynomials and the resultant solution of polynomial equations, for it differs from the present usual procedure only in a simple translation of axes. It is interesting to notice, however, that the method is not emphasized by the author. He appears to have been less interested in graphical representation as a means of solving equations than as a device for determining the range of values of the constant term for which some of the roots of the equation became imaginary. In general, L'Hospital

[24] *Ibid.*, p. 247. Tropfke, *Geschichte*, VI, 164 ff., incorrectly ascribes this statement to Euler and Lacroix.
[25] *Traité analytique*, p. 346.

(like Descartes) was more interested in analytic geometry as a means of expressing loci algebraically than as a method of deriving the properties of a curve from its equation.[26] This latter aspect he seems to have felt belonged more properly to the work in the calculus. His popular *Analyse des infiniment petits* of 1696, the first textbook on the subject, included much material on the properties and singularities of plane curves, using the differential method of Leibniz.

In 1708 there appeared the *Analyse démontrée* of Charles René Reyneau, a work in part closely resembling the far more popular treatises of L'Hospital. The first volume of the *Analyse démontrée* is a complete treatment of algebra in the Cartesian manner. Volume two is on analytic geometry and the calculus, and in this one finds the use of a single axis; a clear statement on negative ordinates and abscissas (but the interpretation nevertheless of $y = bx/a$ and $-y = -bx/a$ as supplementary rays); and second degree equations handled in the manner of Craig, together with a statement of the property of the characteristic. Reyneau uses curves not only to construct equations, but also to solve many physico-mathematical problems. The preface (of the second volume) contains an element of novelty in its insistence on "The perfect accord of analysis [algebra] with geometry. . .. If two solutions are indicated by the solution of an equation, then two lines are represented in the geometric construction. When analysis discloses that the values are impossible, one finds a contradiction in the geometric solution." Polar coordinates are used, in the manner of Varignon, in connection with the spiral $cx = ry$ and also with those of higher order; and a system of mixed coordinates is used in studying the cycloid.

The *Analyse démontrée* appeared in a second edition in 1736–1738, but it seems not to have been well known. Meanwhile, the texts of Guisnée and L'Hospital appeared in numerous editions (1705, 1733, 1753 for the former; 1707, 1720, 1740, 1770 for the latter) and they may be regarded as generally representative of analytic geometry during the first half of the eighteenth century.[27] An English translation by Edmund Stone (†1768) of the *Treatise on Conics* appeared in 1723, and it is of interest to note the influence of L'Hospital in Stone's *A New Mathematical Dictionary* three years later. Under articles on

---

[26] In unpublished work of 1672, Gregory illustrated graphically the roots of a sixth degree equation by sketching the corresponding polynomial curve; but this was not specifically for purposes of solution. See James Gregory, *Tercentenary volume* (ed. by Turnbull, London, 1939), p. 213–216.

[27] A summary of the literature of this period is given by Felix Müller, "Zur Literatur der analytischen Geometrie und Infinitesimalrechnung vor Euler," *Jahresbericht, Deutsche Mathermatiker-Vereinigung*, XIII (1904), p. 247–253.

"Biquadratic Equation" and "Cubic Parabola," the *Dictionary* not only includes graphs of cubic and quartic polynomials, but also a lengthy article on the traditional problem of the "Construction of Equations."

The irrepressible Michel Rolle (1632–1719), in the *Mémoires* of the Académie des Sciences for 1708–1709, raised doubts about the correctness of the Cartesian graphical solution of equations, as he had about the validity of the calculus of L'Hospital. He pointed out that to solve $f(x) = 0$, one arbitrarily chooses a curve $g(x, y) = 0$, and, on combining it with $f(x) = 0$, one obtains new curves, $h(x, y) = 0$, the intersections of which with $g(x, y) = 0$ furnish the solutions of $f(x) = 0$; and he realized that in this way extraneous solutions may be introduced. Imaginary branches further complicated the problem, and although Rolle saw the difficulties, he was unable to solve them. Moreover, his criticism failed to influence general opinion and so did not undermine seriously the prevailing interest in graphical constructions. Incidentally, in Rolle's discussion of this problem the term "analytic geometry" appeared in print, perhaps for the first time, in a sense analogous to that of today.[28]  By analytic geometry Rolle understood research of one or the other of two types: in the one, geometric questions are transformed into problems of algebra, and in the other, one is occupied only with the graphical solution of these problems.[29]  That is, his view was thoroughly Cartesian. His name for the subject, however, did not meet with favor. The designations used by Guisnée and L'Hospital were preferred, and through the books of these men the subject continued, at least as presented in textbooks, to be largely under the influence of Descartes. This situation is confirmed by the fact that more French editions of *La géométrie* were published from 1705 to 1730 than had appeared before or have been seen since.[30] Moreover, it was during this period that coordinate geometry entered the didactic mathematical *Sammelwerke*. One of the earliest collections to take this step was the *Elementa matheseos universae* of Christian von Wolff (1679–1754). This work includes a long section on the application of algebra to "more sublime geometry"—i. e., "that part of it

[28] Several years before, however, there appeared in 1698 at Cadiz a work by Antonio Hugo de Omerique with the title *Analysis geometrica, sive nova et vera methodus resolvendi tam problemata geometrica quam arithmeticas quaestiones.*  See P. A. Berenguer, "Un geometra español del siglo XVII," *El Progreso Matemático*, V (1895), p. 116–121.  The word analysis here seems to have been used, however, in the old Platonic sense.  On analysis and synthesis in Greek thought see J.-M.-C. Duhamel, *Des méthodes dans les sciences de raisonnement* (part I, 3rd ed., Paris, 1885), p. 39–68.

[29] Rolle, "De l'evanoüissement des quantitez inconnuës dans la géométrie analytique," *Académie des Sciences, Mémoires*, 1709, p. 419–450.

[30] See Gustav Eneström, "Über die verschiedenen Auflagen und Übersetzungen von Descartes' 'Géométrie,'" *Bibliotheca Mathematica* (3), IV (1903), p. 211.

which treats of curve-lines and solids generated from them."[31]   This
portion of the *Elementa* is virtually a textbook on plane analytic
geometry, in the sense both of Descartes and Fermat.   As in Fermat,
curves are defined by means of equations.   For example, "a parabola
is a curve in which $ax = y^2$"; and "circles of higher kind" are given by
$y^{m+1} = ax^m - x^{m+1}$.   Numerous algebraic higher plane curves are
introduced through Cartesian equations; and there is some treatment
of transcendental curves.[32]   The quadratrix is given, in the manner of
Newton, by the mixed polar and rectangular equation $ay = bx$.   The
treatment of the general quadratic equation is in the manner of DeWitt,
Craig, and L'Hospital.   The section on coordinate geometry closes
with the traditional Cartesian "construction of the higher equations,"
an art which Wolff ascribes, however, to Sluse.[33]   In the matter of
classifying degrees of simplicity, Wolff places the parabola in secon´
place, the circle third, the equilateral hyperbola fourth, the ellips
fifth, and the "hyperbola within its asymptotes" sixth.   The works of
Wolff enjoyed quite a vogue in the first half of the eighteenth century
and appeared in several languages.   However, in contrast to the textbooks and *Anfangsgründe*, original memoirs of the time tend to follow
Fermat and Newton in placing greater emphasis on the sketching of
curves other than conics, and also to make some contribution to the
much neglected field of solid analytic geometry.   In the latter connection one is tempted to recall the complaint in Plato's *Republic*[34] of the
"ludicrous state of solid geometry", in contrast to that in the plane.

In 1705 Antoine Parent (1666–1716) published in his *Essais et recherches de mathématique et physique* a memoir on Cartesian geometry
in three dimensions which he had presented to the Académie des
Sciences five years before.   Since the time of Fermat and Descartes it
had been known that for space three coordinates are required, and that
an equation in three unknowns represents a surface locus.   Lahire had
even found the equation of such a locus, but he did not sketch or study
it as a surface.   Other work of the seventeenth century on surfaces,
such as that of Wren and Wallis on the lines of the single-sheeted
hyperboloid, was carried out without the use of coordinate geometry.
Consequently, Parent's paper of 1700 on *Des affections des superficies* represents essentially the first analytic study of a curved surface.
It is an awkward treatment of the surface of a sphere, but it shows a

---

[31] Christian Wolff, *A Treatise of Algebra; with the Application of it to a Variety of Problems in Arithmetic, to Geometry, Trigonometry, and Conic Sections* (transl. from the Latin,
London, 1739), p. 227.
   [32] *Ibid.*, p. 268 ff.
   [33] *Ibid.*, p. 300–340.
   [34] Jowett translation, section 528.

full knowledge of space coordinates.   He took a base plane $HQ$ and in
it a fixed line $IMQ$ as axis, with $Q$ the origin of abscissas (Fig. 29).
Assuming a point $O$ above the plane to be the center of a sphere of
radius $r$, he dropped a perpendicular $OH = a$ to the plane, drew $HI =$
$c$ perpendicular to $IQ$, and let $IQ = b$.   The quantities $a$, $b$, and $c$ are

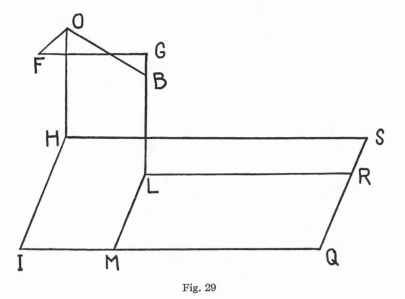

Fig. 29

then the rectangular Cartesian coordinates of the center of the sphere
with respect to the given plane and axis and origin.   Now he chose
any point $B$ on the surface of the sphere and designated the coordinate
lines $BL$, $LM$, and $MQ$ as $z$, $y$, and $x$, respectively.   Lacking a specific
formula for the distance, Parent was compelled to make further geometrical constructions in order to find the equation of the spherical
surface.   He passed through $O$ a plane parallel to $HIQ$ cutting $LB$ in
$G$; and through $O$ and $G$ he drew lines parallel to $HI$ and $IQ$, intersecting in $F$.   Then $OB$ is the diagonal of a rectangular parallelepiped
of sides $a - z$, $b - x$, and $c - y$, and setting the square of OB equal to
$r^2$, Parent obtained $c^2 + y^2 - 2cy + b^2 + x^2 - 2bx + a^2 + z^2 -$
$2az = r^2$ as the equation of the sphere.   This shows that Cartesian
geometry had not yet been sufficiently formalized to dispense with frequent reference to geometrical figures, even in the most elementary
situations.

Parent next showed how to determine a tangent plane to the sphere
at a point, but one notes here a wide divergence from the modern

point of view.   As in plane geometry one never asked at that time for the equation of a line, but only for its construction, so Parent did not feel it necessary to give the equation of the plane, but only to determine it by means of two intersecting lines.   He therefore took through the point in question two planes perpendicular to the coordinate plane and parallel to the ordinates and the axis of abscissas.   The directions of the two lines going through the point and tangent to the circular sections thus determined were found through partial differentiation rather than through coordinate methods alone.   The desired tangent plane was then uniquely determined by these two tangent lines.

The three-dimensional work of Parent includes not only the determination of the equation of a given locus, but also the converse study of a surface determined by a given equation.   That is, he made use of both the Cartesian and the Fermatian aspect of algebraic geometry. Beginning with the "equations superficielles" $y = (b + x)\sqrt{(z - x)/z}$ and $y = z^3/(x^2 + az)$, he considered curves on these surfaces. However, his discussion was more concerned with problems of the calculus than with analytic geometry, and included the determination of the points of inflection of sections by planes parallel to the coordinate plane or perpendicular to either the ordinates (tiges) or the line of abscissas (neuds).   He did not give in either case a diagram of the surface as a whole.   In a memoir read two years later, Parent discussed the one-sheeted hyperboloid of revolution along the lines given by Wren and Wallis.   He described the lines on it, and later also the elliptical, hyperbolic, and parabolic sections, and the asymptotic cone; but his treatment here was not analytic and no equations are given in connection with the surface and curves.[35]

The *Essais et Recherches* of Parent appeared in a second edition in 1713, but his work on solid analytic geometry seems nevertheless to have left little impression.   Jean Bernoulli, too, in correspondence with Leibniz in 1715, showed familiarity with space coordinates,[36] using perpendiculars dropped from a point to three mutually perpendicular planes; but this first use of *three* coordinate planes long went unpublished and unnoticed, so that later writers reverted to the use of a *single* coordinate plane.

While contributions to solid analytic geometry were sporadic and aimless, new research in plane coordinate geometry of the time was de-

---

[35] For Parent's work on solid analytic geometry see his *Essais et recherches de mathématique et de physique* (2nd ed., 3 vols., Paris, 1713), II, p. 181–200, 645–662; III, p. 470–528.   An account of this will be found in Cantor, *Geschichte*, III, p. 417 f.

[36] An excellent account of the early history of solid analytic geometry is found in Coolidge, "The Beginnings of Analytic Geometry in Three Dimensions," *The American Mathematical Monthly*, LV (1948), p. 76–86.

voted largely to a single topic—the study of higher plane curves along the lines suggested by Newton.   In the *Enumeratio linearum tertii ordinis* Newton had not proved the results he gave, and so in 1717 James Stirling (1692–1770) published the *Lineae tertii ordinis Neutonianae* in which the material in the *Enumeratio* is demonstrated and considerably amplified.   The Stirling edition is, in fact, virtually a new work, for the original Newtonian treatise makes up only the first thirty-six pages in a volume of close to two hundred pages.   Stirling showed, among other things, that a curve of order $n$ cannot have more than $n - 1$ differently directed asymptotes, and that an asymptote of the curve cannot cut the curve in more than $n - 2$ points.   Moreover, if the $y$-axis is an asymptote, then the equation of the curve cannot contain a term in $y^n$.   Stirling added four new cubics to Newton's list of 72, and he showed that a curve of order $n$ is in general determined by $n\,\dfrac{(n + 3)}{2}$ points.   There is a noticeable lacuna in that Stirling failed to prove possibly the most difficult theorem in the *Enumeratio*—that on the projective generation of cubic curves from five primitive types.

One of the important additions in Stirling's work is the formal analytic treatment of general second-degree equations.   Ever since the time of Descartes geometers had given indications of how to transform the equations of conics to normal forms, and Wallis had asserted that from the coefficients of the equations alone one could determine the characteristics of the curves.   De Witt, Craig, L'Hospital, and Wolff had shown how to determine the shape and position of a conic from its equation; but Stirling was perhaps the first one to complete in analytic detail the program of reducing the general quadratic to canonical forms.   Beginning with the equation $y^2 + Axy + By + Cx^2 + Dx + E = 0$ in general oblique coordinates, he showed this is reducible to $y^2 = Ax^2 + Bx + C$ where $A$ is less than, equal to, or greater than zero according as the figure is an ellipse, parabola, or hyperbola.   Then by a translation of the origin on the axis of abscissas he reduced the first and last cases to the forms $y^2 = B - Ax^2$ and $y^2 = Ax^2 + B$.   This was not really new, for similar work along these lines had been given frequently in the seventeenth century; but Stirling went further and calculated *analytically* from these forms the characteristic properties of the conics with respect to axes, vertices, asymptotes, and parameter—aspects which previous writers had derived geometrically or had taken over from Apollonius.   For rectangular coordinates such calculations are a simple matter, but the program of Stirling for oblique coordinates called for ingenuity.   In the case of the ellipse, for example, he proceeded as follows: First find the $x$-intercept $CL = \sqrt{B/A}$ (Fig. 30). For oblique coordinates $L$ is not a vertex of the ellipse. Stirling found the vertices by taking a circle with center C and radius $CL$ and finding the other intersection $E$ of the circle with the ellipse. Then the bisector of the angle $ECL$ intersects the ellipse in a vertex $H$, the coordinates of which Stirling calculated in terms of $A$ and $B$. Similarly the extremities of the minor axis can be determined. For the parabola $y^2 = Ax + B$ Stirling substituted for the auxiliary circle a line through $L$ perpendicular to the axis and intersecting the

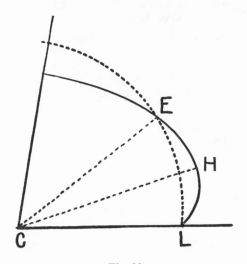

Fig. 30

curve in $E$. Then the perpendicular bisector of $EL$ intersects the parabola in the vertex $H$. For the hyperbola $y^2 = Ax^2 + B$ he found the asymptotes $y^2 = Ax^2$ and bisected the angle between the axis and an asymptote to get a vertex.[37]

This calculation by Stirling is significant for the analytic nature of the treatment, anticipating somewhat similar work later by Euler; but it is noteworthy also for its concern with those arithmetic aspects of the conics which play such a prominent role in modern textbooks. At the time of Stirling, equations were still largely decorative, as far as the conic sections were concerned, and the central problem was to recognize or construct the conic. Today one seldom asks for the *construction* of a conic section but rather for the *calculation* of important

[37] For an excellent account of this work and its significance, see Heinrich Wieleitner, "Zwei Bemerkungen zu Stirlings 'Linea tertii ordinis Neutonianae,'" *Bibliotheca Mathematica* (3), XIV (1914), p. 55–62.

magnitudes, such as the lengths of certain lines or the coordinates of points.

Stirling said that through such calculations as he gave for second-degree equations the analogies between the conics and curves of higher order are more apparent. Thus his commentary on the *Enumeratio* served also as a good introduction to curve tracing in general, for numerous and varied graphs are given. For graphs of rational functions $y = f(x)/\phi(x)$ he found the vertical asymptotes by setting $\phi(x)$ equal to zero.[38] Newton necessarily had included cubic polynomials

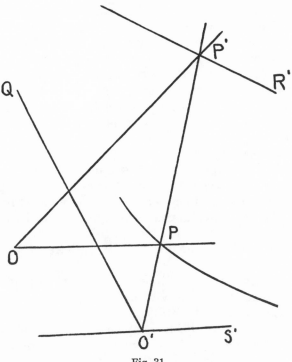

Fig. 31

in his *Enumeratio*, but he had illustrated these only for the special case $y = ax^3$. Here Stirling went further and gave what appears to be the first really systematic exposition of the modern graphical representation and solution of polynomial equations. He drew a series of graphs of general quadratic, cubic, and biquadratic polynomial functions, both with and without imaginary roots. In connection with

[38] For a summary of this work and a biography of the author, see C. T. Tweedy, *James Stirling, a Sketch of His Life and Works, Along with Scientific Correspondence* (Oxford, 1922).

these Stirling pointed out that the roots of $x^4 + bx^3 + cx^2 + dx + e = 0$, for example, are given by the intersections $A$, $B$, $C$, and $D$ of $y = x^4 + bx^3 + cx^2 + dx + e$ with the $x$-axis; and similarly for quadratic and cubic polynomial equations.[39] It will be noticed that, as in many other works of that day, the $y$-axis is not drawn; but this is not essential to the method, and its omission does, in fact, emphasize the functional relationship more effectively than would the use of two axes. The abandonment of homogeneity in his equations is a modern step. However, Stirling did not indicate clearly on his axis either the zero-point for $x$ or the scale used. In fact, one looks in vain for the application of his method to numerical examples. Apparently graphs here, as in L'Hospital, were not used to determine the values of the roots but only to indicate whether they were real or imaginary. Graphical solutions were not regarded as a practical method for the approximate solution of specific polynomial equations, an attitude which seems to have persisted throughout much of the eighteenth century. In the physical and social sciences the situation was much the same. Huygens in 1669 had plotted graphically the mortality statistics of Graunt, Plot in 1684 had sketched a sequence of barometric readings, and Halley in 1686 had drawn a curve illustrating Boyle's law; but such examples were isolated cases until Watt and Playfair, just about a century later, began the systematic practice of graphical representation.[40] One is indeed amazed at the failure of Stirling's contemporaries to make effective practical use of graphical methods. Perhaps this failure was the result of too thoroughgoing a separation at the time of algebra and geometry, in line with the advice of Newton in the *Arithmetica*.

The work of Newton and Stirling on the theory of plane curves was continued in England, especially by Colin Maclaurin (1698–1746) who in 1720 published the *Geometria organica sive descriptio linearum curvarum universalis*. In this book, completed at the age of twenty-one,[41] he varied the Newtonian organic construction in various ways. For example, he kept the vertex $O$ of the one rotating angle fixed, but al-

[39] In the *Novi Commentarii Academiae Petropolitanae* for 1758–1759 (1761), Segner simplified this construction of polynomial curves by using the addition of ordinates, a method used earlier by Newton in the *Enumeratio*. Still later Rowning in *Philosophical Transactions* for 1770 gave a mechanical construction for such curves.

[40] See M. C. Shields, "The Early History of Graphs in Physical Literature," *The American Journal of Physics*, V (1937), p. 68–71; VI (1938), p. 162. Cf. also C. B. Boyer, "Note on an Early Graph of Statistical Data," *Isis*, XXXVII (1947), p. 148–149; and "Early graphical solutions of polynomial equations," *Scripta Mathematica*, XI (1945), p. 5–19.

[41] For a brief summary of his life and work see H. W. Turnbull, "Colin Maclaurin," *American Mathematical Monthly*, LIV (1947), p. 318–322. This is based upon a longer account appearing in Maclaurin's posthumous *Account of Sir Isaac Newton's Philosophical Discoveries* (London, 1748).

lowed the vertex $O'$ of the second angle to slide along a straight line $O'S'$ while the one side of the angle passed through a fixed point $Q$ (Fig. 31). Then if the intersection point $P'$ of the free side of the second angle with one side of the first angle moves along a straight line, $P'R'$, then the intersection point $P$ of this free side with the other side of the first angle will describe a cubic. Later Maclaurin generalized this construction by substituting a curve for one or both of the lines along which $O'$ and $P'$ move. He proved, for example, that if $O'$ moves along a line while $P'$ moves along a curve of order $n$, then $P$ will describe a curve of order $3n$; but if $O'$ moves along a curve of order $m$ while $P'$ moves along one of order $n$, then $P$ will describe a curve of order $3mn$.

Maclaurin's proofs generally are in geometrical form, although he did incidentally give the equations of curves in special cases.[42]  Of particular importance in analytic geometry, however, is a theorem which he stated on the number of points in which curves intersect. Newton had interpreted the degree of the equation of a curve in terms of the number of possible intersections of the curve with a straight line, and presumably the generalization for intersections of the given curve with curves of higher order was known to him and his successors. It is implied, for special cases, by the early graphical constructions of polynomial equations. The tedious listing of curves of minimum degree needed in these solutions—such as that of Lahire—is tantamount to a recognition of the general rule given by Jacques Bernoulli[43] and L'Hospital in terms of the square root of the degree of the equation. Maclaurin, however, gave a clear-cut statement of the theorem that a curve of order $m$ in general intersects a curve of order $n$ in $mn$ points. This often is known as the theorem of Bézout, in recognition of the man who later first gave a satisfactory proof. In this connection Maclaurin came across the difficulty which usually is known as Cramer's paradox. A curve of order $n$ generally is determined (as Hermann in 1716 and Stirling in 1717 had indicated) by $[n(n+3)]/2$ points, which for a cubic is nine; but two curves of order $n$ intersect each other in general in $n^2$ points, and for two cubics this is also nine. The one theorem implies that a cubic is uniquely determined by nine points, the other implies that it is not uniquely determined. The answer to this paradox was not given until almost precisely a century later.

The work of Newton, Stirling, and Maclaurin was continued in England by William Braikenridge (c. 1700–1759), but it seems finally to

[42] Chasles, Aperçu historique, p. 162–170, gives a good account of the geometry of Maclaurin.

[43] Opera, I, p. 343; II, p. 677–679.

have influenced Continental Europe only toward the close of the third decade of the century. Beginning with 1729, papers dealing with higher plane curves began to appear regularly not only in the *Philosophical Transactions* of London, but also in the Paris *Mémoires*. Here François Nicole (1683–1758) in 1729–1731 filled in details missing from Newton's classification of cubics; Christophe Bernard de Bragelogne (1688–1744) in 1730 and 1731 attempted a systematic examination of curves of fourth order; and Bernard de Fontenelle (1657–1757), editor of the *Mémoires*, added general historical accounts for these years. Such work on higher plane curves was closely associated with contemporary developments in the calculus, illustrated by the 1729 memoir of P. L. M. de Maupertuis (1698–1759) on the singularities of higher plane curves. However, it seems to have had little immediate influence on the elementary Cartesian geometry of the time. In 1730 there appeared posthumously the *Commentaires sur la géométrie de M. Descartes* of Claude Rabuel (1669–1728), a prolix traditional treatment which enjoyed quite a vogue. Rabuel regarded Descartes' *Géométrie* as of an "almost insurmountable difficulty" and felt that Schooten also was concerned more with fame than simplicity of exposition. His *Commentaires* is therefore a large volume of 590 pages—in contrast to the brevity of Descartes—and it consists of an amplification in explanatory detail of the original material, rather than of a contribution of new results. Even the errors of Descartes on the normals to a space curve are left uncorrected. Rabuel retained the Cartesian classification of curves, although he mentioned the possibility of arranging them by degrees; and he followed the rules of Descartes on the canonical constructions of equations, while pointing out the objections Fermat, Lahire, Bernoulli, and others had raised against these. Rabuel did, however, depart from his master in the conception of coordinates, for he used two axes and pointed out more clearly than had anyone heretofore that they are on an equal footing. He showed that from given points one can, if he prefers, draw parallel abscissa lines to the $y$-axis and then from the ends of these measure the ordinates along this axis from the origin. For the two branches of the conchoid, however, he gave two distinct equations, indicating a failure to grasp the significance of negative coordinates. In the matter of space coordinates Rabuel went no further than Descartes; but solid analytic geometry was being developed at that very time by Leonhard Euler (1707–1783).

The *Commentarii Academiae Scientiarum Imperialis Petropolitanae* of 1728 includes a paper by Euler, "De linea brevissima in superficie quacunque duo quaelibet puncta iungente," the significance of which for the history of analytic geometry has not been fully appreciated, perhaps because its publication was delayed for four years. Lahire had given, quite incidentally, the equation of a surface, without describing it; Parent had discussed a few surfaces given by equations; and Jean Bernoulli had added a further illustration. Nevertheless, it was Euler who presented for the first time a reasonably systematic analytic treatment of whole classes of surfaces. His language is surprising in its implication that the analytic representation of surface loci was practically unknown. In spite of the fact that his researches on geodesics on surfaces were suggested by Jean Bernoulli, Euler at first adopted a single coordinate plane and a single axis, probably not being acquainted with Bernoulli's use of three mutually perpendicular coordinate planes. Euler found the coordinates of a point $M$ in the manner of Lahire and Parent—by dropping a perpendicular $y = MP$ to the coordinate plane,

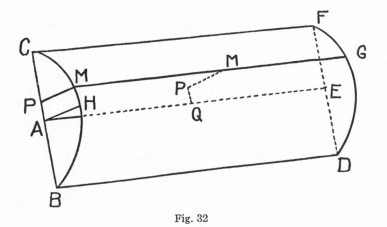

Fig. 32

then in the plane drawing an ordinate $x = PQ$ perpendicular to the axis, and then taking as abscissa the distance $t = QA$ from a fixed point $A$ on the axis (Fig. 32). After giving the equation of the sphere in simplest form as $a^2 = t^2 + x^2 + y^2$, Euler applied his analysis to three broad classes of surfaces—cylinders, cones, and surfaces of revolution. In all of these classes he refers not to special cases, as had always been the case heretofore, but to general types. Thus by cylinder he means not only common cylinders with circular bases, but "any body of which sections perpendicular to an axis are similar and equal to each other." He did not define the word axis, but any line parallel to elements of the surface would do. It is not clear whether the surface is necessarily closed, but the word "body" suggests stereometric considerations and seems to imply that Euler regarded the surface as bounding a solid.   Many years later Euler asked, in a famous paper on developable surfaces of 1771, whether there are "solids" other than the cylinder and cone whose surfaces can be unfolded upon a plane, his phraseology again indicating a view of surfaces as necessarily linked with volumes.[44]   Kaestner, too, referred to "the nature of surfaces which bound bodies, expressed by equations."   The definitive work on surfaces considered as truly independent entities generally is ascribed many years later to Gauss;[45] but Euler's work was the most advanced for his day.

In finding the equation of his cylindrical surface Euler takes as origin a fixed point $A$ not on the surface, as axis a line $AQE$ parallel to an element of the cylinder, and as coordinate plane one through the axis and two elements $BD$ and $CF$ of the cylinder.   Euler then observed that the differential equation of the surface will be $P\,dx = Q\,dy$, in which $P$ and $Q$ do not involve the letter $t$; and this is the equation also of the "base" $ABHC$.   For the special case in which the base is a circle with center at $A$, the equation will be $x\,dx = -\,y\,dy$.   The fact that the equations are here given in differential form is not significant for the history of analytic geometry inasmuch as they are easily seen to be equivalent, respectively, to $F(x, y) = 0$ and $x^2 + y^2 = a^2$.

With the same degree of generality Euler defined a cone as "a solid bounded by straight lines drawn through the points of any curve and a fixed point taken outside of the plane of the curve."   One is surprised that he should have restricted himself in this definition to plane curves, for Henri Pitot (1695–1771) in the Paris *Mémoires* for 1724 had studied the helix on a cylinder and had prophetically said that perhaps some day "curves of double curvature," a name here coined for the first time, would be the object of researches in geometry.   Pitot also pointed out an interesting connection between the ellipse and the sine curve:  If one traces on a right circular cylinder an ellipse cut from it by a plane inclined at an angle of 45° to the base, and if the cylinder is then rolled along a plane, the ellipse will trace out a sine curve on this plane.   He did not carry such work further, however, and the systematic study of skew curves really began about half a dozen years later.

In connection with the cone Euler took the origin as vertex, a line through this as axis, and a plane through this cutting two elements of the cone as coordinate plane.   Then he said that all sections by planes perpendicular to the axis are similar, so that the equation in $t$, $x$, $y$ is such that if two coordinates are increased or diminished in a given

[44] See Cajori, "Generalizations in Geometry as Seen in the History of Developable Surfaces," *American Mathematical Monthly*, XXXVI (1929), p. 431–437.
[45] See Cantor, *Vorlesungen uber Geschichte der Mathematik*, IV, p. 457.

ratio, the third will be increased or diminished in the same ratio. Hence if for $t$, $x$, $y$ one substitutes $nt$, $nx$, $ny$, the equation will remain unchanged; and so "conoidical bodies," i. e., conical surfaces, have the property that the equation with respect to the vertex is homogeneous and may be written in a form in which "$\frac{t}{x}$ is equal to a homogeneous function of $x$ and $y$ of degree zero." Euler then put this in the language of differential equations in order to find geodesics on the surface. In this latter connection he gave a hint of later work on developable surfaces in his statement that the shortest curve between two points on the cone would become the straight line between these points if the surface were flattened out into the form of a plane.

The third general type of surface which Euler studied consisted of those of revolution. Special cases of these had been considered frequently since antiquity, but Euler was the first one to give a whole class by means of an equation. If the axis of revolution is taken as the $t$-axis, Euler expressed the equation of the surface in the form $x^2 + y^2 = T$, where $T$ is any function of $t$. This is virtually the modern form of expression for such surfaces.

The work of Euler on surfaces in three-space was almost simultaneous with the more famous contribution by Alexis Claude Clairaut (1713–1765) on skew curves. His *Recherches sur les courbes à double courbure* was presented to the Académie des Sciences in 1729 when Clairaut was only sixteen years old, but it was published two years later. The *Recherches*, like the *Géométrie* of Descartes, appeared without the name of the author on the title page, although this was generally known. It carried out for space curves the program that Descartes had suggested almost a century before—their study by a consideration of the projections on two coordinate planes. From the preface of his book one gathers that he knew only of the work of Descartes, Bernoulli, and Pitot in three dimensions, and not of the far more important contributions of Lahire, Parent, and Euler. It was apparently a memoir by Jean Bernoulli which first drew his attention to the study of surfaces.

One might have expected Clairaut to open with the straight line, or at least to consider first the theory of surfaces; but he began *in medias res* with a study of a space curve with reference to a single axis. He calls his coordinates $x$, $y$, $z$ and says that if the $z$'s are connected with the $x$'s and with the $y$'s by non-linear equations, then they determine a space curve. He calls it a "curve of double curvature," following Pitot, because its curvature is determined by that of two curves obtained by projection of the original curve upon two perpendicular

planes.   Clairaut said that he would give only algebraic curves, but he
asserted that transcendental curves are just as easily handled.   He
promised, in the preface, to give later a treatise on "curves given by
coordinates from a point"—i. e., polar coordinates in three dimensions.
This would have been the earliest use of such a system, but unfortunately the work never appeared.

Clairaut implicitly corrected the error of Descartes on normals by
pointing out that such a curve has infinitely many normal lines;
and he saw that, for a given space curve, any two projecting cylinders
$y = f(x)$ and $z = g(y)$ determine a third $z = F(x) = g[f(x)]$.   This suggested the use of three mutually perpendicular coordinate planes instead of the single plane and axis with which he had begun.

Clairaut's discussion of surfaces is surprisingly similar to that of
Euler, but less systematic.   He first gave as examples the sphere
$a^2 = x^2 + y^2 + z^2$, the cone $(n/m)z = \sqrt{x^2 + y^2}$, and the paraboloid
$y^2 + z^2 = ax$.   Then he went on to give the equations of other surfaces of revolution, such as the ellipsoid, the one-sheeted hyperboloid,
and the surface obtained by revolving the parabola $ay = x^2$ about the
tangent at the vertex.   Clairaut gave a treatment of general conical
surfaces, using as generating curves the higher parabolas, ellipses, and
hyperbolas—$x^r = a^{r-1} y$ and $ax^{r+q}/c = y^r (a \pm y)^q$.   Like Euler, he
knew that the equation of a cone with vertex at the origin is homogeneous.   Clairaut also gave numerous examples of curves defined not by
projecting cylinders but through the intersection of other surfaces;
and he determined whether or not these curves lie on a given surface.
He considered both positive and negative coordinates, although only
portions of the figures are given in the diagrams.   He showed how to
sketch a surface by considering plane sections, using Bernoulli's surface $xyz = a^3$ as an illustration.   He knew that the equation of a plane
is linear in the variables, and in the Paris *Mémoires* for 1731 (the very
year in which his *Recherches* was published) he wrote this in the inter-
cept form $(ax/c) + (ay/b) + z = a$; but he did not go further in his
study of linear equations.   Clairaut included in book II some results
on tangent and normal lines to a space curve, making use of the calculus; and the *Recherches* closes with book III in which the integral
calculus is applied to curves and surfaces.   This work is carried
through in geometrical form dependent upon diagrams, indicating that
the development of differential geometry waited upon further work
by Euler and Monge.

There is one point in plane coordinate geometry where Clairaut
made an innovation of some interest.   Historical accounts of analytical
geometry place considerable emphasis upon the introduction of the

distance formula, generally thought[46] to have taken place in 1797 and 1798. However, the familiar formulae, for both two and three dimensions, are found in the *Recherches* of Clairaut. They are, of course, incidental to other matters, but they are clearly stated in connection with the determination of the equation of a spherical surface. Let the center $C$ have coordinates $AB = \pm a$, $BD = \pm b$, and $DC = \pm c$ with respect to the axis $AB$ and the base plane $ABD$ (Fig. 33); and let $N$ be any point on the sphere with coordinates $AP = x$, $PM = y$, and $MN = z$. Then Clairaut wrote[47] that $EN = MD = \sqrt{x \mp a^2 + y \mp b^2}$. He followed this with the analogous formula for three-space— $f = \sqrt{x \mp a^2 + y \mp b^2 + z \mp c^2}$, where $f = CN$. This is possibly the

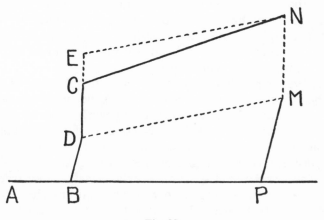

Fig. 33

first time that either of these formulae appeared in print. Consequently, credit for them should go, pending further evidence, to Clairaut. His forms differ but slightly from the modern equivalents, mainly in the failure to regard the literal quantities $a$, $b$, and $c$ as indifferently either positive or negative—a point which Hudde had made as early as 1659.

The contribution of Clairaut in this connection should not, however, be exaggerated. The distance formulae are, after all, obvious analytical expressions of an ancient theorem named for Pythagoras but known to the Babylonians of some four thousand years ago. There can be little doubt but that their equivalents were known to the earliest

[46] See, e. g., Tropfke, *Geschichte*, VI, p. 124; Loria, "Da Descartes e Fermat a Monge e Lagrange," p. 840–842; Wieleitner, *Geschichte*, II (2), p. 42; Coolidge, *History of Geometrical Methods*, p. 134.
[47] *Recherches, sur les courbes à double courbure* (Paris, 1731), p. 98.

analytic geometers, including Descartes and Fermat.   The equations
of circles and spheres, given long before Clairaut, are tantamount to
distance formulae; and the rectification of curves, known since 1659,
is dependent upon some such equivalent.   The formulae for distance
in infinitesimal analysis—$ds = \sqrt{dx^2 + dy^2}$ and $ds = \sqrt{dx^2 + dy^2 + dz^2}$
—appear frequently in the *Recherches*, but these are not to be ascribed to him.   It would not be surprising, in fact, if further research
would reveal explicit, as well as implicit, anticipations of Clairaut's
distance formulae, not only for the calculus, but also for analytic
geometry.   Nevertheless, it is important to note that, in formalization, infinitesimal analysis had at that time far outstripped Cartesian
geometry, even though the invention of the latter preceded that of the
former by about half a century.   Formulae had been a natural outgrowth of the algorithms of Newton and Leibniz, but the coordinate
geometry of Descartes and Fermat still leaned heavily upon auxiliary
diagrams.   Consequently, the distance formulae did not appear systematically until the time of Lagrange.

   In 1731, the year of his *Recherches*, Clairaut published in the *Mémoires* of the Académie a paper relating his solid analytic geometry to
the theory of higher plane curves.   In this work, "Sur les courbes que
l'on forme en coupant un surface courbe quelconque, par un plan donné
de position,"[48] he proved Newton's well-known theorem on the projective transformation of cubic curves.   Using the equation of the
conical surface $xyy = ax^3 + bxxz + exzz + dz^3$, whose traces in the
planes $x = k$ are the divergent parabolas, Clairaut showed that the
plane sections of this surface have as equations the various species of
cubics in the *Enumeratio*.   An interesting case of the simultaneity of
ideas is found in the fact that Nicole presented a similar analytic proof
of the theorem in the very same volume of the *Mémoires*.[49]   The corresponding theorem for the conics was proved, also in this volume, by
Charles-Marie de la Condamine (1701–1774).   He showed that the
conic sections are derivable as plane sections of the cone $nnxx = yy + zz$,
apparently the first instance in which solid analytic geometry was
applied to this ancient theorem.

   The study of the plane and of other surfaces in space was resumed in
the following year by Jacob Hermann (1678–1733).   In a paper,
"De superficiebus ad aequationes locales revocatis, variisque earum
affectionibus," published in the Petersburg *Commentarii* for 1732–

[48] *Académie des Sciences, Mémoires*, 1731, p. 483–493.
[49] "Maniére d'engendier dans un corps solide toutes les lignes du troisième ordre," *Ibid.*,
p. 494–510.   See also another paper by Nicole, "Sur les sections coniques," in the same
volume, p. 130–143.

1733, he said that up to that time the geometry of surfaces other than planes and surfaces of revolution had scarcely been considered.[50] This would indicate that he did not know of the *Recherches* of Clairaut, and that he was unaware of the work of his own associate, Euler. Hermann attributed the neglect of solid analytic geometry to the prolixity caused by the use of three unknowns, whereas for the plane two suffice.   Hermann used the coordinates $x$, $y$, $z$ defined in terms of a single coordinate plane and one directrix [or axis]; and he limited himself entirely to the first four octants, generally to the first alone. His study of the equation $az + by + cx - e^2 = 0$ is better than any previously given, for he determined the position of this plane by finding its traces and intercepts; and then he showed conversely that every point in this plane satisfies the given equation.   He made a beginning in the important matter of directional properties and metric considerations by finding that the sine of the angle between this plane and the coordinate plane is $\sqrt{b^2 + c^2} : \sqrt{a^2 + b^2 + c^2}$.   The choice of the sine instead of the cosine is of no great consequence, but the use of only one coordinate plane concealed the symmetries which later encouraged the further use of analytic methods in the metric study of three dimensions.   Hermann saw that solid analytic geometry could be applied in spherical trigonometry; but this idea was not effectively exploited until over a century later, when Cesàro revived it.

Hermann's work in solid analytic geometry seems to have been carried out independently of that of his predecessors.   It is surprising how slowly ideas in this field spread, even among members of the same academy.   His study of curvilinear surfaces was less general than that of Euler and covered somewhat the same ground as that of Clairaut.   He identified successfully the parabolic cylinder $z^2 - ax - by = 0$ and the cones $z^2 - xy = 0$ and $az^2 - bxz - cyz + cy^2 = 0$; but two somewhat more general quadrics, $z^2 - ax^2 - bxy - cy^2 - ex - fy = 0$ and $az^2 + byz + cy^2 - exz + fx^2 + gz - bx = 0$, he described only as "conoidal surfaces" whose sections are conics.   His most general statement on surfaces is that $u^2 - x^2 - y^2 = 0$ represents a solid of revolution if $u$ is "any quantity involving $z$ and constants." For surfaces such as the above, Hermann gave maxima and minima, tangent planes, and geodesic lines, but his methods, making use of partial derivatives, resemble those of Parent and are less elegant than those of Euler.   He added also a study of the ruled surface which Wal-

---

[50] In the same volume (VI, 1732–1733) p. 13–27 of the *Commentarii* of the Petersburg Academy there appeared a paper by G. W. Krafft, "De ungulis cylindrorum varii generis," in which cycloidal cylinders, cissoidal cylinders, and other types of cylindrical surfaces are considered; but they are not studied analytically in terms of equations in three coordinates.

lis had described as a cono-cuneus.   Here for the first time the analytic
equation is given, appearing as $(b - z) \sqrt{a^2 - y^2} = bx$.

Hermann's work is not well organized, but it shows a genuine enthusiasm for the study of solid analytic geometry.   He promised further studies on surfaces, but he died before they could be completed;
and so it remained for Euler to give the first general treatment of surfaces of second degree.   Before passing on to this, however, it will be
well to refer briefly to Hermann's analysis of the general equation of
second degree in plane analytic geometry.   In 1729 he published a
paper in which he recalled and extended the method Descartes had
used to identify conics.[51]   Beginning with the general equation
$\alpha yy + 2\beta xy + \gamma xx + 2\delta y + 2\epsilon x + \phi = 0$, a form which had appeared
essentially in L'Hospital's *Conics*, he solved this for $y$ and showed that
the curve is an ellipse, parabola, or hyperbola according as $\beta^2 - \alpha\gamma$ is
less than, equal to, or greater than zero; this result was known earlier
to Craig and L'Hospital, to whom Hermann refers, along with Debeaune, De Witt, and Schooten.   He indicated further that if, in the
solution for $y$ in terms of $x$, the radical sign disappears—i. e., if $(\alpha\epsilon - \beta\delta)^2 = (\delta^2 - \alpha\phi)(\beta^2 - \alpha\gamma)$—then the equation represents a pair of intersecting straight lines (for $\alpha \neq 0$).   Descartes in this case had
thought of the result as a single line, for he had failed to use the double
sign before the radical.   The work by Hermann on conics and quadrics
shows that the analytic study of second-degree plane curves had
reached maturity while that of second-degree surfaces was still in its
infancy.

The eighteenth century in numerous respects is noteworthy for its
elaboration and perfection of implications inherited from the earlier
periods, and the case of polar coordinates, to which Hermann made an
important contribution, is an apt example of this.   Newton's use of
polar coordinates was as yet unpublished, but Jacques Bernoulli, in
continuing the study of Fermat's parabolic spirals, specifically proposed the use of vectorial lines in 1691 and 1694.[52]   Varignon about a
decade later suggested that from known curves new types might be obtained by the simple expedient of interpreting the variables in the rectangular Cartesian representation of the former as polar coordinates
for the latter.[53]   For example, the higher parabolas became parabolic
spirals, and the higher hyperbolas became hyperbolic spirals.   A hint

[51] "De locis solidis ad mentem cartesii concinne construendis," *Commentarii Academiae Petropolitanae*, IV (1729), p. 15–25.
[52] See Jacques Bernoulli, *Opera* (2 vols., Genevae, 1744), I, p. 431 f.
[53] See *Académie des Sciences, Mémoires*, 1704 (1722).   Such coordinate transformations are analytic equivalents of the geometric transformations so popular in the preceding century.

of space polar coordinates is found in Clairaut's preface, where he refers to curves of double curvature "dont les coordonées partent d'un
point"; but this idea, like those of other anticipators, was not amplified.

Far more significant than these somewhat equivocal adumbrations
was the clear-cut proposal, made by Hermann in another paper[54] of
1729, that polar coordinates are as appropriate for the study of geometric loci as are Cartesian. "But with equal right it [the doctrine of
loci] can be explained through the relationship which vectorial radii
bear to the sine or cosine of the angles of projection, from the consideration of which the properties of curves flow just as elegantly as they are
brought out in the usual manner." Hermann gave equations for

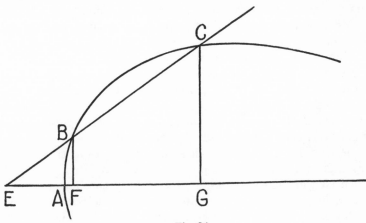

Fig. 34

transforming equations from rectangular to polar coordinates, using
the letters $m$ and $n$ for the sine and cosine of the vectorial angle and $z$
for the radius vector. The form he proposed is more general than that
now ordinarily used, for the pole did not necessarily coincide with the
origin. If $AF$ (or $AG$) is the abscissa, and $BF$ (or $CG$) is the ordinate of
a point on the curve $ABC$ (Fig. 34) and if the pole is taken as $E$, where
$EA = a$, then the equations of transformation are $x = nz - a$ and
$y = mz$. As an illustration Hermann converted the Cartesian form
of the parabola $y^2 = px$ to the polar equation $m^2z^2 = npz - ap$—perhaps the first instance of the application of polar coordinates to a conic
section, or to curves other than spirals.[55] Among other examples,

[54] "Consideratio curvarum in punctum positione datum projectarum, et de affectionibus
earum inde pendentibus," *Commentarii Academiae Petropolitanae*, IV (1729), p. 37–46.
[55] Tropfke, *Geschichte*, VI, p. 169, incorrectly ascribes the polar equation of the conic to
Lacroix.

Hermann gave the transformation of the folium of Descartes $y^3 = bxy - x^3$ to $z = bmn/(m^3 + n^3)$, where the pole and origin coincide. He suggested also the study of equations in mixed forms, such as that involving the variables $y$ and $z$.

In view of the clarity and generality with which Hermann presented the case for polar coordinates, it is difficult to see why the major portion of credit for this system should be given by historians to others much later. Smith, for example, reports that "the idea of polar coordinates seems due to Gregorio Fontana (1735–1803), and the name was used by various writers of the 18th century."[56] The idea—and more—must obviously be ascribed to Hermann, if not earlier to Newton or Bernoulli. Nevertheless, Hermann's work in this connection seems not to have been widely known, and hence the definitive use of polar coordinates may perhaps with justification be ascribed to Euler about a score of years later.

For fifteen years following the work of Clairaut and Hermann there were few significant new treatises touching upon analytic geometry. One might except the work of Jean Paul de Gua de Malves (1713–1785), *Usages de l'analyse des Descartes pour découvrir, sans le secours du calcul différentiel, les proprietés, ou affections principales des lignes géométriques de tous les ordres*, which appeared in 1740. This implies in the title that De Gua felt, with some justification, that Cartesian geometry was overshadowed at the time by the calculus, and he therefore presented the theory of plane curves along the lines of Newton's *Enumeratio*, having recourse to infinitesimal methods only to shorten the calculations. His work is noteworthy for its thorough treatment of curve sketching, especially with respect to singular points, and for the use of Newton's parallelogram in a new form, known as De Gua's analytical triangle. In this "algebraic triangle," as he calls it, all sides are on the same footing, a situation which facilitates the study of the infinite branches of curves. De Gua added new results in the theory of curves, such as the theorem (implied in Clairaut's paper of 1731), that if a cubic has three points of inflection, these lie on a straight line. He showed, in a general way, that singularities are compounded of ordinary points, cusps, and points of inflection. He gave also, like Clairaut and Nicole, a proof of Newton's theorem on the divergent parabolas, and he added two new species of cubics to the 76 recognized by Stirling. De Gua used translations and rotations of axes in various forms, but without trigonometric symbolism. In general his work is more significant in establishing curve theory as a subject in

[56] *History of Mathematics* (2 vols., New York, 1925), II, p. 324. Cf. also Cantor, *Geschichte*, IV, p. 513, and Encyklopädie, III, p. 596, 656.

itself, and for new results outside the scope of elementary analytic geometry, than for any influence on the methods of coordinate geometry. Moreover, his book seems to have been little known until referred to ten years later by Cramer in a more popular work on algebraic curves.[57]

The year of De Gua's *Usages* saw also the appearance of the *De lineis curvis* of J. B. Caraccioli. Italy had contributed little to the development of analytic geometry during the century following Descartes' *Géométrie*, but in 1738 there appeared at Rome a book which enjoyed quite a vogue—the *Institutiones analyticae earumque usus in geometria cum appendice de constructione problematorum solidorum* of Paolino Chelucci. This contributed nothing essentially new to the ideas of analytic geometry, and it strikes the modern reader as very tedious and highly unanalytic; but it appeared in at least four editions by 1761 and so undoubtedly aided in making algebraic methods better known in Italy. Caraccioli's *De lineis curvis* is also analytic only in part, but it is far more modern in treatment than the work of Chelucci. It is unusual especially in the great variety of curves presented, transcendental as well as algebraic. Considerable space is devoted to the generalized ellipses and hyperbolas—$y^{m+n} = (a \mp ax/b)^m x^n$. In spite of the predominantly synthetic form of the book, polar coordinates (using the then customary symbols $x$ and $y$) are applied to the higher spirals $b^m x^n = a^n y^m$. Mixed coordinates are used, as in Newton, for the generalized quadratrices $a^n y^m = b^m x^n$; and, as in Reyneau, three interdependent coordinates are used in the analytic equation of the cycloid.

The middle years of the eighteenth century produced a number of unusually popular works bearing on analytic geometry, and 1748 in particular was responsible for three of international significance. The three books were from widely separated regions, and each one was translated later into other languages. One, *A Treatise of Algebra* by Maclaurin, appeared (posthumously) in England; another, the *Instituzioni analitiche* of Maria Gaetana Agnesi (1718–1799), came from Italy; and the third, the *Introductio in analysin infinitorum* of Euler, was by a Swiss who wrote in Latin, spoke in French, lived in Germany and died in Russia. Each one of these three treatises includes a section on "The Application of Algebra to Geometry," the title then preferred for what is now called analytic geometry. Maclaurin had planned his volume in 1729 as a commentary on the *Arithmetica* of

---

[57] For an account of de Gua's work, see Paul Sauerbeck, "Einleitung in die analytische Geometrie der höheren algebraischen Kurven nach den Methoden von Jean Paul de Gua de Malves," *Abhandlungen zur Geschichte der mathematischen Wissenshaften*, XV (1902), p. 1–166.

Newton, but he showed less reluctance to combine algebra and geometry. Of the three main divisions in the book, the third is devoted to "the use of algebra in the resolution of geometric problems; or reasoning about geometric figures; and the use of geometric lines and figures in the resolution of equations." As in the *Geometry* of Descartes, quantities are represented by lines, and equations are "constructed" geometrically; but Maclaurin gives more careful consideration to the linear equation. Given $ay - bx - cd = 0$, he finds the locus as follows: With reference to a pair of perpendiculars $APE$ and $PNM$, draw a line $AN$ with an angle of inclination $NAP$ such that the cosine is to the sine as $a$ is to $b$ (Fig. 35). Then draw $AD$ parallel to $PM$ and equal to $cd/a$, taking $AD$ on the same side of $AE$ as $PN$ if $bx$

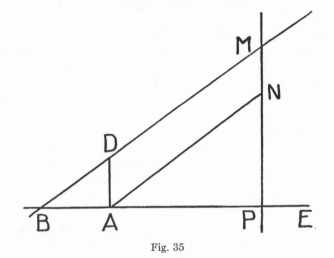

Fig. 35

and $cd$ have the same sign, otherwise on the opposite side. Then through $D$ draw a straight line $BDM$ parallel to $AN$. The line $BDM$ is, of course, the required locus with respect to the lines $AP$ and $AD$ as axes. Here one can see the gradual development of the idea of the slope-intercept form of the straight line; but it was another fifty years before the modern forms made their definitive appearance. Moreover, the old hesitation in the face of negative coordinates is evident in Maclaurin's omission, reminiscent of that of De Witt and L'Hospital, of the form $y = -ax - b$ in his enumeration of the various types of linear equations: $y = ax + b$, $y = ax - b$, and $y = -ax + b$.[58]

Maclaurin gives constructions for quadratic equations and discussions of curves of second order, but these are unexceptional. In an

[58] See *Algebra* (London, 1748), p. 305 f, and also appendix.

appendix to the *Algebra*,[59] "Concerning the general properties of geometrical lines," the author acknowledges the continuing influence of Newtons's *Enumeratio*. In this work, one of the last Maclaurin wrote, curves are not constructed but only traced in the modern manner. Maclaurin gave also an application of the method of Bernoulli, L'Hospital, and Stirling on the graphical solution of polynomial equations. In this connection he used the graph of a cubic polynomial to illustrate the occurrence of imaginary roots. The use of graphical methods in algebra was unusual for that time. For over a hundred years the most popular textbooks on elementary algebra—by Clairaut, Saunderson, Simpson, Euler, Bossut, Lacroix, Loomis, Davies—show a complete lack of graphical methods. A statement by Clairaut indicates that such omission was not due to lack of acquaintance with graphical work. In the preface to his *Elemens d'Algèbre* of 1746 he remarked, apropos of equations above the fourth degree, that, except for special cases, one is here reduced to simple approximations; "and as these approximations are often simplest when one is aided by geometry, I propose to treat these equations when I shall expound the theory of curved lines." This projected treatise by Clairaut seems not to have been completed, but his plan indicates how clearly the fields of algebra and analytic geometry were separated. Maclaurin's *Algebra* is exceptional in this respect, but this book (which appeared in at least half a dozen editions, including a French translation of 1753) shows also the lack in England of a clear-cut analytical program to take the place of the Cartesian tradition.

A perfect example of balance between the points of view of Descartes and Fermat is seen in the *Instituzioni analytiche* of Maria Agnesi,[60] written for the instruction of a younger brother. This is not only the first important surviving work written by a woman; it is also one of the few early contributions to analytic geometry to come from Italy. It does not include any essentially new material; but it is significant for its clarity of exposition and its widespread influence as a textbook,[61] as well as for the picture it gives of the state of the subject at that time. The first book of this is devoted to the "Analysis of Finite Quantities," which, the author says, "is commonly called the Algebra of Cartesius."

---

[59] *Op. cit.* The appendix contains both the Latin original, *De linearum geometricarum proprietatibus generalibus*, and an English translation.

[60] The *Éloge historique de Marie-Gaetane Agnesi* (transl. from Italian, Paris, 1807) by A.-F. Frisi supplies biographical details but not an adequate analysis of her work.

[61] The preface of the English edition, *Analytical Institutions* (London, 1801), indicates that John Colson learned Italian for the sole purpose of translating this work which was "well known on the Continent." An extensive commentary on Book I is added by the translator.

In it one recognizes indeed *La géométrie*, somewhat rearranged and amplified. Even Descartes' method of tangents is described, although this is supplemented by an alternative in which a line is substituted for the circle. The geometric construction of algebraic expressions, equations, and loci comes in for all the emphasis Descartes had given to it—with an additional section on the construction of first-degree equations along the lines given by De Witt in the Latin editions of *The Geometry*. One sees in Agnesi's work some of the old errors with respect to negative coordinates, for $y = ax/b$ is regarded as a ray lying entirely in the first quadrant, with $y = -ax/b$ as its complementary half-line. The coefficients of equations are not general as to sign, so that quadratic equations in one unknown are divided into four types. Not one, but six different forms of linear equations are considered, and these lines (or rays) are constructed in geometric style somewhat as in Maclaurin's *Algebra*. Right and oblique coordinates are used more or less indiscriminately; and a single axis is usually adopted, although a second is sometimes implied. The study of the conics and their application to the solution of equations is in the usual tradition. The equations of the circle and the general conic sections are given with respect both to the center and to a vertex. Other cases are handled by the equivalent of a translation of axes. A long section is devoted to the construction of determinate equations, with rules determining the simplest possible loci required; but the Cartesian manner of constructing polynomial equations is supplemented briefly by a description of the method of L'Hospital, in which $x^5 - bx^4 + acx^3 - aadx^2 + a^3cx - a^4f = 0$, for example, is solved through the intersections of the curves $z = \dfrac{x^5 - bx^4 + acx^3 - a^2dx^2 + a^3cx}{a^4}$ and $z = f$. The influence of the Newtonian curve-tracing is seen in her recognition of two ways of constructing a locus or indeterminate equation: "The first manner of tracing them is by finding an infinite number of points. The second is by means of other curves of an inferior degree which are already described." The first [or Fermatian] method is explained at some length, presumably because it was not so well known, after which the author adds, "This method of describing curves by an infinite number of points may perhaps be reduced to a greater perfection by making use of geometrical constructions." The old Cartesian hierarchy of curves is here well illustrated by the "second" manner of tracing $a^2y = x^3$ through the motion of lines with respect to an Apollonian parabola. Let the parabola be given in rectangular coordinates by $x^2 = az$, where $OR = a$, $OB = z$, $QP' = y$, and $BP = OQ = x$ (Fig. 36). Draw $PQ$ perpendicular to $OR$ and draw the line $BR$. Then if $OP'$ is drawn parallel to $RB$, the locus of $P'$ will be the curve desired, as one sees immediately from the proportion $OR:OB = OQ:QP'$. This curve is now in turn used in the construction of others of higher degrees.

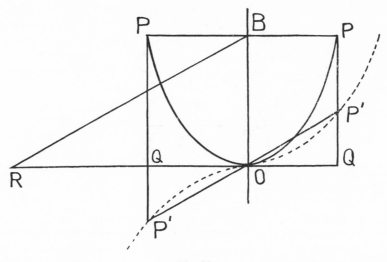

Fig. 36

The well-known "witch of Agnesi" is an excellent example of the author's "double standard" for curves. This curve had been rediscovered in 1703 by Guido Grandi (1671–1742) who gave it the name *versiera*.[62] Maria Agnesi first plotted it from the equation (which Fermat had given more than a century earlier) and then constructed it in a strictly Cartesian manner through the well-known locus of lines with respect to a circle (given half a century before by Newton). Using a notation similar to that above, where $OR = a$, $OB = y$, $BP = z$, $BP' = RQ = x$, and the circle is $z^2 = ay - y^2$ (Fig. 37), we have $RQ/RO = BP/BO$. Squaring and substituting, this becomes $x^2/a^2 = (ay - y^2)/y^2$ or $y = a^3/(a^2 + x^2)$.

The works of Maclaurin and Agnesi in 1748 show the tendency of authors to compromise with the Cartesian tradition on constructions,

[62] The name "witch," customarily used in English, apparently is due to a mistranslation. The word "versiera" which Grandi coined in 1718 to indicate the manner in which the curve is generated, has also the meaning "witch," in Italian, but this has no connection with what Grandi and Agnesi had in mind. See Gino Loria, *Spezielle algebraische und transcendente ebene Kurven* (Leipzig, 1902), p. 75. On the origin and properties of this and other curves, and also for numerous bibliographic references to sources, the work of Loria is most valuable. See also R. C. Spencer, "Properties of the Witch of Agnesi," *Journal of the Optical Society of America*, XXX (1940), p. 415–419.

but Euler's contribution of this same year marks a complete victory for the attitude of Fermat.    Loria has well said that in the history of analytic geometry the year 1748 is almost as important as 1637, for the subject then became "a robust structure" affording material for a college course.[63]    The *Introductio* of Euler is referred to frequently by historians, but its significance generally is underestimated.    This book is probably the most influential textbook of modern times.    It is the

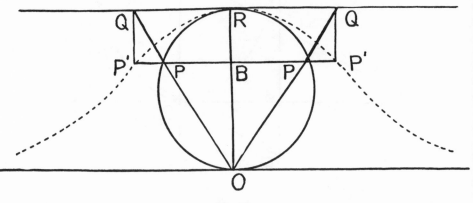

Fig. 37

work which made the function concept basic in mathematics.    It popularized the definition of logarithms as exponents and the definitions of the trigonometric functions as ratios.    It crystallized the distinction between algebraic and transcendental functions and between elementary and higher functions.    It developed the use of polar coordinates and of the parametric representation of curves.    Many of our commonplace notations are derived from it.    In a word, the *Introductio* did for elementary analysis what the *Elements* of Euclid did for geometry.    It is, moreover, one of the earliest textbooks on college-level mathematics which a modern student can study with ease and enjoyment, with few of the anachronisms which perplex and annoy the reader of many a classical treatise.    It can be read, however, only in Latin, French, or German,[64] whereas Maclaurin's *Algebra* can be read in English as well as French, and Agnesi's *Instituzioni* appeared in Italian, French, and English.

[63] "Da Descartes e Fermat a Monge e Lagrange," p. 825–827.
[64] A Russian translation was once planned, but seems not to have appeared.    For a complete bibliography of the works of Euler, see Gustaf Eneström, "Verzeichnis der Schriften Leonhard Eulers," *Jahresbericht der Deutschen Mathematiker-Vereinigung*, Ergänzungsbände, IV, 1910–1913.    This contains 866 entries, excluding multiple editions!

The first part of the *Introductio* is devoted to "pure analysis," the second to the "application of algebra to geometry." The latter is a systematic treatise on analytic geometry in the sense of Fermat. Both sides of the fundamental principle are clearly stated by Euler: With Descartes, he recognizes that "the nature of any curve whatever is given by an equation between two variables $x$ and $y$ of which $x$ is the abscissa and $y$ is the ordinate"; and with Fermat he holds that "any function whatever of $x$ gives rise to a continuous curved line which can be described by plotting." Note that Euler here uses the word "plotting," and not the Cartesian word "constructing." The *Introductio* is one of the first treatises to give numbers of graphs of specific curves with *numerical* coefficients, clearly indicating the units used on the axis of abscissas. Euler does not take up systematically the Cartesian derivation of equations of loci; for him analytic geometry was more Fermatian in aspect and centered about the sketching of curves as given by their equations. By 1748 the old Cartesian classification of curves had been virtually abandoned, so that it is not given by Euler, Agnesi, or Maclaurin. Euler did bow to the Cartesian tradition to the extent of including a section on "The construction of [determinate] equations"; but this is one chapter out of twenty-two, whereas with Descartes this topic had occupied two books out of three. Moreover, Euler preferred to follow Newton in his selection of intersecting curves for their simplicity of description rather than according to Descartes' rule with respect to order; and he closes the brief chapter on constructions with the apologetic statement that he "has stayed over long on this question, more curious than useful."

The most noteworthy feature of the *Introductio*, from the point of view of the development of plane analytic geometry, is undoubtedly the generality of Euler's treatment. One of the chief advantages of modern analytic methods over the synthetic approach of the ancients is that many special cases can be included within one comprehensive formulation; but this aspect, to some extent understood by Fermat and Descartes, had been largely overlooked during the first century of the new analysis. Halley in 1694 had well expressed this advantage:

> The excellence of modern geometry is in nothing more evident, than in those full and adequate solutions it gives to problems; representing all cases in one view. . . .[65]

but neither he nor his successors adequately appreciated this remark. Descartes himself had laboriously worked out variations of sign be-

---

[65] *Philosophical Transactions*, 1694, p. 960. For a list of Halley's writings see *Correspondence and Papers of Edmond Halley* (ed. by E. F. Mac Pike, Oxford, 1932), p. 272f.

longing to different cases; and as late as 1748 the works of Maclaurin and Agnesi continued to splinter the linear equation into numerous different cases. Distance and direction formulae had but rarely been introduced; transformations of coordinates remained on an *ad hoc* basis; and dependence upon the use of diagrams continued to be the rule. Euler did not change *all* of this, but he did more than any other single individual to generalize analytic methods and to make the use of coordinates a point of departure for a systematic exposition of algebraic geometry. To begin with, before introducing the conic sections, "which up to that time had been almost the sole object of this branch of mathematics," he gave a theory of curves *in general*, based on the idea of function which had been developed in the first volume of the *Introductio*. The Cartesian distinction between geometrical and mechanical curves is given in modern terminology by the names algebraic and transcendental, and this is supplemented by a subdivision of functions into continuous and discontinuous, single-valued and multivalued.

Following the brief introduction to curves in general, Euler turned in order to curves of various degrees. His treatment of the linear equation is characteristic for its generality, but it is startingly abbreviated. To cover all forms of the straight line, Euler gave the single general equation $\alpha x + \beta y - a = 0$, remarking incidentally that the intercepts are $a/\alpha$ and $a/\beta$ (for $\alpha \neq 0 \neq \beta$). He mentioned specifically the cases $\alpha = 0$, $\beta = 0$, and $\alpha = a = 0$, but not $\beta = a = 0$, possibly because he used a single axis. It is noteworthy that the geometrical construction of lines is completely abandoned and that no use is here made of diagrams. Euler remarked that a line is uniquely determined by two points, evidently implying that its equation can be found by means of undetermined coefficients; but he went no further in the study of equations of first degree for the reason that "the geometry of the straight line is well-known." This is greatly to be regretted, especially since he was composing an elementary textbook. Had he amplified the treatment of the linear equation, the history of analytic geometry might have been advanced by almost precisely half a century, for not until 1797–1798 did equations of first degree become an integral part of textbooks in the subject. The situation with respect to the circle is similar. Presumably Euler felt much as Newton did—that analytic geometry is not to be applied to elementary problems involving the straight-edge and compasses. How different was the situation to be just half a century later!

What Euler might have done for the line and circle he did effectively accomplish for the conic sections. Here again one finds the treatment truly analytic—general, and free from reference to diagrams.  Wallis had freed the conics from the cone, but Euler went further.  Pointing out that previous writers had derived the properties of the curves either from the cone or from a geometric construction, he adds, "We will be satisfied by examining here what one can deduce from their equation, without recourse to other means."[66]  Beginning, as had Hermann, with the general equation $\zeta yy + \epsilon xy + \delta xx + \gamma y + \beta x + \alpha = 0$, Euler solved the equation for $y$ in terms of $x$ and found diameters much as Newton and Stirling had earlier—through the sum of the roots. Finding the diameter which bisects all chords parallel to the ordinates, first for rectangular coordinates and then again for oblique coordinates at any angle, the intersection of these diameters gave him the center of the conic.  Then starting with the rectangular Cartesian equation of the central conic with respect to its principal axes, Euler easily found the usual points, lines, and ratios associated with the curve, completing the analytic study begun by De Witt and continued by Wallis, Craig, L'Hospital, Stirling, and Hermann.  He was, of course, familiar with the characteristic, stating that the conic is an hyperbola if $\epsilon\epsilon > 4\delta\zeta$, and he knew that for the equation of the hyperbola with respect to its axes, the asymptotes are obtained by equating to zero the terms of highest degree.  The looseness with which he handled infinitesimals in the calculus is here paralleled by the statement that the parabola is nothing but an ellipse of which the major axis has been increased to infinity.   In contrast to the modern treatment, Euler derived the properties of the parabola from those of the ellipse.  He first considered the equation of the ellipse with respect to its center and axis—$yy = \dfrac{bb}{aa}(aa - xx)$—and then with respect to a vertex and an axis—$yy = 2cx - [c(2d - c)xx/dd]$, where $c = bb$ is half the parameter or latus rectum, and $d = a - \sqrt{aa - bb}$ is the distance between a vertex and its focus.  In the language of Euler, when $2d = c$ the ellipse becomes a parabola and the semiaxes $a$ and $b$ become infinite.

Euler's treatment of conics is thorough, and carried out in terms of both rectangular and oblique coordinates.  The use of a single axis did not prevent him from giving perhaps the first analytic treatment of the transformation of coordinates.  A single pair of equations, for example, suffices to cover the transformations from rectangular to oblique axes, without recourse to geometrical figures:

$$x = nr - (nv - m\mu)s - f$$
$$y = -mr + (\mu n + vm)s - g,$$

[66] *Introductio*, vol. II (1797), p. 40.

where $f$ and $g$ are the rectangular coordinates of the new origin, $m$ and $n$ are the sine and cosine of the angle between the old and new axes, and $\mu$ and $\nu$ are the sine and cosine of the angle of obliquity of the new ordinates; and where the old and new coordinates of any point are $x$, $y$ and $r$, $s$, respectively. These equations are easily reconciled with the modern forms for rectangular coordinates by noting that in this case $\mu = 1$, $\nu = 0$, and the sign of the angle between the axes is reversed.

Euler abandoned the Cartesian arrangement of curves by classes for the Newtonian classification by degrees, although he still felt it necessary to justify such a step. Following his study of quadratic curves, he then started with the general cubic and subdivided these curves into types, with their principal properties. Next he gave a similar treatment of quartics in which 146 normal forms are included. Following this, Euler reverted to the properties of curves in general— their tangents, asymptotes, diameters, curvature, and singular points. He seems not to have been aware of the earlier work of Stirling and De Gua. An elementary illustration of his greater generality is found in the fact that whereas Lahire and De Gua had noted that $y^3 = x^3$ represents three lines, two of which are imaginary, Euler enunciated the theorem that $f(x, y) = 0$ represents a system of $m$ lines [real or imaginary] through the origin if $f$ is a homogeneous algebraic function of degree $m$. After a chapter on the intersections of curves, Euler added one, previously mentioned, on the construction of equations. Here the statement that by means of two curves of orders $m$ and $n$, respectively, one can construct the roots of equations of degree not greater than $mn$, was but a reformulation of Fermat's correction of Descartes and of Maclaurin's theorem on the number of possible intersections of algebraic curves.

One of the unusual features of Euler's work on analytic geometry is the inclusion of a chapter on transcendental curves. The Newtonian school followed Descartes in considering algebraic curves primarily; and although logarithms and trigonometric functions had been long and widely used, their graphs seldom appeared in print. The forms of these curves had been given about a century before the *Introductio*, but Euler's work seems to have been definitive in bringing them into books on an elementary level. During the latter part of the eighteenth century, in fact, the analysis of the trigonometric functions was freely ascribed to Euler, and with not a little justification. Euler listed systematically all the usual formulae of goniometry, with special reference to the multiple-angle formulae; he treated the circular functions as ratios rather than as geometric lines; he emphasized the periodic nature of the functions and drew their graphs; he correlated circular and exponential functions through the use of imaginary quantities; and he differentiated and integrated the direct and inverse trigonometric expressions logarithmically rather than in terms of geometric properties. In short, he made the study of trigonometric and other elementary transcendental curves part of coordinate geometry and the calculus. Curves of the usual elementary functions were not the only ones which Euler included in his analytic geometry. The chapter on transcendental curves includes such oddities as $y = x^{\sqrt{2}}$, $y = x^x$, $y^x = x^y$, and $y = (-1)^x$, some of which he undoubtedly owed to his teacher, Jean Bernoulli.

The *Introductio* includes also two accounts of polar coordinates which are so thorough and systematic that the system frequently is ascribed to Euler.[67] Whereas Newton, Bernoulli, and Varignon had applied vectorial coordinates only to transcendental curves, and Hermann had used them only for algebraic curves, Euler devoted a section to each of these classes. In connection with the latter case,[68] he gave the equations of transformation $x = z \cos \phi$, $y = z \sin \phi$, introducing modern trigonometric symbolism into polar coordinates. He gave general consideration to $z$ as a function of $\sin \phi$ and $\cos \phi$, and he noted in more detail the limaçons $z = b \cos \phi \pm c$ and the conchoids $z = (b/\cos \phi) \pm c$. Like Hermann, Euler, too, suggested the transformation of equations of conics into polar coordinates.

One is surprised that neither Hermann nor Euler referred to those curves which now are so prominent in the elementary use of polar coordinates—the curves $r = a \sin n\phi$ and $r = a \cos n\phi$. These had been described by Grandi as early as 1713 in letters to Leibniz, and they were published in England[69] in 1723 and in Italy[70] in 1728. However, Grandi had not handled his "roses" analytically, but had described them in words. The curve $r = \sin \frac{a}{b}\phi$, for example, was indicated as follows: In the circle with center $C$ and radius $CA$ (Fig. 38) take angles $ACD$ and $ACG$ in the given ratio, $a$ to $b$; and along $CD$ take $CI$ equal

[67] See, e. g., E. Müller, "Die verschiedenen Koordinatensysteme," *Encyklopädie der mathematischen Wissenschaften*, III (1), 596–770, especially p. 656–657. Cf. *Encyclopédie des sciences mathématiques*, III (3), 1, p. 47. On the other hand, Coolidge (*History of Geometric Methods*, p. 171–172, mentions no contributions to plane polar coordinates between 1691 and 1794.

[68] *Introductio* (1748), II, p. 212 ff.

[69] Guido Grandi, "Florum geometricorum manipulus," *Philosophical Transactions*, XXXII (1723), p. 355–371. This account includes a score of beautifully drawn figures—bifoliate, trifoliate, etc. Grandi waxes eloquent on the role of geometry in the beauties of nature.

[70] Guido Grandi, *Flores geometrici ex rhodonearum* (Florentiae, 1728).

to *HG*, the sine of angle *ACG*.  The locus of the point *I* is then the curve in question.  That is, "the rose curves are those which are formed by segments, cut off from the infinite number of branches going out from the center, equal to the sines of those angles which are to the cor-

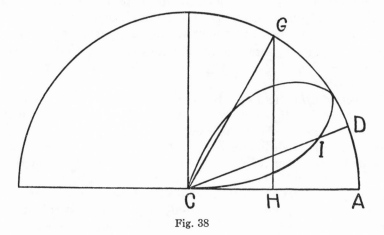

Fig. 38

responding angles of the branches in a given ratio."[71]    What tongue-twistings Grandi would have been spared had he been familiar with polar coordinates!

In the polar treatment of transcendental curves Euler adopted a somewhat different notion and notation for the independent variable.[72] Here he studied curves of the form $z = f(s)$, where the argument *s* is the arc of a unit circle which measures the angle $\phi$.  Ostensibly even Euler, more than a century after the appearance of *La géométrie*, felt that coordinates must of necessity denote lengths.  In connection with the spiral curves which he drew, Euler made use of the general angle, allowing *s* to increase indefinitely, both positively and negatively.  As a consequence of his use of negative polar coordinates, the spiral of Archimedes here appeared, perhaps for the first time, in its dual form.[73]

It is probably not too much to say that although Newton may have originated polar coordinates, it was the work of Euler which was the

[71] *Flores geometrici*, p. 2.
[72] *Introductio*, II, p. 284 ff.
[73] Loria mistakenly ascribes negative radii and the dual form of this spiral to Cournot about a century later.   See Gino Loria, "Perfectionnements, évolution, métamorphoses du concept de coordonnées," *Mathematica*, XVIII (1942), p. 125–145; XX (1944), p. 1–22; XXI (1945), p. 66–83, in particular, p. 139.   This important paper appears also in *Osiris*, VIII (1948), p. 218–288.

decisive factor in making the system a traditional part of elementary analytic geometry. Jean-le-Rond d'Alembert (1717–1783), in the article "Géométrie" in the *Encyclopédie*, took cognizance of the system in writing that the equation of a curve can be given in terms of parallel coordinates, or of the radius vector and either the abscissa or the ordinate, or of the radius vector and the sine, secant, or tangent of the angle of inclination. The *phrase* "polar equation" was used in 1784 by Gregorio Fontana (1735–1803) in connection with polar curves of the form $z = f(u, \sin u, \cos u)$, where $z$ is the radius vector and $u$ is the arc of a unit circle;[74] and this fact seems to have led Smith unwarrantedly to ascribe to him the *idea* of polar coordinates.[75] Certainly the system of plane polar coordinates is not to be attributed to any one later than the time of Euler, for it began then to appear in representative didactic works of the century. For example, the *Instituzioni analytiche* of Riccati and Saladini (1765–1767) included the formula for radius of curvature in polar coordinates, using the Newtonian scheme with $(x, y)$ for $(r, r\theta)$.

The method of parametric representation of curves, systematized largely by Euler, again was a development of implications of a hundred years before. In presenting the fundamental principle of analytic geometry, Descartes had pointed out that, for a curve, one must have one more unknown than there are conditions or equations. Parametric equations of a curve are only a special case of this principle. In fact, the composition of movements, so popular in the seventeenth century but in reality going back to the quadratrix of Hippias, is also an anticipation of the parametric representation of curves. But it was Euler who pointed out specifically the advantages of such forms. Where $y$ and $z$ are related by an implicit algebraic function, such as $y^{10} = 2ayz^6 + byz^3 + cz^4$, he suggested the substitution $y = xz$, as the result of which $z$ and $y$ are each expressible algebraically in terms of the parameter $x$. Similarly the curve $y = x^x$ is given, after the substitution $y = tx$, in the parametric form $x = t^{1/(t-1)}$ and $y = t^{t/(t-1)}$, or, if $t - 1 = u$, as $x = \left(1 + \frac{1}{u}\right)^u$ and $y = \left(1 + \frac{1}{u}\right)^{u+1}$.

Where Leibniz represented the cycloid by a differential or integral equation, Euler wrote it parametrically in the now more usual manner[76]

[74] "Sopra l'equazione d'una curva," *Memorie di Matematica e Fisica della Società Italiana*, II (part i, 1784), p. 123–141, especially p. 128. Cf. also his *Disquisitiones physico-mathematicae* (Papiae, 1780), p. 184–185. Here Fontana uses the letters $x$ and $y$ instead of $u$ and $z$. Earlier (in 1763), Fontana had given the formula for radius of curvature in polar coordinates.
[75] Smith, *History of Mathematics*, II, 324. The work of Fontana gives one the impression of having been influenced by Euler.
[76] See *Introductio*, I, p. 39–46; II, p. 294.

as

$$\begin{cases} x = b - b \cos \dfrac{z}{a} \\[2mm] y = z + b \sin \dfrac{z}{a}. \end{cases}$$

Possibly the most important general contribution of the eighteenth century to mathematics was the development of the notion of function. This idea again had been familiar to earlier men, from Descartes and Fermat to Newton and Leibniz; but it was Euler who established it firmly in his *Introductio*. Improving on the definition given fifty years earlier by Jean Bernoulli, Euler wrote: "A function of a variable quantity is an analytic expression composed in any manner whatever of this variable and constants." In this definition Bernoulli and Euler had in mind the chief means of defining plane curves—by means of "analytic expressions" or equations in two variables; but the emphasis, as in the case of Descartes, was on the equation rather than on the curve. On at least one occasion, nevertheless, Euler inverted the situation and used the term function to designate the relationship between $x$ and $y$ implied by any curve "drawn free-hand" in the $xy$ plane. This generalization of the Cartesian mechanical motions, however, was not exploited until a century later.

The *Introductio* closes with a long and systematic appendix on solid analytic geometry. This is perhaps the most original contribution of Euler to Cartesian geometry, for it represents in a sense the first textbook exposition of algebraic geometry in three dimensions.[77] Inasmuch as Clairaut's well-known book had presented the theory of curves of double curvature, Euler devoted most of his appendix to the study of surfaces and touched but lightly upon skew curves. As in his plane analytic geometry Euler had used a single axis, so in three dimensions he continued his early use of a single coordinate plane as basic. Nevertheless, he expressly remarks that three planes can be used, and he himself makes frequent use of this scheme, especially in illustrations. Moreover, he pointed out the signs of the coordinates for the eight octants associated with such a trihedral of reference, and tests for symmetry and extent are given, even though Euler limited his diagrams largely to the first octant. The order of treatment of the material is

---

[77] There is surprisingly little material on the history of surfaces. The catalogue of the New York Public Library, for example, includes a trayful of references on surfaces; yet not more than three or four of these are in any real sense historical, and even these refer only to recent developments. It is interesting to note in this connection that the first decade of the twentieth century produced a greater number of papers dealing with surfaces than did any comparable period before or since. There are at least half a dozen extensive treatments of the history of curves (see the bibliography at the close of this work), but one looks in vain for analogous works on the history of surfaces.

similar to that given above for two dimensions.  Surfaces are first considered in general, and are divided into two classes, algebraic and transcendental.  They are studied in the usual manner by a consideration of the traces in various planes.  Cones, spheres, cylinders, and conoids serve as illustrations.  Paralleling his contribution to plane analytic geometry, Euler gave the first formulae for translation and rotation of axes in three dimensions—the non-symmetric form still known by his name:[78]

$$\begin{cases} x = t \cos \zeta + u \sin \zeta \cos \eta - v \sin \zeta \sin \eta - a \\ y = -t \sin \zeta + u \cos \zeta \cos \eta - v \cos \zeta \sin \eta - b \\ z = u \sin \eta + v \cos \eta. \end{cases}$$

This has been a classic portion of solid analytic geometry ever since.

For three dimensions, as for two, Euler's treatment of the linear equation is characteristically general but disappointingly brief.  He gives the plane in the form $\alpha x + \beta y + \gamma z = a$, and finds its traces and intercepts with the coordinate plane and axis.  Like Hermann, he found the angle between the given plane and the coordinate plane, but he wisely gave the cosine of this angle, $\gamma / \sqrt{\alpha\alpha + \beta\beta + \gamma\gamma}$, instead of the sine.

Euler's classification of quadric surfaces is significant as the first unified treatment of the subject.  It is surprising to note that here one finds apparently the first conception of the surfaces of second degree as a family of quadrics in space analogous to the plane conic sections. Euler begins with the general quadratic equation with ten terms and points out that the aggregate of terms of highest degree furnishes the equation of the asymptotic cone, real or imaginary.  He indicates that the general equation can be reduced by transformation to the canonical forms $App + Bqq + Crr + \kappa = 0$, $App \pm Bqq = ar$, and $App = aq$, from which he derives a classification of quadric surfaces.  He includes the five fundamental types—the ellipsoid, the hyperboloids of one and of two sheets (*elliptico-hyperbolica* and *hyperbolico-hyperbolica*), and the elliptic and hyperbolic paraboloids (*elliptico-parabolica* and *parabolico-hyperbolica*)—but he does not list all of the degenerate types.  We have seen above, however, that he was thoroughly familiar with cones and cylinders, general as well as quadric. Euler's work here represents the first attempt at a unified treatment of the general quadratic equation in three unknowns—over a century after Fermat and Descartes had done a similar thing for binary quadratics.  Such a classification has remained ever since a part of standard courses in analytic geometry.

[78] Wieleitner, *Geschichte*, II (2), p. 53, inadvertently credits this to Meusnier in 1785.

Euler did not undertake to give a full discussion of the curves of intersection of surfaces, but he showed not only how to study these in the Cartesian manner by means of projection curves, but also how to write the equation of a plane section of a surface in terms of two co-ordinates in the plane. He investigated in particular the sections of the sphere, elliptic cylinder, and quadric cone, and pointed out that the order of a plane section of a surface is not higher than the order of the surface.

The work of Euler may be said to mark a turning point in the de-velopment of analytic geometry in several respects. In the first place, it marks the close of the period in which the strict Cartesian point of view was dominant. The subject no longer had for its chief end the solution of problems in geometrical construction. It was not, in fact, concerned exclusively with two dimensions. Analytic geometry for Euler meant essentially but one thing—the sketching and study of curves and surfaces by means of their equations, i. e., the graphical representation of functions. In retrospect one can easily appreciate the service he rendered in releasing the subject from the rigidity of the older tradition.[79] In a sense, one may say that Euler freed the subject from geometrical fetters, even though he did not go beyond three di-mensions. Under the Cartesian view, analytic geometry began from a geometrical problem and ended with a geometrical construction; the algebraic study of curves and equations constituted the connecting link from the initial to the terminal stage. For Euler, on the other hand, analysis was not the application of algebra to geometry; it was a subject in its own right—the study of variables and functions—and graphs were but visual aids in this connection. To some extent this represents a return to the use of the word *analysis* by Viète; but whereas formerly it referred to algebraic calculation in which unknowns were operated upon as if they were known, it now dealt with continuous variability based upon the function concept. It is, of course, possible to read this meaning into the works of Fermat and Descartes, or even into that of Viète; but only with Euler did it take on the status of a conscious program.

Euler's analytic theory of functions was invaluable for the develop-ment of mathematics, but it seems to have over-shadowed the develop-ment of one other aspect of plane analytic geometry. He and most of his successors failed to appreciate the advantage of algebra as a fitting language for *elementary* geometric concepts. As Loria has pointed

---

[79] Hermann Hankel, *Die Entwickelung der Mathematik in den letzten Jahrhunderten* (2 ed., Tübingen, 1884), p. 12. One should bear in mind, nevertheless, that the increased general-ity often was accompanied by a corresponding loss of vigor.

out,[80] analytic geometry now generally includes four things: (1) generalities on the method of coordinates; (2) examples of equations of loci and the tracing of curves; (3) formulae for solving fundamental questions on points, lines, and planes; (4) the study and classification of curves and surfaces of second degree.   All but the third of these are found clearly developed in the *Introductio*, so that this work sometimes is regarded as the first textbook containing the essence of analytic geometry.[81]   The one thing missing is the plane and solid algebraic geometry of rectilinear configurations.   The variables $x$, $y$, and $z$ and the functional relationships $\alpha x + \beta y = a$ and $\alpha x + \beta y + \gamma z = a$ were of interest as aspects of analysis, not as algebraic counterparts of points, lines, and planes.   This may well be the reason that such fundamental formulae as those for distance, midpoint, slope, angle, and area, appeared so late in the history of analytic geometry.   The background for the formula-building stage of algebraic geometry arose very gradually during the half-century following the *Introductio*, and it is paradoxical to note that the program for two dimensions did not in general precede, it followed, the introduction of corresponding formulae in *solid* analytic geometry, of which branch Euler was the effective founder.

[80] "Qu'est-ce que la géométrie analytique," *L'Enseignement Mathématique*, XIII (1923), p. 142–147.
[81] See D. J. Struik, *A ConciseHistory of Mathematics* (2 vols., New York, 1948), II, p. 138, 169.

# CHAPTER VIII

# The Definitive Formulation

*A thorough advocate in a just cause, a penetrating mathematician facing the starry heavens, both alike bear the semblance of divinity.*

—GOETHE

THE second half of the eighteenth century was a period of conflicting tendencies in the history of analytic geometry. In a broad sense it is possible to distinguish three trends: there was one which may be called the continuation of the Cartesian tradition of constructions and graphical solutions; a second direction emphasized the Fermatian aspect as formulated in Euler's theory of functions; and finally there was the gradual development of a third movement which might be characterized as foreshadowing the formula-building of modern algebraic geometry. The books of Guisnée and L'Hospital, so characteristic of the Cartesian tradition of the first half of the century, continued to be popular during the second half and appeared in new editions. In the famous *Encyclopédie*, D'Alembert devoted several pages to an article on "Construction," in which equations are solved graphically by means of circles and parabolas; and it was natural for him here to cite the textbooks of Guisnée and L'Hospital. These texts were supplemented by a number of new books of similar character, such as the *Conics* of La Chapelle, which appeared in 1750. The full title of the work is *Traité des sections coniques et autres courbes anciennes*, an interesting commentary on the Cartesian lack of interest in the host of new curves which analytic geometry might have suggested.[1] La Chapelle held that at first there were only two men in Europe—de Beaune and van Schooten—who understood the geometry of Descartes, and that for a century it was the object of commentaries by the strongest mathematicians. His own work is, in a sense, one of these commentaries. It is based upon Lahire and, more

---

[1] A work with a very similar title and in much the same spirit as La Chapelle's appeared two years later—*Les sections coniques, et autres courbes anciennes traités profondement* (Paris 1752), by Jean Edme Gallimard (1685–1771).

especially, Guisnée and L'Hospital, who, the author says, were about the only writers on the conics known in his day. La Chapelle's book is not at all like a modern textbook, or even like that of Euler. It makes use of the older language of proportions, and, except incidentally in connection with the conics, it does not give equations of the curves considered, such as the cissoid, conchoid, quadratrix, spiral of Archimedes, and cycloid. There is no plotting of curves and little of the Eulerian analytic approach to geometry.

The older tradition, so clearly represented by La Chapelle, had become intrenched in the didactic *Sammelwerke* of the period. During the seventeenth century analytic geometry had not generally been included in mathematical compendia and works of instruction; but with the opening of the next century one found the subject in the textbooks of Wolff and others. The mathematical program of the time seems to have become more or less standardized in the French *Cours de mathématiques* (such as those by Sauri or Bézout), the German *Anfangsgründe zur höheren Analysis* ( e. g., by Kästner or Wolff), and the Italian *Instituzioni analytiche* (Riccati and Saladini). These invariably contain a section on "the application of algebra to geometry," following the pattern of L'Hospital and Guisnée. The first American textbook on analytic geometry, in 1826, was derived from the *Cours* of Étienne Bézout (1764–1769), and one can still hear echoes of Descartes in the ingenuous justification for the study of the conic sections: "When questions one wishes to solve do not exceed the second degree, one does not need these curves, but beyond this they become necessary."[2]

The multi-volumed collections of the later eighteenth century, nevertheless, betray clearly a tendency to reach a compromise between the points of view of Descartes and Fermat. The works of Euler did not enjoy an impressive multiplicity of editions, but his all-pervasive influence is quite evident in the literature of the age. Practically every one of the authors of the ubiquitous continental compendia expresses indebtedness to the great analyst, but no one of them forged ahead to the next stage in coordinate methods. Emphasis upon the notion of functions and long sections devoted to curve-tracing are found in all the works; but there was a natural tendency for the material on curves to be merged with that on the calculus, and hence analytic geometry in this sense sometimes lost its identity. One of the best-known single-volume textbooks of the period, however, valiantly and effectively maintained, free from reference to the cal-

---

[2] Lacroix and Bézout, *An Elementary Treatise on Plane and Spherical Trigonometry, and on the Application of Algebra to Geometry* (Cambridge, Mass., 1826), p. 110.

culus, the tradition of Fermat, Newton, and Euler against that of Descartes, L'Hospital, and Guisnée. This was the *Introduction a l'analyse des lignes courbes* of Gabriel Cramer (1704–1752) which appeared in 1750, the same year as La Chapelle's very different work on "conic sections and other *ancient* curves."

Cramer's volume resembles strongly in aim and content the work of de Gua[3] and Euler. The author, however, says that his book had been almost completed before the appearance of the *Usage de l'analyse* in 1740, and that the *Introductio* of 1748 was published too late for him to make use of it. The first half of this statement is belied to some extent by the effective use Cramer made of the analytical triangle, for the introduction of which he compliments de Gua; but the latter half is borne out by his failure to employ the analytical trigonometry of Euler. Apparently Cramer was influenced primarily by Newton's *Enumeratio* and Stirling's *Commentary*. He did not apply the calculus, and so his book rivals that of Euler as a résumé of the application of analytic geometry to the study of curves. It opens with a general theory and classification of algebraic curves. Transcendental curves are illustrated by the cases $y = ba^x$ and $y^{\sqrt{2}} + y = x$, but these are not systematically considered. A whole chapter is devoted to the transformation of axes, but trigonometric symbolism is not here used. Cramer was among the first to make formal use of two axes and to define the two coordinates simultaneously and symmetrically in terms of these. The coordinates are designated by the terms *coupée* and *appliquée*, as well as by the more modern names abscissa and ordinate.

One finds an intruding chapter on the old subject of the construction of equalities, with a table like that of Lahire indicating the curves of minimum degree necessary for the graphical solution of equations in the traditional manner. The table runs from quadratic equations to those of degree one-hundred, for the last of which two tenth-degree curves are indicated as necessary. Cramer even quotes the rule given in L'Hospital's *Conics* to cover this favorite Cartesian topic of the simplest possible curves; but he adds hesitantly that he doesn't know whether or not this restriction is for the best. He suggests as an alternative the Newtonian idea that it is less the algebraic simplicity than the geometric facility of description which one should seek. This being so, he doesn't see why one should reject the type of graphical solution of polynomial equations which is found in L'Hospital's *Conics* and which Cramer ascribes to Jacques Bernoulli. This method, described above, makes use of the intersections of the line $x = a$ and the curve $x = by + cy^2 + dy^3 + ey^4 + \ldots$ to solve the

---

[3] Sauerbeck, *op. cit.*, refers to Cramer's work as in a sense an improved edition of de Gua.

polynomial equation $a = by + cy^2 + dy^3 + ey^4 + \ldots$. A score of cubic and quartic polynomial equations of this form are plotted graphically.[4] Cramer pointed out that "this construction is simple and of an easy practicality," but he himself used it, following L'Hospital and Stirling, largely to illustrate the character of the roots of equations as determined by the coefficients. No examples are given in which the coefficients are simple numbers. Cramer did, however, express a well-warranted surprise that such graphical solutions should be rejected[5] (presumably by the algebraists of his day).

Like Euler, Cramer devoted much space to the properties of plane curves, both elementary and higher—their diameters, singular points, and infinite branches. He made extensive use of the Newton-de Gua triangle and of series developments for singularities, both at the origin and at infinity. Specific cases of equations with numerical coefficients are given frequently, an exceptional practice in that day. The general equation of the circle, rarely given by writers of the time, appears in the form $(y - a)^2 + (b - x)^2 = rr$; but neither of the curves of elementary geometry—the line and the circle—was studied systematically. The straight line appears in the general form $a = \pm by \pm cx$, and various cases (cutting across each of the four quadrants, respectively) are geometrically constructed. The equations of the axes, $x = 0$ and $y = 0$, are specifically mentioned. In determining a curve of order $n$ through $\dfrac{n(n + 3)}{2}$ points, Cramer suggested the method of undetermined coefficients. Thus to find the conic through five points, the coordinates of the points are substituted in the equation $A + By + Cx + Dyy + Exy + xx = 0$; and by means of the five linear equations thus obtained one calculates the five unknown coefficients. Admitting that the calculation is fairly long, Cramer adds the comment, "I believe I have found for this a rule which is quite convenient and general when one has any given number of equations and unknowns none of which exceeds the first degree."[6] Using exponents as distinguishing indices instead of powers, Cramer wrote the equations as

$$A^1 = Z^1 z + Y^1 y + X^1 x + V^1 v + \ldots$$
$$A^2 = Z^2 z + Y^2 y + X^2 x + V^2 v + \ldots$$
$$A^3 = Z^3 z + Y^3 y + X^3 x + V^3 v + \ldots$$

. . . . . . . . . . . . . . . . . . . . . . . . . . . . . . . . . .

[4] Gabriel Cramer, *Introduction a l'analyse des lignes courbes algebriques* (Genevae, 1750), plate opposite p. 108; cf. also p. 92.

[5] An exception to this is found in Andreas Segner, "Methodus simplex et universalis, omnes omnium aequationum radices detegendi," *Novi Commentarii Academiae Petropolitanae*, v. VII, 1758–1759 (1761), p. 211–226. In this the graphical solution of polynomial equations is freely used in the modern manner (except for the continued use of a single axis of coordinates).

[6] *Introduction*, p. 60, 657–659.

"Then to find the values of the unknowns one forms fractions as follows: in the denominator one writes all possible products of a $Z$, a $Y$, an $X$, a $V$, etc., always written with letters in order and with signs determined according as the number of inversions in superscripts is odd or even. In the numerator one substitutes the $A$'s for the coefficients of the unknown desired." Through this device, now known as Cramer's rule, the author is entitled to credit as a rediscoverer of the value of the determinant notation of Leibniz. Leibniz in a letter to L'Hospital had shown[7] how to take advantage of a harmonious notation in eliminating two unknowns from three linear equations, boasting that this "little pattern" showed that Viète and Descartes were not aware of all the mysteries. Determinant patterns were occasionally used in the latter half of the eighteenth century, but it was almost a hundred years after the time of Cramer before mathematicians generally realized the role that determinants can play in analytic geometry.

Among the properties of curves cited in the *Introduction* is one which has become known as "Cramer's paradox," although it had been referred to in a memoir by Euler two years before,[8] and was mentioned also by Maclaurin. This dilemma pointed out that in general two algebraic curves of degree $n$ intersect in $n^2$ points, but that the number $n^2$ usually is greater than $\dfrac{n(n+3)}{2}$, the number of points by which each of the curves presumably should be *uniquely* determined. The only exceptions are the conics $\left(\text{for which } n^2 \text{ is 4 and } \dfrac{n(n+3)}{2} \text{ is 5}\right)$ and the cubics (for which both numbers are 9). Cramer seems to have realized, in a vague sort of way, that the question of independence of points was involved here; but it remained for the nineteenth century to give a clear explanation of this paradoxical situation.[9]

Cramer's *Introduction* is the work of an expert on the subject. It includes almost 700 pages of exposition and hundreds of illustrations—a worthy successor to Newton's *Enumeratio*.

Half a dozen years after Cramer's popular treatise on curves, there appeared anonymously a far less known and much less extensive

---

[7] See *Leibnizens mathematische Schriften* (ed. by Gerhardt), v. II (Berlin, 1850), p. 239–240.

[8] "Sur un contradiction apparente dans la doctrine des lignes courbes," *Mém. de Berlin* v. IV, 1748.

[9] Gergonne, "Sur quelques lois qui régissent les lignes et les surfaces algébriques," *Ann. de Math.*, v. XVII (1826–1827), *Plücker, Analytische geometrische Untersuchungen* (1828), v, I, p. 41. For an excellent historical account of the paradox see Charlotte A. Scott, "On the Intersections of Plane Curves, *American Mathematical Society Bulletin*, v. IV (1897–1898,) p. 260–273.

work, a *Traité des courbes algébriques* which was significant for a somewhat new orientation. The authors of this book were in reality M.
P. Goudin (1734–1817) and A. P. Dionis du Sejour (1734–1794).[10]
They opened the subject of curves, as did Euler and Cramer, with the
translation of axes. This led them then to the first systematic treatment of the derivation of the equation of the straight line, as distinct
from the earlier constructions of lines of which the equations were
given. It is interesting to note, however, that even here, in 1756, the
linear equation was not studied for its intrinsic interest, but because
it was found to be of value in disclosing the properties of higher plane
curves through the determination of the number of times various
secants cut a given curve.

Goudin and Dionis du Sejour first wrote the rectangular Cartesian
equation of a line through the origin as $mz - nu = 0$, where $u$ and $z$
are the abscissa and ordinate, respectively, and $m : n$ is "the ratio of
the cosine to the sine" of the angle of inclination. Loria exaggerates[11]
in saying that this marks "the first elementary metric question which
is found in the literature of coordinates," for Euler earlier had given
the intercepts of the line (in two dimensions) and he and Hermann had
given a direction angle for a plane (in three dimensions). Nevertheless, it should be emphasized that this is perhaps the first treatise
to give specific formulas for finding equations of straight lines directly
from given data without reference to geometric diagrams. The form
of presentation for the general line, based as it is upon transformations
of axes, differs somewhat from the modern equivalent. Upon translating the axes either $p$ units to the right or $q$ units upward, the equation $mz - nu = 0$ becomes either $ms - nr - mq = 0$ or $ms - nr - np = 0$,
where $r$ and $s$ are the new abscissa and ordinate, respectively. Consideration is given to the special cases for which $m = 0$ or $n = 0$.
The book is full of examples of linear equations, derived from the
above forms and used in the study of curves. This shows that the
general impression given by historians of analytic geometry that the
equations of straight lines did not enter until the *end* of the eighteenth
century is inaccurate. It is true that the form in which linear equations are given by Gaudin and Dionis du Sejour is less convenient and
less rigidly formalized than that in modern textbooks, but this is
relatively a minor matter. One could, in fact, impute to the authors
by implication the point-slope form of the straight line, for the equa-
tion $mz - nu = 0$, coupled with the transformation $\begin{cases} x = u + r \\ y = z + s \end{cases}$ which they

[10] See *Académie des Sciences, Histoire*, 1756, p. 79.
[11] "Da Descartes e Fermat a Monge e Lagrange," p. 829.

suggest, leads directly to the form $\dfrac{y-s}{x-r} = \dfrac{n.}{m}$ This latter form, however, does not explicitly occur in their work. It seems first to have
been given a quarter of a century later by Monge, a point again overlooked by historians of the subject.

Except for a somewhat more pronounced attention given to the
linear equation, the *Traité des courbes algébriques* is in the manner of
Euler and Cramer. It is what the name implies—a study of higher
plane curves, rather than an analytic geometry in the modern sense.
The authors indicate, for example, that a curve of order $n$ can have not
more than $n(n-1)$ tangents with a given direction. This observation was later developed synthetically by Poncelet into the important
idea of the *class* of a curve. Gaudin and Dionis add also that the
curve can have no more than $n$ asymptotes; and they point out, as had
Maclaurin, that an asymptote cannot cut the curve in more than
$n-2$ points. In England similar work on plane algebraic curves was
carried out by Edward Waring (1743–1798) in his *Miscellanea analytica de aequationibus algebraicis et curvarum proprietatibus* of
1762. The most unusual part of this work is a section on surfaces in
which Waring treats these from a general point of view. Among other
things, he gave the number of independent coefficients for an algebraic
surface of degree $n$ as $\dfrac{(n+1)(n+2)(n+3)}{1 \cdot 2 \cdot 3} - 1,$ an extension by
analogy of the corresponding result for algebraic plane curves. He
indicated that most of the theorems on plane curves can be extended
to surfaces and curves of double curvature; but he did not develop
this idea extensively. Waring, too, seems to have been more interested
in the volumes of bodies than in the properties of surfaces. The
rivalry of the time between the adherents of the method of fluxions
and those who espoused the differential calculus seems to have caused
a gulf in geometry as well, for Waring does not cite a single Continental
mathematician. His work, however, was more a part of the method of
fluxions than analytic geometry, to which British mathematicians
contributed little of significance during the second half of the century.

The nature of analytic geometry did not change appreciably in the
twenty-five years following Euler's *Introductio*. Mich. Hube (1737–
1807), in his *Versuch einer analytischen Abhandlungen von den Kegelschnitten*[12] of 1759, sought to give wider circulation to Euler's general

---

[12] I have not seen this work, but have depended upon the account given by Cantor,
*Vorlesungen über Geschichte der Mathematik*, v. IV, p. 453f, and Wieleitner, *Geschichte der
Mathematik*, v. II p. 21, 40. I have seen K. C. Langsdorf, *Ausführung der Erläuterungen
über die Kästnerische Analysis des Unendlichen, nebst Anmerkungen zu Hubens analytischer
Abhandlung von den Kegelschnitten* (Giessen, 1781), but this is virtually unintelligible with-
out access to Hube's work.

analytic treatment of conics. Hube's *Versuch* has therefore been referred to as the "first German testbook of analytic geometry." A long introduction to this by Kaestner, on the advantages of analysis as compared with synthesis, shows that there was a tendency to distinguish more sharply between the two approaches. The treatises of Vincenzo Riccati and Girolamo Saladini (1765–1767) and of the Abbé Sauri (1774) show strongly the influence of Euler, especially in trigonometry and in "géométrie sublime ou géométrie des courbes"; but if anything they were less courageous in breaking away from the older dependence upon Descartes and the geometric background. The spirit and arrangement of the first six chapters in the *Instituzioni* of Riccati and Saladini is clearly Cartesian—the geometric construction of determinate equations and problems, first of degrees one and two, then of degree three or four, and finally for those beyond four. The *Cours* of Sauri is somewhat less Cartesian, but it also includes a long section on constructions of determinate and indeterminate equations. It is interesting to note that in at least one respect virtually all compendia of the time depart from the older tradition—in the use, to a limited extent, of polar coordinates. Riccati and Saladini give "the equation of a curve related to a focus,"[13] including the logarithmic spiral $u = ly$; and Sauri refers to "curves of which the ordinates issue from a point called focus,"[14] including the hyperbolic spiral $z = rc/y$, where $z$ is the arc of a fixed circle of radius $r$.

It should be noted, however, that in both cases these applications are made in connection with the calculus rather than with coordinate geometry. A few years later Goudin composed a *Traité des propriétés communes à toutes les courbes* (Paris, 1778) in which are listed 371 formulas (mostly from the calculus) expressing the properties of curves; and among these one finds the equivalent of transformations from rectangular coordinates $(x, y)$ to polar coordinates $(t, z)$—$t^2 = x^2 + y^2$, $ry = x \tan z$, $rx = t \cos z$, $ry = t \sin z$, where $r$ is a constant. Bézout's *Cours* (third edition) and Kaestner's *Anfangsgründe* also contain brief reference to polar coordinates. One gets the impression, nevertheless, of a certain timidity in the use of polar coordinates in the eighteenth century. As late as 1797 there appeared a paper by S. Gourieff[15] in which the equations $x = z \cos w$, $y = z \sin w$ for transformation from rectangular to polar coordinates, as well as the polar formula for radius of curvature, are derived as though they were novel. Here, in spite

[13] *Instituzioni analitiche*, v. II, p. 176, 255.

[14] *Cours complet*, v. III, p. 70f.

[15] "Mémoire sur la résolution des principaux problèmes qu'on peut proposer dans les courbes dont les ordonnées partent d'un point fixe," *Nova Acta Academiae Petropolitanae*, v. XII (1794), p. 176–191. The paper was presented May 22, 1797, but the volume containing it was published in 1801.

of the earlier work of Euler and Hermann (former members of the same Academy), one finds an account of polar coordinates more repetitious and less expert than that of Newton more than a century before.    In some respects the development of coordinate geometry was surprisingly halting.

From Descartes and Fermat to Euler and Cramer, analytic geometry had been concerned almost exclusively with constructions and loci other than those covered by the line and circle.    Rarely had analytic methods been used in connection with problems of elementary geometry, for which synthesis had afforded solutions.    Euler had freed the analysis of conics, higher plane curves, and quadric surfaces from geometric considerations; but his analytic geometry did not invade the domain in which synthesis reigned supreme—the study of lines and circles, spheres and planes.    In the article on "Géométrie" in the *Encyclopédie*, D'Alembert adopted Euler's view when he said, "Algebraic calculation is not to be applied to the propositions of elementary geometry because it is not necessary to use this calculus to facilitate demonstrations, and it appears that there are no demonstrations in elementary geometry which can really be facilitated by this calculus except for the solution of problems of second degree by the line and circle."

The first decisive steps in the application of algebra to the problems of elementary geometry were taken in three-space rather than in two dimensions.    The history of the point and line in solid analytic geometry differed, for obvious reasons, from that in the plane.    The straight line in space does not lend itself readily to construction in the Cartesian sense or to plotting point by point in the sense of Fermat, Newton, and Euler.    Moreover, given two points or two lines in a plane, their mutual relationships in terms of distance and direction are obvious on the face of it, irrespective of a coordinate system; but this is not true of these elements seen in three-space or in a plane perspective drawing of them.    Their relative positions are made clear by referring them to some familiar configuration, such as a coordinate system.    Inasmuch as visualization and pictorial representation are more difficult in three dimensions than in two, it is here desirable, even for relatively simple situations, to express geometric elements in algebraic terminology.    When Lagrange, the greatest mathematician of the age, in 1773 suggested the analytic treatment of certain aspects of *elementary* geometry, he naturally chose an illustration in three dimensions.

Joseph Louis Lagrange (1736–1813) reminds one of Euler in the international character of his life.    He was born in Italy and died in

France, living for over a score of years in each of these countries and for a comparable period in Germany. His work, too, resembles Euler's in its elegance and generality. In a paper on "Solutions analytiques de quelques problemes sur les pyramides triangulaires,"[16] Lagrange set himself an old and familiar simple problem—the determination of the surface area, center of gravity, and volume of a tetrahedron, as well as the centers and radii of the inscribed and circumscribed spheres. The significance of the work lay more in the point of view than in the substance, as Lagrange realized: "I flatter myself that the solutions which I am going to give will be of interest to geometers as much for the method as for the results. These solutions are purely analytic and can even be understood without figures."[17] And true to his promise, there is not a single diagram throughout the work. The paper is characteristic of its author for generality and elegant symmetry. Beginning with the four vertices $(0, 0, 0)$, $(x, y, z)$, $(x', y', z')$, and $(x'', y'', z'')$, he found the six edges of the tetrahedron by means of the distance formula. It is interesting to note that this formula had appeared long before in Clairant's work on skew curves. That the distance formulas did not reappear in the interval is, in a sense, symptomatic of the lack of appreciation before Lagrange of the value of analysis as a geometrical medium of expression for elementary geometry. Where diagrams are readily available, there is less need for formulas. In three dimensions, however, geometrical figures often are not at hand and perspective drawings of them are not so easily interpreted. Conclusions in this case are derived with greater precision from analytic formulas than from synthetic forms. To this situation is probably ascribable the fact that the definitive steps in the analytic geometry of rectilinear figures were made first in three dimensions rather than two.

To determine the altitude of the tetrahedron from the origin of coordinates, Lagrange wrote the equation of the plane of the opposite face as $u = l + ms + nt$, where $l$, $m$, and $n$, the coefficients, are found by means of the coordinates of the three vertices determining the plane. The normal form of the plane not being known, he found the altitude by the methods of the calculus, getting the minimum distance from the origin to the plane. The volume, $V = \frac{1}{3} Bh$, was then determined; but Lagrange went further and expressed this with beautiful elegance in various forms (equivalent to determinants) in terms of the sides, faces, and vertices. In another connection[18]

[16] *Oeuvres*, v. III, p. 658–692. Although delivered in 1773, the paper was published in 1775.

[17] *Oeuvres*, v. III, p. 661.

[18] *Oeuvres*, v. III, p. 585–586.

he made use of these formulas in expressing the condition that four points be coplanar through the fact that the points determine a tetrahedron of volume zero. This work of 1773 represents one of the earliest associations of linear algebra with analytic geometry. Again displaying the advantages of his method, Lagrange solved the problem of finding in the tetrahedron a point such that, on connecting it with the four vertices, the volumes of the four pyramids with the point as vertex and the faces of the given tetrahedron as bases shall be in a given ratio. Then he found the center of the circumscribed sphere by means of undetermined coordinates, equating the distances from the center of the sphere to the vertices of the tetrahedron. For the center of the inscribed sphere (and the centers of the escribed spheres) he again used the calculus to find the distance from a point to a plane and then equated the distances to the faces of the tetrahedron. For the center of gravity of the tetrahedron, he found the intersection of planes through an edge and the midpoint of the opposite side. It should be emphasized that the choice of general coordinates for three of the vertices (the fourth being at the origin) made the beautifully neat and symmetric results of Lagrange possible and freed algebraic geometry from constant reference to special axes and from frequent appeal to geometric diagrams and theorems. Cartesian geometry, at least for three dimensions, was at length being truly arithmetized.

Lagrange closed his paper with the apologetic remark that he has presented this work only to give an example of the application of analysis to this type of research; but his interest in solid analytic geometry is apparent also in other memoirs of the same year. In one, "Sur l'attraction des spheroides elliptiques,"[19] he presented again a problem which Maclaurin had given before in elegant synthetic form— to show "the detractors of analysis that it furnishes a solution which is simpler, more direct, and more general." The distance formula, for two dimensions as well as three, is used freely, and the general equation of the sphere is given both in rectangular Cartesian and in spherical coordinates. This seems to be the first application of polar coordinates to solid analytic geometry, although Clairaut had long before promised to carry out such work. The novelty of this method in 1773 seems to be implied by the unusually detailed explanation with which it is presented—yet Lagrange refers to it as "ordinary," possibly to indicate that he himself had used it frequently:

> One of the most useful and ordinary transformations is to introduce, in place of the rectangular coordinates $x$, $y$, and $z$, a radius vector issuing from

[19] *Nouveaux Mémoires de l'Académie Royale des Sciences et Belles-Lettres de Berlin*, 1773. See *Oeuvres*, v. III, p. 617–658.

a fixed point which is called the center of the radii, together with two angles $p$ and $q$ which determine the position of this radius, of which $p$ is that angle which the radius makes with one of the axes, such as that with the $z$-axis, or rather with an axis parallel to this, but passing through the center of the radii; and of which the other $q$ is the angle which the projection of the radius $r$ on the plane of the $x$, $y$ coordinates makes with the $x$-axis, or, which is the same thing, with an axis parallel to the latter and passing through the center of the radii. If one denotes by $a$, $b$, $c$ the rectangular coordinates which determine the arbitrary position of the center, it is clear that one will then have

$$r = \sqrt{(x - a)^2 + (y - b)^2 + (z - c)^2}$$

from which one easily finds

$$\sin p = \frac{\sqrt{(x - a)^2 + (y - b)^2}}{r}, \text{ and } \sin q = \frac{y - b}{\sqrt{(x - a)^2 + (y - b)^2}};$$

and from these one concludes that[20] $x - a = r \sin p \cos q$,

$$y - b = r \sin p \sin q, \quad z - c = r \cos p$$

Perhaps the most significant feature of the paper for the development of analytic geometry is the use, generally overlooked by historians, of the rotation of axes in modern symmetric form:[21]

$$\begin{cases} x = \lambda x' + \mu y' + \nu z' \\ y = \lambda' x' + \mu' y' + \nu' z' \\ z = \lambda'' x' + \mu'' y' + \nu'' z' \end{cases}$$

where the nine coefficients are connected by the six familiar relationships

$$\begin{array}{lll} \lambda^2 + \lambda'^2 + \lambda''^2 = 1 & & \lambda\mu + \lambda'\mu' + \lambda''\mu'' = 0 \\ \mu^2 + \mu'^2 + \mu''^2 = 1 & \text{and} & \lambda\nu + \lambda'\nu' + \lambda''\nu'' = 0 \\ \nu^2 + \nu'^2 + \nu''^2 = 1 & & \nu\mu + \nu'\mu' + \nu''\mu'' = 0. \end{array}$$

This form of the equations of transformation, so much more convenient than that given twenty-five years before by Euler (although sometimes incorrectly ascribed[22] to Meusnier in 1785), is eminently characteristic of the precision, elegance, and generality of the work of Lagrange.

In point of view, the analytic geometry of Lagrange comes closer to the modern form of the subject than that of any of his predecessors. It was elementary geometry in analytic language, quite independent

[20] *Oeuvres*, v. III, p. 626–627. For the case in which $r$ is constant, these equations are a parametric form for the spherical surface; but this does not seem to have been generally recognized at the time.

[21] See *Oeuvres*, v. III, p. 646–648.

[22] See Wieleitner, *Geschichte der Mathematik*, v. II (2), p. 53.

of reference to geometric diagrams.[23] Analytic formulas replaced geometric entities, and the calculations were carried out with full generality. Unfortunately, however, Lagrange was not a geometer, and so he never wrote a textbook on the subject. He did not, in fact, digress long enough to give a treatment for two dimensions analogous to his papers on solid analytic geometry. He turned instead to physics and in the *Mécanique analytique* of 1788 he did for mechanics what he might have done for geometry—developed the subject from first principles without reference, as he specifically boasted, to a single diagram. In this book he remarked that mechanics may be looked upon as the geometry of a four-dimensional space (time being the fourth dimension); but he did not take steps to build up a multidimensional coordinate geometry.

For about a score of years, Lagrange's suggestion to geometers of the possibility of a complete analytic form went largely unnoticed until finally a student of his, by the name of Lacroix, undertook to put the idea into textbook form. But Lacroix had the inspiraton not only of Lagrange, the analyst, but also of Gaspard Monge (1746–1818), the greatest geometer of the century.[24] The analytic geometry of Monge, like that of Lagrange, was limited pretty largely to three dimensions. In an important paper on developable surfaces delivered in 1771 (two years earlier than that of Lagrange) but published in 1785, he considered a number of problems in coordinate geometry. The first one was to find the plane through a point $(x', y', z')$ and perpendicular to the line given by $ax + by + cz + d = 0$ and $a'x + b'y + c'z + d' = 0$. This he solved in the usual modern manner by taking the plane to be found as $A(x - x') + B(y - y') + C(z - z') = 0$ and then determining[25] the ratios of the coefficients $A$, $B$, and $C$ from the proportions $\dfrac{A}{\alpha} = \dfrac{B}{\beta} = \dfrac{C}{\gamma}$, where $\gamma = ab' - a'b$, $\beta = ac' - a'c$, and $\alpha = bc' - b'c$. The quantities $\gamma$, $\beta$, and $\alpha$ are, of course, the two-rowed determinants of the matrix $\left\| \begin{matrix} a & b & c \\ a' & b' & c' \end{matrix} \right\|$, but such notations, the natural results of the symmetries appearing in the work of La-

[23] Brunschvicg in *Les étapes de la philosophie mathématique* (Paris, 1912), p. 293, sees in the Geometry of Descartes the prototype of such work as that of Lagrange; but Brunschvicg, like Comte, overemphasizes the part pure calculation played in the Cartesian point of view.

[24] For a good biographical account, see Louis de Launay, *Un grand français. Monge. Fondateur de l'école polytechnique* (Paris, 1933). For further details on his work, see Charles Dupin, *Essai historique sur les services et les travaux scientifiques de Gaspard Monge* (Paris, 1819). An edition of Monge's works is badly wanted. An excellent summary of his work, however, is available in René Taton, *L'oeuvre scientifique de Monge* (Paris, 1951).

[25] "Mémoire sur les développées, les rayons de courbure, et les différens genres d'inflexions des courbes a double courbure," *Mém. d. math. et de physique, présentés a l'Académie Royale des Sciences*, v. X (1785), p. 511–550.

grange and Monge, were not used until almost the middle of the nineteenth century.

Another problem of the "Mémoire sur les développées" involved the perpendicular distance from the point to the line. Extending the notation above, Monge expressed the distance in a compact form somewhat similar to the work of Lagrange of 1773:

$$\frac{\sqrt{\lambda^2 + \mu^2 + \nu^2}}{\alpha^2 + \beta^2 + \gamma^2}, \text{ where } \begin{array}{l} \lambda = \alpha y' - \beta z' + \delta \\ \mu = \beta x' - \gamma y' + \zeta \\ \nu = \gamma z' - \alpha x' + \xi \end{array} \text{ and } \begin{array}{l} \delta = ad' - a'd \\ \xi = cd' - c'd \\ \zeta = bd' - b'd. \end{array}$$

Monge here seems to have anticipated Lagrange not only in symmetries of notation but also in the complete avoidance of diagrams in connection with solid analytic geometry. The distance formula for three dimensions also appears here, a couple of years before it was used by Lagrange; but it is to be noted that this paper by Monge was not published until a dozen years after that of Lagrange had appeared in print.[26]   A third problem in the paper by Monge belongs largely to the calculus: to find the normal plane to the curve $y = \phi(x)$, $z = \psi(x)$ passing through a given point. This has significance for algebraic geometry, however, in that it definitively cancelled the error Descartes had made almost a century and a half before when he had thought that for such a space curve a unique normal *line*, rather than a normal plane, was determined.

The examples above show how thoroughly modern the work of Monge is, both in notation and in method of attack. The only deficiency is in a corresponding treatment for two dimensions. Here the lack is not so pronounced as in Lagrange, for ten years later Monge presented another paper[27] in which the modern point-slope form of the plane equation of the straight line is given explicitly, perhaps for the first time.[28]   The passage in question[29] opens with the statement that "the equation of the line is generally of the form $y = ax + b$." This is not new, for it had appeared in Monge's paper of 1771 and in numerous works as far back as Fermat. But Monge continued, "and if one wishes to express the fact that this line passes through the point M of which the coordinates are $x'$ and $y'$, which determines the quantity $b$,

[26] Loria, "Da Descartes e Fermat a Monge e Lagrange," p. 836–837, has overlooked this delay in publication (and also the work of Clairaut), incorrectly reporting that the distance formula (for three dimensions) "here is found for the first time in the literature."

[27] "Mémoire sur la théorie des déblais et des remblais," *Mémoires de l'Académie des Sciences*, 1781 (pub. 1784), p. 666–704.

[28] This form generally is incorrectly ascribed to Lacroix. See Wieleitner, *Geschichte der Mathematik*, v. II (2), p. 42, 53; Tropfke, *Geschichte der Elementar-Mathematik*, v. VI, p. 123; Loria, "Da Descartes e Fermat a Monge e Lagrange," p. 841.

[29] *Op. cit.*, p. 669.

this equation becomes $y-y' = a(x-x')$, in which $a$ is the tangent of the angle which this straight line makes with the line of $x'$s."

This passage is not of great significance as far as substance is concerned. Mathematicians from the very earliest days of analytic geometry (and probably long before that) were familiar with the property which this equation expresses—that "the straight line lies evenly between its ends." It is not, in fact, greatly different from the forms given by Goudin and Dionis du Sejour. The importance of Monge's statement lies in the tendency to formalize the straight line in analytic symbols, the direction which Lagrange had suggested to geometers. The works of Monge on coordinate methods, in spite of the fact that he was the greatest synthetic geometer of the age, remind one strongly of the analyst Lagrange in the virtual absence of diagrams. Lagrange and Monge seem to have realized more fully than their predecessors how useful might be a complete alliance between analysis and geometry. Analytic geometry was rapidly approaching a new stage.

The point-slope equation of the straight line appears frequently in Monge's paper of 1781, but there is little besides in the way of plane analytic geometry. The interest of Lagrange and Monge lay especially in three dimensions, and it is here that contributions were made also by their contemporaries. The decade from 1771 to 1781 was, in fact, perhaps the most significant of all in the development of coordinate geometry in three-space. Whereas previously solid analytic geometry had lagged about a century behind that in the plane, it now was taking the lead. Journals of the time, in Germany, France, and Russia, give evidence of a sustained interest in at least one aspect of the subject—the transformation of axes. Even the aged Euler, whose definitive work belonged to an earlier period, took part in the program.

In 1771, the year of Monge's first paper (as yet unpublished), Euler wrote an article on developable surfaces in which he expressed these surfaces parametrically. In this work[30] he used the now customary notation for the perpendicularity of two lines in space—$l\lambda + m\mu + n\nu = 0$, where $l^2 + m^2 + n^2 = 1$ and $\lambda^2 + \mu^2 + \nu^2 = 1$. That the sum of the squares of the direction cosines of a line is equal to unity has been ascribed to later men,[31] but it is implicit in the work of Lagrange and Monge, as well as in that of Euler. In another paper a few years later,[32] Euler gave the familiar formulas for finding the

[30] "De solidis quorum superficiem in planum explicare licet," *Novi Commentarii Academiae Petropolitanae*, v. XVI (1771), p. 3–34, especially p. 6 and 22.
[31] Wieleitner, *Geschichte der Mathematik*, v. II (2), p. 52, ascribes this result to Tinseau, as does also Kommerell in Cantor, v. IV, p. 544.
[32] "Nova methodus motum corporum rigidorum determinandi," *Novi Commentarii Academiae Petropolitanae*, v. XX (1775), p. 208–238. See especially p. 219, 230, 235.

direction cosines of the line through the points $(f, g, h)$ and $(x, y, z)$ in the form $x - f = s \cos \zeta$, $y - g = s \cos \eta$, $z - h = s \cos \theta$, where $s$ is the distance between the points. Here he specifically stated that the three direction angles $\alpha$, $\beta$, $\gamma$ associated with a given line are equivalent to two conditions inasmuch as $\cos^2 \alpha + \cos^2 \beta + \cos^2 \gamma = 1$. His familiarity with the work of Lagrange is made evident by his praise of the Lagrangian formulas for the rotation of axes; and Euler used these in a form which was especially popular in the early nineteenth century:

$$\cos zA = \cos ZA \cos aA + \cos ZB \cos aB + \cos ZC \cos ac$$
$$\cos zB = \cos ZB \cos bB + \cos ZC \cos bC + \cos ZA \cos bA$$
$$\cos zC = \cos ZC \cos cC + \cos ZA \cos cA + \cos ZB \cos cB.$$

Analogous formulas were given in the same year by Euler's colleague at St. Petersburg, A. I. Lexell (1740–1784), who added also equations for the translation of axes. Lexell not only remarked that the sum of the squares of the direction cosines of a line is unity, but added also that the sum of the squares of the sines is equal to two.[33] The relationship $\cos^2 \alpha + \cos^2 \beta + \cos^2 \gamma = 1$ had been given also in 1774 by Ch. Tinseau (1749–1822), but his memoir, like that of Monge, was not published until 1785.[34] In this article Tinseau gave an interesting generalization of the Pythagorean theorem for space of three dimensions: the square of the area of a plane surface is equal to the sum of the squares of the projections of this surface upon three mutually perpendicular coordinate planes. De Gua in 1783 claimed that in 1760 he had anticipated Tinseau in this discovery, and for the special case of a trirectangular tetrahedron the theorem had been known earlier to Faulhaber and Descartes; but even as late as the end of the nineteenth century the theorem was not well known.[35] To Tinseau it appears that the use of the word "conoid" in the modern sense is due. He adopted it for the locus of a line which moves so as to remain parallel to a given plane and to cut a given curve and a given line perpendicular to the plane. As an example, the hyperbolic paraboloid $ky = xz$ here appeared as a locus of a line instead of (as in Euler) the locus of points satisfying an equation. Tinseau also gave the equation of the tangent plane to a surface, but this was a familiar result of the eighteenth century and is traceable back to Parent.

J. B. Meusnier (1754–1793), in his work on the curvature of surfaces

[33] A. I. Lexell, "Theoremata nonnulla generalia de translatione corporum rigidorum," *Novi Commentarii Academiae Petropolitanae*, v. XX (1775), p. 239–270. See especially p. 244, 246–247, 250, 261, 270.

[34] *Mém. de math. et de physique, présentés a l'Académie Royale des Sciences*, v. X (1785), p. 593–624.

[35] See G. Eneström, "Note historique sur une proposition analogue au théorème de Pythagoras," *Bibliotheca Mathematica* (2), v. XII (1898), p. 113–114.

(read in 1776 and published in 1785, along with the memoirs of Monge and Tinseau), again made use of Euler's form for the equations of rotation;[36] and Monge in 1784 gave again[37] the symmetric form of Lagrange, showing that Cartesian geometry for three-space was taking on its definitive modern form.    Plane analytic geometry, however, had not changed appreciably since 1748.    The textbooks of Bézout appeared in several editions throughout the later eighteenth century, and these contained the usual part on "the application of algebra to geometry" in which a balance is maintained between Euler's geometry of curves and Descartes' construction of problems.    When in 1795 Laplace taught at the École Normale, he summarized the status of the subject by citing Cramer's *Lignes courbes* and volume II of Euler's *Introductio* as supplying "all the details one can desire in this respect," adding that one should also read "the two original works which gave birth to them"—the *Geometry* of Descartes and Newton's *Enumeration of Cubic Curves*.[38]    During the closing years of the century, however, this situation was radically changed.

The establishment in 1794 of the École Polytechnique was decisive in changing the form of analytic geometry, for it brought together the three men who engineered the change—Lagrange and Monge, and their student, Lacroix.    Instruction at the school included two chief branches, one devoted to mathematics, the other to physics and chemistry.    The former included two parts: mathematical analysis (with applications to geometry and mechanics) and descriptive geometry (with three subdivisions—stereotomy, architecture, and fortification).    Students took mechanics only in the second and third years after a first year including analysis and the applications of algebra to the geometry of space.[39]    Admission requirements included application of algebra to [plane] geometry, as well as arithmetic, algebra (including the solution of equations through quartics), geometry (including trigonometry), and conics.[40]    Lagrange taught the analysis, presumably along the lines indicated in the textbook on *Théorie des fonctions analytiques* which he later published in 1797.    Monge was in charge of the solid analytic geometry, and he too found no suitable text available.    He referred students tentatively to his classic paper of 1781 and he set to work to supply them with notes.    These he called *Feuilles d'analyse*

---

[36] Wieleitner, *Geschichte der Mathematik*, v. II (2), p. 53, says that Meusnier first gave the transformation of space coordinates in full generality, but one finds the equivalent of this in Euler, Lagrange, Monge, and, especially, in Lexell.

[37] "Mémoire sur l'expression analytique de la génération des surfaces courbes," *Mém. de l'Acad.*, 1784 (1787), p. 85–117.    See p. 112–114.

[38] See *Journal de l'École Polytechnique*, cahiers 7–8, 1796, p. 122–123.

[39] *Journal de l'École Polytechnique*, cahiers 1–2, 1795–1796.

[40] *Op. cit.*, cahier IV, p. ix.

*appliquée à la géométrie*, the first edition appearing in 1795. They were reprinted in 1801 and reappeared with modifications—also with new title, *Application de l'analyse à la géométrie* —in 1807, 1809, and 1850.

The *Feuilles d'analyse* are characteristic of analytic geometry as Lagrange and Monge had visualized it. No figures or diagrams are used until the author is well into differential geometry in the last third of the work. Unfortunately, however, it includes only a few brief introductory paragraphs on the analytic geometry of two dimensions, and even these were omitted in editions after the second.[41] The book opens with the equations of the line given in his paper of 1781—first the slope-intercept form, $x = az + b$, and then the more general point-slope form, $x - x' = a(z - z')$. Then he gives the condition that the two lines $x = az + b$ and $x' = a'z + b'$ be perpendicular, writing this in the form $aa' + 1 = 0$. This is perhaps the first time that this familiar result appeared in print. The bulk of the work is then devoted to three dimensions, so that it is virtually the first textbook on solid analytic geometry and differential geometry. The equations of a line through two points $(x', y', z')$ and $(x'', y'', z'')$ are given as

$$x(z' - z'') = z(x' - x'') + x''z' - x'z''$$
$$y(z' - z'') = z(y' - y'') + z'y'' - z''y'.$$

The distance formula for three-space appears in the usual form. The direction cosines of the angles between the plane $Ax + By + Cz + D = 0$ and the coordinate planes are given in the ordinary radical form, as is also the formula for the cosine of the angle between two planes, given previously by Euler. The method of undetermined coefficients is employed to find the plane through three points. For the line through a point and perpendicular to a plane, the projection form $\begin{cases} x = az + \alpha \\ y = bz + \beta \end{cases}$ is adopted. The condition that this line be perpendicular to the line $\begin{cases} x = a'z + \alpha' \\ y = b'z + \beta' \end{cases}$ is given as $1 + aa' + bb' = 0$.

The usual problems on points, lines, and planes are included—to find a line through a point and perpendicular to a line; given two planes, to find the projections of their intersection (this is essentially the symmetric form of the line, although this does not enter explicitly); to find the distance between two parallel planes; to find the angle between two lines, or between a line and a plane; to find the shortest distance between two lines, and to determine the equations of the

---

[41] This omission probably accounts in part for the failure of historians correctly to ascribe the point-slope equation of the line to Monge.

common normal. These problems are handled algebraically in a thoroughly modern manner. The remainder of the *Feuilles d'analyse* (and by far the largest part) is devoted to a study of surfaces and skew curves by means of the calculus.

It is interesting to note the difference in emphasis in the analytic geometries of Euler and Monge. The former had stressed the study of curved lines and surfaces by algebraic means, omitting almost entirely the line and plane; the latter gave the algebraic geometry of lines and planes, but relegated the study of curves and surfaces (even those of second degree) to the calculus. Only after Lacroix had united these points of view in the first truly modern textbook on plane analytic geometry did Monge and Hachette collaborate to issue a solid analytic geometry in the sense of Euler—an *algebraic* study of the quadric surfaces.[42]

Monge had a fertility of imagination and of geometrical innovation which seems to have been the envy of Lagrange himself, but he was distracted by political interests.[43] An ardent republican, and later an equally enthusiastic Bonapartist, he lacked the time or patience to compose systematic introductory treatises in the manner of Euler. His papers on solid analytic geometry, differential geometry, and descriptive geometry are classics; but he wrote almost nothing on plane analytic geometry, and even the little he did publish has been generally overlooked. The need for an introductory work on plane algebraic geometry for students in the newly formed French schools was nevertheless clear and this was soon met by S. F. Lacroix (1765–1843), student and colleague of Monge.

The *École Normale* was founded in the same year as the *École Polytechnique*, and among its faculty were Lagrange, Laplace, Monge, Hachette, and Lacroix. The last named was undoubtedly the most prolific textbook writer of modern times, if allowance is made for multiple editions;[44] and among his numerous works are two which mark the definitive stage in the history of elementary plane analytic geometry: the *Traité de calcul* of 1797 and the *Traité élémentaire de*

[42] *Traité des surfaces du second degré*, Paris, 1813. This appeared in 1801 in cahier 11 of the *Journal de l'École Polytechnique;* in 1805 under the title *Application de l'algèbre à la géométrie*, by Monge and Hachette; and in the 1807 edition of Monge's *Application de l'analyse à la géométrie*. The 1807 edition includes also the transformation of coordinates, especially the equations of rotation in both the Euler and the Lagrange form.

[43] See D. E. Smith, "Gaspard Monge, Politician," *The Poetry of Mathematics and Other Essays* (New York, 1934), p. 71–90. This paper is found also in *Scripta Mathematica*, v. I.

[44] In 1848 there appeared at Paris the 20th edition of his *Traité élémentaire d'arithmétique* and the 16th edition of his *Élémens de géométrie*. The 20th edition of his *Élémens d'algèbre* was published at Paris in 1858 and the 9th edition of the *Traité Élémentaire de calcul* in 1881. In 1897 there appeared a 25th edition of his work on trigonometry and analytic geometry! And these figures do not take into account the large number of translations into other languages.

*trigonométrie rectiligne et sphérique et application de l'algèbre à la géométrie* of 1798–1799. Here Lacroix did for two dimensions what Lagrange and Monge had done for three-space. His object is clearly stated in the preface of his *Traité de calcul*.

In carefully avoiding all geometric constructions, I would have the reader realize that there exists a way of looking at geometry which one might call *analytic geometry*, and which consists in deducing the properties of extension from the smallest possible number of principles by purely analytic methods, as Lagrange has done in his mechanics with regard to the properties of equilibrium and movement.

Lacroix goes on to describe the program further and to give some hint of its inspiration. He points out that Adrien-Marie Legendre (1752–1833), in notes to his famous geometry of 1794, had given the equivalent of the equation of the straight line and had suggested an analytic treatment of certain parts of geometry.[45] In view of the fact that Lacroix began assembling material for his *Traité* in 1787 and that the printing began in 1795, it is doubtful that Legendre's reference, more suggestive than helpful, exerted an appreciable influence on Lacroix. The inspiration went back much earlier. As Lacroix said, Lagrange's work in 1773 on pyramids is a chef-d'oeuvre of the type of geometry he had in mind. He adds, however, that he believes Monge "was the first one to think of presenting in this form the application of algebra to geometry." This statement is interesting for its inference that the work of Monge was part of a conscious program and that this view of analytic geometry had occurred to Monge before Lagrange made his suggestion to geometers. The early works of Monge and Lagrange were composed at practically the same time, and it would be interesting to know whether there existed mutual influences. J. B. Biot (1774–1862) and L. Puissant (1769–1843), writing a few years after Lacroix, ascribe the new program jointly to Lagrange and Monge; but in 1810 J. B. Delambre (1749–1822) writes that the "resurrection of the alliance of algebra and geometry, contracted by Descartes, was due to the works of Monge," and that his influence extended also to elementary works such as those of Biot and Lacroix.[46] Whatever the relative weight of inspiration may have been, it is probably fair to speak of the new program as "analytic geometry in the sense of Lagrange, Monge, and Lacroix." Lagrange suggested the new orientation and illustrated it for a number of problems; Monge applied it systemati-

[45] Cf. Legendre, *Élémens de géométrie* (Paris, 1794), p. 287 f.

[46] Biot, *Essai de géométrie analytique* (Paris, 1802). L. Puissant, *Recueil de diverses propositions de géométrie résolues ou démontrées par l'analyse algébrique* (2nd ed., Paris, 1809), avant-propos. The first edition appeared in 1801. Delambre, *Rapport historique sur les progrès des sciences mathématiques depuis 1789 et sur leur état actuel* (Paris, 1810), p. 39–42.

cally to the geometry of three dimensions; and Lacroix first explicitly formulated it and presented it for two dimensions in textbook form.

In 1798 there appeared an odd little book by L. A. O. de Corancez, *Précis d'une nouvelle méthode pour réduire à de simples procédés analytiques la démonstration des principaux théorèmes de la géométrie, et la dégager des figures & constructions qu'on y a employées jusqu'à présent,* in which credit for the analytic point of view is ascribed largely to Lagrange and Legendre; but this work is not analytic geometry in the usual sense. The emphasis is upon the avoidance of figures through the use of algebra, especially in elementary Euclidean geometry, rather than upon the explicit introduction of coordinate systems. This seems to be essentially what Legendre had had in mind in 1794— the succinct demonstration of the fundamental propositions of geometry (in particular, those on similar figures) by the consideration of functions[47]—but this program was not strictly identifiable with that of Lagrange, Monge, and Lacroix.

Algebra and geometry had been associated in various ways even before Descartes. The Greek geometric algebra was an early example of this; Descartes had emphasized the value of algebra as an intermediary between the formulation and the construction of geometric problems; Fermat had stressed the study of geometric curves as given by algebraic equations; and Euler had thought of curves as geometric representations of the theory of algebraic functions. Lacroix complained, in the introduction to his *Traité de calcul,* of the manner in which almost all books mix up geometric considerations with algebraic calculations. This indictment may well have been directed in particular against the form of analytic geometry as found in Descartes, with its emphasis upon the construction of equations. Even as late as 1791 one finds a whole book by G. C. F. de Prony (1755–1839) on "Exposition d'une méthode pour construire les equations indéterminées qui se rapportent aux sections coniques." Lacroix held that algebra and geometry "should be treated separately, as far apart as they can be; and that the results in each should serve for mutual clarification, corresponding, so to speak, to the text of a book and its translation."[48] (Sophie Germain later expressed this idea when she said that algebra is but written geometry and geometry is but figured algebra.) He adds in other passages that "algebra is a language appropriate for propositions," and again that "the application of algebra to geometry is not restricted to the use of algebra in the study of extension" [the Cartesian view], "one sees in it also all the properties that an algebraic expression sig-

[47] *Élémens de géométrie,* p. 287 f.
[48] *Traité de calcul,* v. I, p. xxvi.

nifies"[49] [the view of Euler]. The result of the point of view of Lacroix was an analytic geometry remarkably like that of present-day textbooks. The chapter on "Théorie des lignes courbes" (occupying over 100 pages in the first edition[50] of the Traité de calcul) presents explicitly—in many cases for the first time—most of the material commonly found in the early chapters of any modern text; and about a year later Lacroix published much the same material in his textbook on trigonometry and the application of algebra to geometry. The distance formula, for example, is given in the familiar form $\sqrt{(\alpha'-\alpha)^2 + (\beta' - \beta)^2}$. The point-slope equation of the line is presented systematically,[51] together with the associated two-point form,

$y - \beta = \dfrac{\beta' - \beta}{\alpha' - \alpha}(x - \alpha)$. The tangent to the circle $x^2 + y^2 = r^2$ through

a point $(\alpha, \beta)$ on it is given as $y - \beta = -\dfrac{\alpha}{\beta}(x - \alpha)$. The area of the tri-

angle with vertices $(0, 0)$, $(\alpha, \beta)$, and $(\alpha', \beta')$ is given as $\dfrac{\alpha\beta' - \alpha'\beta}{2}$.

The perpendicular distance from a point $(\alpha, \beta)$ to the line $y = ax + b$ is given as $\dfrac{\beta - a\alpha - b}{\sqrt{1 + a^2}}$, the equivalent of the use of the normal form of a line. There are formulas for the sine, cosine, and tangent of the angle $\theta$ between two lines: $\sin\theta = \dfrac{r(a'-a)}{\sqrt{1 + a^2}\sqrt{1 + a'^2}}$, $\cos\theta = \dfrac{r(1 + aa')}{\sqrt{1 + a^2}\sqrt{1 + a'^2}}$,

$\tan\theta = \dfrac{a'-a}{1 + aa'}$, where $a$ and $a'$ are the slopes and $r$ is the "radius." (The trigonometric functions were still taken as lines rather than ratios.) The transformation of coordinates is given in simple formal manner. Also given is the general equation of the circle. This had been known, of course, as far back as Roberval, Fermat, and Descartes, and it had been given incidentally by Cramer; but analytic geometry had been preoccupied with conics and higher plane curves, so that a systematic treatment of the circle had not appeared before. No significant part of Lacroix's material represents new discoveries; it is novel only in the form of exposition used. The continued emphasis

[49] Traité élémentaire de trigonométrie (10th ed., Paris, 1852), p. 87, 120.
[50] In the preface to the 2nd edition (Paris, 1810), Lacroix wrote that he had suppressed the preliminary work on plane elementary geometry inasmuch as it had appeared in his Traité d'application de l'algèbre à la géométrie and thence had passed into many other textbooks. He explains further that he has sought to make the treatment of three dimensions still more independent of geometric considerations.
[51] Wieleitner ("Zur Erfindung...," in Zeitschrift für math. Unterricht, v. XLVII (1916), p. 414–426) misleadingly says the equation of the straight line first appeared in a textbook on analytic geometry in Biot's work of 1802.

upon the almost automatic application of formulas made the subject resemble an algorithm, in which independent reference to the geometrical properties of figures is dispensed with. For pedagogical reasons, however, Lacroix's work does include diagrams to about the extent now customary in elementary textbooks.

The usual elementary treatment of curves and loci, with emphasis upon conics, is much like that presented today. An elementary treatment of conjugate diameters is included in rectangular coordinates; and the significance of the characteristic $\beta^2 - 4\alpha\gamma$ is noted.[52] The fundamental principle of analytic geometry is clearly stated: "the equation of a curve is obtained by expressing analytically one of its properties"; and, reciprocally, an equation "gives rise to a curve, the properties of which are made known by the equation." The equation of the parabola with focus at the pole is given in polar coordinates as $z = \dfrac{2c'}{1 + \cos \phi}$, where $z$ is the radius vector, $\phi$ the vectorial angle, and $c'$ is the distance from focus to vertex; the analogous form for the central conics is $z = \dfrac{c'(1 + e)}{1 + e \cos \phi}$, where $e$ is the eccentricity. (By substituting $c' = \dfrac{ep}{1 + e}$ one obtains the now usual polar form.)[53]

The plane analytic geometry which is found in Lacroix's *Traité de calcul* of 1797 is found also in his *Traité élémentaire de trigonométrie* of about a year later. Less than half of the latter small volume is on trigonometry, the larger portion being devoted to the remainder of the title— *et application de l'algèbre à la géométrie*. There are, of course, some things in this analytic geometry which are not found in modern texts, and conversely. There is, for example, a hangover of Cartesianism in short sections on the construction of the roots of the equation $x^2 - ax = \pm b^2$ by means of line and circle, and on the construction of quartic equations by conics. On the other hand, there is nothing on transcendental plane curves, one-parameter families of curves, or curves given by parametric equations. But, nevertheless, the work of Lacroix is the first textbook treatment which could serve (with some relatively slight modification) as a basis for a modern course in plane analytic geometry.

The *Traité de calcul* included also a chapter on curved surfaces and curves of double curvature, but this is omitted in the earliest ed-

[52] Tropfke (*Geschichte der Elementar-Mathematik*), v. VI, p. 164, incorrectly ascribes the general statement of the characteristic to Lacroix, whereas in reality it goes back at least to L'Hospital.

[53] Tropfke, *op. cit.*, p. 169, ascribes the polar equation of the conic to Lacroix, but various forms had appeared earlier in Euler and Hermann.

itions of the *Traité élémentaire de trigonométrie*[54]   The theory of this chapter, which comprises the material for a course on elementary solid analytic geometry, Lacroix says is almost entirely due to Monge, "who in a way rediscovered the work of Euler and Clairaut and gave it a new and considerably amplified form."   The preliminary work on points, lines, planes, angles, distances, projections, and transformation of coordinates is practically in modern form.   Transformations are given from rectangular coordinates to spherical coordinates, in the manner of Lagrange: $r = \sqrt{x^2 + y^2 + z^2}$, $z = r \sin p$, $y = r \cos p \sin q$, $x = r \cos p \cos q;$   and also in the more symmetric form[55] now generally known as polar coordinates: $z = r \cos \phi$, $y = r \cos \psi$, $x = r \cos \pi$, where $\cos^2 \phi + \cos^2 \psi + \cos^2 \pi = 1$.   The treatment of surfaces of second degree follows Monge in making early use of the methods of the calculus rather than those of algebra.

Lacroix did not write a textbook devoted solely to analytic geometry. This is not of great significance and probably was merely the result of the fact that he wrote a series of texts to cover the usual program of mathematics.   The sequence of courses in Paris seems to have become more or less standardized in the following order: arithmetic, algebra, geometry, trigonometry, applications of algebra to geometry, complement of algebra, descriptive geometry, and calculus.   In composing his series of textbooks for this "Cours de mathématiques" at the École Central des Quatre Nations,[56] it was convenient to group together trigonometry and coordinate geometry.   This combination of subjects is occasionally found at the present time.   It is surprising that, although Lacroix in 1797 referred to this new point of view in geometry as analytic, he did not use this in the title of the textbook he wrote. He preferred instead to retain the century-old designation used by Guisnée.   The phrase analytic geometry evidently was suggested to Lacroix by the title of Lagrange's *Mécanique analytique*, but in reality it goes back at least as early as 1709 when Rolle used it.   During the second half of the eighteenth century the word "analytic" was much discussed and frequently used in titles.   D'Alembert, in the article on "Analysis" in the *Encyclopédie*, wrote that it "is properly the method of solving mathematical problems in reducing them to equations." Hence, he said, the words "analysis" and "algebra" often were re-

---

[54] The third edition of 1803 included an appendix (p. 233–259) on solid analytic geometry. This edition, the earliest I have had an opportunity to examine, does not differ greatly from later editions I have seen, including the fourth, eighth, and tenth.

[55] *Traité de calcul*, v. I, p. 464.   Coolidge (*History of Geometrical Methods*, p. 172) ascribes this system to Monge in 1807; and it may well be that Lacroix learned of it from his teacher before 1797.

[56] See S F. Lacroix, *Essais sur l'enseignement en général, et sur celui des mathématiques en particulier* (Paris, 1805).

garded as synonymous.[57] For the article on Cartesian geometry D'Alembert used the title "Application de l'algèbre ou de l'analyse à la géométrie," and in it he held that analysis is just as rigorous as the geometry of the ancients. In the article on "Conique," he said that it was possible to "achieve a truly analytic treatise on conic sections, i. e., in which the properties of the curves are deduced immediately from the general equation"—presumably as Euler had done.

In the titles of books and articles the word "analytic" came to be used with increasing frequency. Müller in 1760 published a *Traité analytique des sections coniques*, a title similar to that used by Hube the year before; Klügel in 1770 issued an *Analytische Trigonometrie* and in 1778 an *Analytische Dioptrik*, fittingly dedicated to Euler. The name "analytic geometry" appeared in 1779 as the title of Newton's *Method of Fluxions* in the *Opera* edited by Horsley, and in the titles of two papers by Fuss in 1780 and 1781.[58] In 1779 a book by a Swedish author, Niels Schenmark, bore the title *Analytische Geometrie*;[59] and in 1782 the words "Algebra et geometria analytica" appear on the title page of volume I of the *Operum* of Paolo Frisi (1728–1784). The introduction to the latter work includes an historical résumé in which the author explained that he had put less emphasis upon the solution of cubic and quartic equations and had treated at greater length that part of algebra in which his age excelled—"the application of analysis to geometry, which for this reason obtains the name analytic geometry". This would imply that the designation "analytic geometry" was by no means new at the time, only a few years before Lacroix began his work.

Analytic methods in the eighteenth century flourished more in France and Germany than in England and Italy. Klügel and Kaestner especially did much in Germany to popularize analytic methods. Besides the volumes mentioned above, Klügel composed a volume on the mutual relationship of analysis and synthesis.[60] In this, however, he does not use the term analysis in the clear-cut algebraic sense but

---

[57] This is indicated also by the *Analyse démontrée* and *Usage de l'analyse* of Reyneau in 1708 and 1738, for these volumes are on algebra, as well as the application of algebra to geometry. For more extensive discussion and references on the use of the word analysis see my paper, "Analysis: Notes on the Evolution of a Subject and a Name," *The Mathematics Teacher*, v. XLVII (1954), p. 450–462.

[58] "Exercitatio analytico-geometrica" and "Disquisitio analytico-geometrica," in *Acta Academiae Scientiarum Imperialis Petropolitanae*, v. I and II (1780–1781). The phrase "analysin geometricam" appeared also in Hermann's work of 1729 in the *Commentarii Academiae Petropolitanae*, v. IV (1729), p. 47.

[59] I have not had an opportunity to refer to this work; I cite it on the authority of Tropfke, *Geschichte der Elementar-Mathematik*, v. VI, p. 154.

[60] *De ratione quam inter se habent in demonstrationibus mathematicis methodus synthetica et analytica* (Helmstadt, 1767). I have not seen this work and have relied here upon the description given in Cantor, *Vorlesungen*, v. IV, p. 455–456.

rather with the older Greek meaning.  Hence the difference was not so much between constructions on the one hand and calculations on the other as it was in the inner nature of truth and the way in which it is to be sought.  He pointed out that analysis not only is better a-dapted to the making of new discoveries, but it also affords greater generality.  Conics, he said, are of three types in synthesis, but only one family in analysis—a statement which seems to imply that he here has in mind the modern rather than the ancient use of the words. A similar discussion of analysis had appeared in Kaestner's preface to Hube's Conics of 1759, and here Kaestner had characterized the an-alytic approach as affording less beauty but more power.

The failure of Lacroix to adopt the phrase "analytic geometry" as a title for his work may have been due to the confusion in meaning evinced by Klügel.  This conjecture is strengthened by some passages which appeared in the *Essais sur l'enseignement* of 1805.  Here La-croix explains at length the difference between analysis and synthesis, using the word in its ancient logical meaning, as described by Pappus. Lacroix points out that his plan in the application of algebra to geom-etry is alternately analytic and synthetic in this sense.  It is prob-able that he did not call his work analytic geometry because he feared this might obscure the earlier use by Plato.  There is nothing alge-braic in the older analysis; it has reference only to the *order* of ideas in a demonstration.

The *Essais* includes also an enlightening critique of the development of coordinate geometry.  Characteristically of the period, he ascribes the invention of the subject categorically to Descartes, while admitting that the first traces are found in Viète.  But Descartes, he points out, applied it only to questions occupying geometers of the time, and not to curves in general.  The theory of curves was begun by Newton and perfected by Euler and Cramer.  Force of habit, however, caused geometers to amalgamate the methods of the ancients with the new, so that they began with curves of second order rather than with the straight line.  Lacroix holds that the views of the eighteenth century on the ap-plication of algebra to geometry should not be followed, inasmuch as Descartes, Euler, Lagrange, and Monge had in mind simply the de-duction of the properties of extension.  In this last assertion, however, Lacroix does not do justice either to his illustrious teachers or to Euler. Whereas earlier Lacroix had given credit to Monge, he seems here to arrogate to himself more than is warranted.  He claims that *his* plan, at the time of the first edition, was without doubt new, as were the means used to fulfil it—at least with regard to its elements.[61]  Unless

[61] *Essais*, p. 381.

one appropriately emphasizes the closing words of this assertion, the claim of Lacroix is too categorical. He did for plane coordinate geometry essentially what Monge and Lagrange had done earlier for solid analytic geometry. In this there is sufficient glory for all three, for they made Cartesian geometry what it is today.

The program launched by the works of Monge and Lacroix in 1795 and 1797 met with such extraordinarily prompt and widespread approval that one might appropriately refer to it as an "analytical revolution," comparable to the very nearly contemporaneous "chemical revolution" initiated by Lavoisier.[62] The early years of the nineteenth century saw the publication of an array of introductory works on plane and solid analytic geometry which are strikingly similar to the pattern set by Lacroix and to modern textbooks, both in point of view and in subject-matter content. In 1801 there appeared the *Essai sur la ligne droite et les courbes du second degré* of F. L. Lefrançais [or Français], and the *Recueil de diverses propositions de géométrie résolues ou démontrées par l'analyse algébrique, suivant les principes de Monge et de Lacroix*, by Puissant. The very titles of these books indicate the shift of emphasis in coordinate geometry from the theory of higher plane curves toward the consideration of elementary geometrical problems. The object of Lefrançais, a former student of the École Polytechnique, was "to present the principles of analytic geometry, or the method which consists of treating questions which are proposed solely by the means which analysis furnishes, drawing from geometry only what is absolutely indispensable to the expression of the conditions of each problem." That of Puissant was similar: "Never to base his calculations upon geometrical constructions . . . so as better to make use of the advantages of algebra." Echoing Lacroix, Puissant contrasts this procedure with "the mixed methods used from Descartes to our day in the solution of most questions in geometry." The arithmetization of geometry was spreading rapidly.

The textbooks of Lefrançais and Puissant illustrate the increased emphasis on the line and circle and the rapid rounding out of the now familiar details. In Lefrançais one finds the formula for the bisectors of the angles formed by the lines $y = ax$ and $y = a' x$, expressed as

$$y = Ax, \text{ where } A = \frac{a\,a' - 1 \pm \sqrt{(1 + a^2)(1 + a'^2)}}{a + a'}. \text{ Puissant gave}$$

[62] It is of interest to note that Monge took part in the chemical revolution also, for in 1783 he performed experiments on the composition of water, not knowing of the earlier work by Cavendish. See Arago, *Oeuvres complètes*, v. II, p. 427–592, for further biographical details on Monge. For work by Monge on the liquefaction of gases see M. G. Beumer "Gaspard Monge as a Chemist," *Scripta Mathematica*, v. XIII (1947), p. 122–123.

the simplified form $\beta y + \alpha x = r^2$ for the tangent to the circle $x^2 + y^2 = r^2$ at the point $(\alpha, \beta)$. ´Proofs of propositions from elementary geometry are given in both books: the concurrency of the perpendicular bisectors of the sides of a triangle appears in Lefrançais; the concurrency of the altitudes is in Puissant; and the concurrency of the medians is given in both. Numerous problems on lines and circles are a feature of these textbooks of 1801, including (in Puissant) an analytic proof that the sum of the squares of the sides of a quadrilateral is equal to the sum of the squares of the diagonals added to the square of twice the segment joining the midpoints of the diagonals. The transformation of coordinates is given by Lefrançais in general form as

$$\begin{cases} x = u \cos q + t \cos p + a \\ y = u \sin q + t \sin p + b, \end{cases}$$

where $p - q = 100°$—an interesting example of the decimal influence of the French revolution. The formula for the removal of the $xy$ term of the general quadratic equation through a rotation of axes is given, except for sign, in strictly modern form, $\tan 2q = b/(c-a)$. The author's avowed intention to avoid negative quantities betrays the contemporary suspicion of all numbers not real and positive.

His textbook, Lefrançais says, was intended to serve as an introduction to the lectures of Monge and Hachette on solid analytic geometry, and so it was limited to two dimensions. Puissant, however, in a greatly amplified second edition of 1809,[63] completed his own book with a section on three dimensions, as well as one on the applications of "transcendental analysis" to geometry. (The book includes also introductory chapters on trigonometry, as in the case of Lacroix.) Monge in his *Feuilles de analysis* of 1795 had given a brief elementary introduction to the solid analytic geometry of lines and planes, and Lacroix had amplified this in a chapter of the *Traité du calcul* of 1797. Monge and Hachette in 1801–1802 had rounded out the textbook material on elementary analytic geometry by presenting a summary of their course, including an algebraic treatment of quadric surfaces.[64] This summary is tantamount to a brief modern course in solid analytic geometry. It opens with points, lines, planes, angles, and the transformation of coordinates (including the transformation from three mutually perpendicular planes to three arbitrary planes). A proof of Tinseau's generalization of the Pythagorean theorem is given. The treatment of the quadric surfaces is in the customary algebraic

[63] The original edition contained only 121 pages, the second edition 442 pages. The New York Public Library has copies of each edition, as well as of the third edition of 1824.

[64] "Application d'algèbre à la géométrie," *Journal de l'École Polytechnique*, cahier XI (1801–1802), p. 143–172.

manner, leading to the determination of principal diametral planes. The notation, phraseology, and methods are virtually the same as those to be found in any textbook of today. The definitive form of analytic geometry finally had been achieved, more than a century and a half after Descartes and Fermat had laid the foundations.

Monge and Lacroix gave analytic geometry its final form, but not its traditional name. Lacroix at one time had used the phrase "analytic geometry" to characterize the sub·ect, but did not adopt it officially. The first of the new textbooks to carry this name in the title [65] seems to be the *Essai de géométrie analytique* (1802) of Biot, a work which rivalled that of Lacroix in popularity. This book was translated into numerous other languages and served for many years as the textbook at the United States Military Academy at West Point.[66] As in the case of Lacroix, the work opens with some of the old-fashioned material on geometrical constructions equivalent to the algebraic operations, but the body of the work is devoted to what Biot regards as the two divisions of analytic geometry: *"determinate geometry*, which consists in the application of algebra to determinate problems"; and *"indeterminate geometry*, which consists in the investigation of the general properties of lines, surfaces and solids, by means of analysis."[67] The material does not differ appreciably from that in Lacroix, Lefrançais, and Puissant, except that plane and solid analytic geometry are integrated in a single volume devoted solely to this one subject. There are a few specific contributions which perhaps deserve to be noted. One is the recognition of the discriminant $F(AC-B^2) + E(BD-AE) + D(BE-CD)$ of the general conic. Another is the use of the equations $\dfrac{xx_1}{a^2} \pm \dfrac{yy_1}{b^2} = 1$ for the tangents to the ellipse and hyperbola. The distance from a point to a line is given for oblique, as well as rectangular, coordinates. As in other textbooks of the time, the use of polar coordinates is limited to conics with a focus at the pole, and a simple treatment of conjugate diameters is given in Cartesian coordinates. One misses the problems on polygonometry which Puissant had emphasized, and one is struck by the frequent use of the form $y = x\dfrac{\sin \alpha}{\sin (\beta - \alpha)}$ for the straight line through the origin, referred to oblique coordinates; but in most respects the modern

[65] Tropfke, *Geschichte der Elementar-Mathematic*, v. VI, mistakenly says the first appearance of the name in the title of a book was in Garnier's *Éléments de géométrie analytique* of 1808.   Cf. also Wieleitner, *Die Geburt*, p. 6.
[66] See J. B. Biot, *An Elementary Treatise on Analytical Geometry* (transl. by F. H. Smith New York and London, 1840), preface.
[67] *Ibid.*, p. 13.

reader would find the book rather conventional.[68] The text by Biot exerted a very wide influence, and it was probably this that led Lefrançais to change the title in his second edition (1804) to *Essais de géométrie analytique*. It was the text of Biot which found greatest popularity in the United States, even though it made its appearance there as late as 1836 in the edition by Davies.[69]

The popularity of the new aspect of analytic geometry grew with amazing rapidity, evidenced not only by a host of new elementary textbooks, but also by articles in scientific periodicals. The numbers of the *Journal de l'École Polytechnique* and of the *Correspondance sur l'École Impériale Polytechnique* included much old and new material in this field; and in books and articles of the time there are to be found further details which have been incorporated into modern textbooks. In 1808 Jean-Guillaume Garnier (1766–1840) issued another well-known text, *Éléments de géométrie analytique*, in which the concurrency of the angle bisectors of a triangle and the collinearity of the centroid, orthocenter, and circumcenter are proved by analytic means. Monge in 1809 gave the formula $\pm\, 1/2(a'b'' + c'a'' + b'c'' - a''b' - c''a' - b''c')$ for the area of a triangle in the plane, showing that the sign is determined by the sense in which the boundary of the triangle is traversed. By equating the area to zero, he gave the condition that three points be collinear. He added also corresponding formulas (including attention to sign) for the volume of a tetrahedron with one vertex at the origin and for the area of a triangle in space, pointing out[70] that these had been given long before by Lagrange. In other analytic studies on the triangular pyramid Monge proved in various ways that the center of gravity is the point of concurrency of the lines joining the midpoints of opposite edges. He gave also the analogue of the Euler line in three-space, showing that for the orthocentric tetrahedron the centroid is twice as far from the orthocenter as from the circumcenter.[71]

Monge added also to the knowledge of the quadric surfaces. He had called attention to the orthoptic circle of a central conic—the locus of points from which the conic subtends a right angle—and hence this has become known as the "circle of Monge," even though this locus had been given earlier in synthetic form by Lahire. Extending this theorem to space of three dimensions, he showed analytically that the

[68] I have seen the second edition of 1805 and the sixth edition of 1823. The latter consists of about 450 pages, half again as many as in the second.

[69] See L. G. Simons, *Fabre and Mathematics and Other Essays* (New York, 1939), p. 65.

[70] "Essai d'application de l'analyse a quelques parties de la géométrie élémentaire," *Journal de l'École Polytechnique*, cahier XV (1809), p. 68–117.

[71] *Correspondance sur l'École Impériale Polytechnique*, (1804–1816), ed. by Hachette, 3 vols., 1813–1816 See especially v. II, p. 1–6, 96–97, 263–266.

point of intersection of three mutually perpendicular planes each tangent to a central quadric generates a sphere—the director sphere—concentric with the quadric surface. For non-central quadrics the locus is a plane.[72]  In his work on surfaces of second degree, Monge collaborated with his colleague J. N. P. Hachette (1769–1834).  These two men showed more rigorously than Euler that sections of a quadric by parallel planes are homothetic; they discovered for the general case the circular sections which had been noted for particular instances by Wallis and D'Alembert; they noted the double generation of the quadric surface by a moving circle,[73] and they determined the umbilics of the quadric.  Monge and Hachette also studied the properties of the rectilinear generators of the ruled quadric surfaces, showing that there are two systems, that through any point of the surface there passes one generator of each system, that two generators of different systems cut each other, and that two of the same system are not in the same plane.[74]  Other contributions of Monge on families of surfaces and lines of curvature belong more particularly to differential geometry.

The École Polytechnique was the center of the development in analytic geometry in the first decade of the nineteenth century, and there Monge, Hachette, and their associates continued the work of Euler and Lagrange on the transformation of coordinates in three dimensions.  The *Application de l'algèbre à la géométrie* of Monge and Hachette naturally included orthogonal transformations.[75]  J.-J. Livet (1783–1812), a former student, in 1806 published his paper on "Formules pour passer d'un système de coordonnées rectangulaires à un système de coordonnées obliques."[76]  He used these formulas to prove that if $a$, $b$, $c$, are the axes of an ellipsoid and if $a'$, $b'$, $c'$ are three conjugate diameters, then $a^2 + b^2 + c^2 = a'^2 + b'^2 + c'^2$, a generalization of the work of Apollonius.  Lefrançais, another former student, introduced greater generality by transforming from one system of oblique coordinates to another oblique frame of reference. First transforming, as Livet had done, from the rectangular system $x$, $y$, $z$ to an oblique system $x'$, $y'$, $z'$, and then from $x$, $y$, $z$ to another oblique system $x''$, $y''$, $z''$, he obtained the desired transformation from $x'$, $y'$, $z'$ to $x''$, $y''$, $z''$ by eliminating $x$, $y$, $z$ from the above two sets of

[72] See Coolidge, *History of Conic Sections and Quadric Surfaces*, p. 173–174, for the proof of this "Theorem of Monge"; or see Monge and Hachette, *Traité des surfaces du second degré* (3rd ed., Paris, 1813), p. 234–239.

[73] See *Journal de l'École Polytechnique*, v. I, p. 5.

[74] See Monge and Hachette, *op. cit.*, p. 34–44.   Cf. also Kötter in *Jahresbericht der Deutsche Mathematiker Vereinigung* (2), v. V, p. 75.

[75] See *Journal de l'École Polytechnique*, cahier 11 (v. 4, 1801), p. 143–169.

[76] *Journal de l'École Polytechnique*, cahier 13 (v. 6, 1806), p. 270–296.

equations.[77]   Hachette then again took up this question and wrote two articles on the transformation from one oblique system to a second oblique system without recourse to the auxiliary tri-rectangular coordinate planes of Lefrançais.[78]

The *Journal de l'École Polytechnique* and the *Correspondance de l'École Polytechnique* during the first decade of the nineteenth century included numerous articles on elementary analytic geometry.   Monge, Hachette, Lefrançais, Puissant, and others contributed new formulas or new proofs relating to points, lines, and planes, or solved innumerable problems, or disclosed new properties of the conics and the quadric surfaces.   During the first years of the century "Cartesian geometry took a satisfying and probably definitive aspect in consequence of the general formulas which could be applied almost automatically."[79]   But one misses, in the host of elementary analytic geometry textbooks and articles published from 1798 to 1808, the now familiar normal forms of the straight line and the plane.   Various equivalents had appeared occasionally, but the standard equations $x \cos \alpha + y \sin \alpha = d$ and $x \cos \alpha + y \cos \beta + z \cos \gamma = d$ are a striking feature of an otherwise unconventional work of 1809—the *Élémens d'analyse géométrique et d'analyse algébrique, appliquées à la recherche des lieux géométriques* of Simon A. J. L'Huilier (1750–1840).   The first third of the book (about 100 pages) is devoted to algebra and to the application of algebra to geometry without the use of coordinates—the word analysis being used in the sense of Pappus and Viète.   Then, finally, the author applies coordinate methods to the line and the circle, saying that the principles had been developed by the "moderns," and naming in particular L'Hospital, Euler, Cramer, Lagrange, Monge, Lacroix, Puissant, Biot, and Garnier.   The first equation of analytic geometry given by L'Huilier is the normal form of the straight line, and this is used practically to the exclusion of other linear forms.   Finding the circle tangent to three lines was but one of the applications he found for the normal form.   In three dimensions, similarly, the normal form of the plane occupies a position of prominence.   In view of the frequency with which the normal forms appear in L'Huilier's work, it is surprising to find that they often are ascribed to later mathematicians,[80] notably to Cauchy in 1826 or to Magnus in 1833 or to Hesse in 1861!

[77] "Mémoire sur la transformation des coordonnées," *Journal del'École Polytechnique* cahier 14 (1808), p. 182.
[78] See Loria, "Perfectionnements...," *Mathematica*, v. XVIII (1942), p. 125–145.
[79] Loria, "Perfectionnements...," *Mathematica*, v. XX (1944), p. 1–22.
[80] See e. g., *Encyclopédie des sciences mathématiques*, v. III (17), p. 1, 26.   Loria, a competent authority, mistakenly places the first appearance of the form for three dimensions in 1861 with the work of Hesse.   See his "Perfectionnements...," *Mathematica*, v. XX (1944), p. 12.

Historical accounts of analytic geometry frequently close with the work of Monge and Lacroix, leaving the impression that the subject had reached maturity.[81] Lagrange himself apparently believed this to be the case. Having correctly predicted of Monge that "with his application of analysis to geometry this devil of a man will make himself immortal," Lagrange made the mistake of underestimating the future of mathematics. He had written to D'Alembert that it appeared that "the mine [of mathematics] is already too deep, and unless new veins are discovered it will have to be abandoned."[82] So discouraged was Lagrange with the prospects of mathematics that for a while he turned to chemistry. Seldom has a great man been more mistaken. In mathematics as a whole, and in analytic geometry in particular, "progress in the seventeenth and eighteenth centuries was almost negligible compared to what was done in the nineteenth."[83] Elementary coordinate geometry, as now usually taught in a first course, had indeed reached its definitive form, with only details to be added here and there. But analytic geometry in a larger sense was about to burst forth in a period of growth far outstripping all previous ages in rapidity and extent. The estimate that "the nineteenth century alone contributed about five times as much to mathematics as had all preceding history"[84] applies to algebraic geometry as well as to any other branch. To give an adequate account of this work is beyond the scope of this volume; but to omit a survey, even though brief, of this period—unquestionably the Golden Age of analytic geometry—would encourage a distorted view of mathematical history. An attempt will therefore be made in the next and concluding chapter to indicate some of the significant lines of development during the ebullient early nineteenth century.

[81] See, e. g., Tropfke, *Geschichte der Elementar-Mathematik*, v. VI; Wieleitner, *Geschichte der Mathematik*, v. II (2); Smith, *History of Mathematics*, v. II.

[82] See Bell, *Men of Mathematics*, p. 157, 187.

[83] See Coolidge in *Osiris*, v. I (1936), p. 231–250. Cf. his *History of Geometrical Methods*, p. 422–423.

[84] Bell, *Development of Mathematics*, p. 15.

# CHAPTER IX

# The Golden Age

*The faculty of resolution is possibly much invigorated by mathematical study, and especially by that highest branch of it which, unjustly, merely on account of its retrograde operations, has been called, as if par excellence, analysis.*

—EDGAR ALLEN POE

ONE reason that the early rise of the calculus had been rapid, as compared with that of analytic geometry, was that an aura of enthusiasm had been fostered by papers on the subject in the periodicals of the time, especially the *Acta Eruditorum* at Leipzig, the *Philosophical Transactions* in London, and the *Mémoires de l'Académie des Sciences* of Paris. These journals were not in existence when Cartesian geometry was in its infancy, and hence the subject had been poorly publicized. And after the journals were established, there naturally was less interest in the gawky teen-age subject of analytic geometry than in the bouncing infant prodigy, the infinitesimal calculus. Perhaps the nearest approach to a sustained program of interest in coordinate geometry during the eighteenth century is found in the papers on solid analytic geometry by Monge, Lagrange, and Euler from 1771 to 1781 in publications of the Academies at Paris, Berlin, and St. Petersburg. The Golden Age in analytic geometry during the nineteenth century undoubtedly was due in no small measure to the buoyant spirit of the contributors to newly organized periodicals. Of these the *Journal de l'École Polytechnique* was the earliest.

One might almost say that the present college course known as analytic geometry was born of the French Revolution and nourished by the Napoleonic interlude, for the École Polytechnique, established by the Republic in 1795 and fostered by Napoleon himself, was the center from which the new spirit spread to the rest of the world. At Paris the geometer Monge, favorite of the emperor, inspired the men who gave elementary analytic geometry its present form and also the disciples who saw the unlimited possibilities of the subject in new directions. Not the least starry-eyed of the converts to the cause was Joseph-Diaz Gergonne (1771–1859), an artillery officer fired by an enthusiasm born of an education at the École Polytechnique. The

225

226    HISTORY OF ANALYTIC GEOMETRY

tempo of analytic geometry had been high throughout the first decade
of the nineteenth century, but it received still further impetus in 1810
through the journal which Gergonne founded and edited—the *Annales
de mathématiques pures et appliquées*, the first periodical to be devoted
entirely to mathematics. This journal included a special section with
the caption, "Géométrie analytique," and the editor seized every opportunity to point out the power and facility afforded by coordinate
geometry, the modern rise of which he attributed to Monge. Synthetic
methods likewise had received a strong impulse in the descriptive geometry of Monge, and in various quarters among the students of the
École Polytechnique there were those who asserted that analytic geometry often failed where synthesis afforded short and elegant solutions.
This challenge was met eagerly by Gergonne who felt confident that if
coordinate geometry had appeared to fail in certain directions, this
apparent failure was due only to lack of a correct method of attack in
handling the problem. He thus began a new aspect in the development of analytic geometry—its application to the classical problems of
elementary synthetic geometry. The problem of Apollonius—to construct a circle tangent to three given circles—had been a favorite topic
of Viéte and Fermat, as well as of many others, but it had been attacked largely through so-called pure geometry. Gergonne, as the result of reading L'Huilier's *Élémens*, set himself the task of solving the
problem by analytic means, and so well did he succeed that the elegant
solution has become known as the "Gergonne construction," even
though the synthetic basis had been given[1] by L. Gaultier.

   In this construction, presented to the Academy of Turin in 1813, the
desired tangent circles are determined by the points of intersection with
the three given circles of lines drawn through the center of the circle
which is orthogonal to the three circles and through the poles, with respect to these given circles, of the axes of similitude of the three
circles. Gergonne hoped thus "to avenge analytic geometry completely of the reproach too often made of not being able to rival pure
geometry in the construction of problems"; and he tried to prove that
"analytic geometry, suitably handled, offered the most direct, the
most elegant, and the simplest solutions of two problems long celebrated, and which pass as difficult"[2]—the circle tangent to three circles,

   [1] "Sur les moyens généraux de construire graphiquement un circle determine par trois
conditions et une sphère determinée par quatre conditions," *Journal de l'École Polytechnique*,
cahier 14 (v. IX, 1813), p. 124–214.
   [2] *Annales de mathématiques*, VII (1816–1817), p. 289–303. Very elegant analytic solutions of the problem of Apollonius were given also in the early nineteenth century by such
men as Hachette, Poisson, and Plücker; synthetic solutions were given by Chasles, Poncelet,
Steiner, and others. There is an extensive history of the problem. See, e. g., J. T. Ahrens.
*Apollonisches Problem* (Augsburg, 1832).

and the sphere tangent to four spheres. Obviously disappointed by the lack of enthusiasm in the reception of his work, Gergonne amplified the exposition in his *Annales* of 1816. He explains that he does this the more willingly in that "the methods used in this connection seem to open a new field of speculation and of research of a type to make analytic geometry take on an entirely new appearance." The element of novelty to which he referred lay presumably in his use of linear combinations of circles. This elementary principle was indeed destined to change the character of analytic geometry about a decade later, but Gergonne failed to exploit it because he neglected to introduce abbreviated notations.

Gergonne's analytic geometry was many-sided, and the year in which he presented his famous construction he contributed also to another relatively new aspect of the subject—the search for new coordinate systems. In his "Essai sur l'expression analytique des courbes indépendamment de leur situation sur un plan,"[3] he pointed out that the number of possible systems is infinite. He cites polar coordinates as being useful for spirals, and also for conics and for the circle $r = $ const. He suggests also the bipolar equations of the conics $t \pm u = $ const. But in such systems he says the equation depends upon the situation of the curve with respect to the coordinates and does not express "the intrinsic nature of the curve." Gergonne therefore proposed a system, the idea of which he had conceived a long time before: take as the coordinates of a point on a curve the radii of curvature $R$ and $R'$ of the curve and its evolute at corresponding points. Here there is nothing arbitrary, and hence one obtains an "absolute expression" of the curve. He pointed out that some curves with very awkward equations in "ordinary" coordinates have very simple ones in his new system. The cycloid, for example, becomes $R^2 + R'^2 = 16a$; the involute of the circle is given by the equation $R' = a$, and the logarithmic spiral is $R' = R$. Gergonne admits that in his scheme it is not easy to construct a curve from its equation; and the equations which he gives for transformation from rectangular to "absolute" coordinates (and vice versa) are quite complicated. This may be the reason that his ideas were not widely adopted at the time.

Gergonne was not the first one to suggest natural or intrinsic coordinates. Attempts to define curves through connections between quantities inherent in a curve, such as radius of curvature, go back as far as some work of Euler in 1740 in which he was looking for curves (such as the logarithmic spiral) similar to their evolutes.[4] Again in

---

[3] *Annales de mathématiques*, IV (1813–1814), p. 42–55.
[4] *Commentarii Academiae Petropolitanae*, XII.

1764 Euler used semi-intrinsic parameters, arc length and angle of inclination, in proving that the evolute of a cycloid is a cycloid.[5]  Lacroix in 1797 had suggested more definitely that "a curve is given not only when its equation is given in coordinates parallel to two fixed lines or in polar coordinates, but also when one has any relation between two quantities determined by its nature."  For the logarithmic spiral he proposed the equation $u = av$, where $u$ is the radius vector and $v$ is the polar subtangent.  Similarly he said that a relation between radius of curvature and arc length can be looked upon as an equation of the curve, and such an equation has this remarkable characteristic, "that one of the variables is entirely inherent in the curve." Arc length, however, is arbitrary inasmuch as it depends upon the choice of an initial point.[6]  Half a dozen years later A. M. Ampère (1775–1836) presented to the Académie des Sciences a paper in which he proposed to remove the arbitrary element in Lacroix's system by substituting for arc length the parameter of the osculating parabola at the point.  He gave equations connecting rectangular coordinates with his "parabolic coordinates," and he wrote the equations of such curves as the cycloid and the involute of the circle in the latter system and also in the coordinates of Lacroix.[7]

The work of Lacroix and Ampère is cited by Gergonne; but similar ideas in a work published during the very year of Ampère's paper seem to have been pretty generally overlooked.[8]  The *Géométrie de position* of L. N. M. Carnot (1753–1823) is a recognized landmark in synthetic geometry, but it contains also a significant brief section on coordinates. Carnot had been a student under Monge in the military school at Mézières,[9] and, later, he was influential in the organization of the École Polytechnique.  It is consequently natural to find in his work the most general view of coordinate systems since the days of Newton. Innumerable transformations are suggested, including those for polar coordinates ($\tan z = \dfrac{y}{x}$, $t = \sqrt{x^2 + y^2}$ and $x = t \cos z$, $y = t \sin z$) and

---

[5] *Novi Commentarii Academiae Petropolitanae*, X (1764), p. 207–242; XI (1765), p. 152–184.

[6] *Traité du calcul* (1797), I, p. 418.

[7] "Sur les avantages qu'on peut retirer, dans la théorie des courbes, de la considération des paraboles osculatrices," *Journal de l'École Polytechnique*, cahier 14 (v. VII, 1808), p. 159–181.

[8] The most extensive account of the history of intrinsic coordinates is that by E. Wölffing, "Bericht über den gegenwärtigen Stand der Lehre von den natürlichen Koordinaten," *Bibliotheca Mathematica* (3), I (1900), p. 142–159; yet this overlooks also the important work in question. A good account which recognizes the work of Carnot is given by Loria. "Perfectionnements . . .," *Mathematica*, XX (1944).

[9] It was largely through the military genius of Carnot, the "Organizer of Victory," that the French Republic was saved in 1793.  For a thorough account of his political activities see Huntley Dupre, *Lazare Carnot. Republican patriot* (Oxford, Ohio, 1940).

bipolar coordinates[10] ($u = \sqrt{x^2 + y^2}$, $v = \sqrt{(a - x)^2 + y^2}$. For the latter system the circle $u = mv$ is given as an illustration. Carnot proposed also angular coordinates $(u, v)$, where $\tan u = \dfrac{y}{x}$ and $\tan v = \dfrac{y}{a - x}$; and he illustrated these by the circle $u + v = m$ and the hyperbola $u - v = m$. Other unusual schemes are presented, such as the following: If $A$ $(0, 0)$ and $B(a, 0)$ are fixed points and $M(x, y)$ is a variable point (all in rectangular coordinates), let $z$ be the ordinate of the orthocenter $K$ of triangle $ABM$. Then $x$ and $z$ are taken as new coordinates of the point $M$. In this system the ellipse $yy = \dfrac{bb}{aa}(ax - xx)$ becomes $zz = \dfrac{aa}{bb}(ax - xx)$. Carnot suggested also as coordinates of $M$ the quantities $u$ and $v$, where these are the areas of triangles $AMK$ and $BMK$. Equations of transformation from $x$ and $y$ to $u$ and $v$ are included.[11]

After proposing, as coordinates of a point on a curve, the radius vector and the radius of curvature, Carnot substituted for the former the length of arc of the curve. Inasmuch as the initial point from which the arc is measured remained arbitrary, several alternatives are suggested: the coordinate $z$ may be taken as the angle between the tangent at the point and the line which bisects secants drawn infinitely close to the tangent and parallel to it; or $z$ may be the angle between the secant bisector and the normal; or one can substitute still other lines or angles.

Attempts to establish systems of natural coordinates appeared sporadically throughout the nineteenth century. As early as 1802 (and again in 1804 and 1835) K. C. F. Krause (1781–1832) published at Jena his *De philosophiae et matheseos notione earumque intima conjunctione* in which he took as coordinates of a point on a curve the arc length $s$ and the angle $\phi$ which the tangent makes with a fixed reference line. Shortly after Krause's last work, A. Peters (1803–1876) used the same coordinates, in his *Neue Curvenlehre* (Dresden, 1838), to define new curves, such as $s\phi = K$. In England William Whewell (1794–1866) recognized the value of such a system in articles published in 1849 and 1851, in connection with which he used the modern name "intrinsic equation." Toward the end of the century the definitive work on natural coordinates was contributed by Ernesto Cesàro (1859–

---

[10] Loria, "Perfectionnements . . .," *Mathematica*, XVIII (1942), p. 138, overlooks the work of both Newton and Carnot in ascribing bipolar coordinates to Cournot in 1847.
[11] *Géométrie de position* (Paris, 1803), p. 458–480.

1906), culminating in his *Geometria intrinsica* (Napoli, 1896). Cesàro used arc length and radius of curvature as his intrinsic coordinates, and since then these have been adopted more widely than other combinations.[12]

Intrinsic coordinates seem to have had but little effect upon the development of analytic geometry as a whole; the influence of Gergonne was stronger in other directions. His contributions of 1813 fortunately were not limited to natural coordinates and the problem of Apollonius alone. He wrote also on subjects which gradually led him toward the important idea of duality. In his program of proving famous theorems analytically, he gave demonstrations of the dual theorems of Pascal and Brianchon on hexagons inscribed and circumscribed to conic sections.[13] A more important paper of that time presented an analytic theory of poles of conics and quadrics.[14] This opens with another defense of coordinate methods.

> Even those who are most familiar with the advantages which are presented by analytic geometry properly so-called, such as the uniformity of its processes, knowing full well that it alone affords the privilege of leading us constantly to the end of our researches without any kind of uncertainty, reproach the subject pretty generally for furnishing in the solution of problems only very complicated constructions, and of demonstrating theorems only by a calculation of which the prolixity often is repulsive. I have always thought that, most often, these inconveniences pertain perhaps less to the nature of the instrument than to the manner in which it is employed.

Gergonne therefore intends to show in this article that "analytic geometry, properly employed, can furnish, for the solution of problems, constructions which concede nothing, for elegance and simplicity, to those which one deduces by purely geometric considerations." The earnestness with which the author continued his plea for coordinate geometry led him, within a few years, into bitter conflict with the foremost French protagonist of synthetic geometry, Jean-Victor Poncelet (1788–1867).

Poncelet, like Gergonne, had been educated at the École Polytechnique, where he had been deeply influenced by Monge and Carnot; but he was attracted more by the synthetic than the analytic geometry of his teachers. He completed his education in 1810, the year in which the *Annales de mathématiques* first appeared; but while Gergonne was writing in praise of analytic geometry, Poncelet was in

---

[12] I have used the German translation by G. Kowalewski, *Vorlesungen über natürliche Geometrie* (Leipzig, 1901).

[13] *Annales de mathématiques*, IV (1813–1814), p. 381–384.

[14] *Annales de mathématiques*, III (1812–1813), p. 293–302.

far-off Russia, a prisoner-of-war after the ill-fated Napoleonic expedition of 1812. During the years of imprisonment, 1813 and 1814, Poncelet composed a long work, *Applications d'analyse et de géométrie, qui ont servi de principal fondement au Traité des proprietés projectives des figures*,[15] which has been largely overlooked—partly because publication was delayed for half a century, chiefly, perhaps, because it has been overshadowed by the more famous *Traité* of 1822 to which it was to have been the introduction. The *Applications d'analyse* shows that it is incorrect to hold to the traditional view that "Poncelet neglected all that was connected with the analysis of R. Descartes."[16] The second cahier of the book is, in fact, a good textbook on analytic geometry, typical of the subject at the time at which it was composed. It seems to be clear from this work that it was analytic geometry which led him a little later to his characteristic principles in synthetic geometry. He believed strongly in the generality of analysis and hence he sought to give geometric interpretations where none were thought possible before. On his return to France, Poncelet published his views in Gergonne's *Annales* for 1818, the very year of Monge's death. Admitting that analytic geometry had acquired a superiority over ordinary geometry and a generality which it was impossible to contest, he felt that it was nevertheless possible to give to ordinary geometry the same degree of perfection. Inasmuch as the source of the power of analysis did not lie, he held, in the use of algebra or of coordinates, but rather in its generality, it was only necessary for synthesis to borrow from coordinate geometry the "principle of continuity or of the permanence of mathematical relations."[17] According to this principle, "The metric properties discovered for a primitive figure remain applicable, without other modifications than those of change of sign, to all correlative figures which can be considered to spring from the first." This view—which he held to be implicit in, and inseparable from, algebraic analysis—he assumed could clearly be extended to pure geometry.[18] As an example of the principle, when applied to synthetic geometry, Poncelet cited the equality of the products of the segments of intersecting chords in a circle. This becomes, when the point of intersection lies outside the circle, an equality of the products of the segments of secants. If one of the lines is tangent to the circle, the theorem remains valid for this figure also on substituting the square of the tangent for the product of the segments of the secant. In a

---

[15] Two volumes, Paris, 1862–1864.
[16] See *Encyclopédie des sciences mathématiques*, III, 3, p. 193.
[17] See *Applications de l'analyse*, II, p. 296, or Gergonne's *Annales* for 1818.
[18] A good philosophical account of the principle of continuity will be found in Ernst Cassirer, *Substance and Function* (Chicago and London, 1923), p. 79 f.

limited sense this idea of a law of continuity had been proposed before by Kepler and Desargues; but whereas they had appealed to it
only in the case of elements at infinity, Poncelet extended it to imaginary points, both finite and at infinity.  All circles, he found, have in
common two imaginary points at infinity—the so-called circular points.

Poncelet boldly applied his principle of continuity to the discovery
of many new and useful theorems, but he never was able to demonstrate it in a manner to satisfy his opponents.  He maintained that it
could be justified, but he omitted a proof because he wished to present
it as a purely geometrical principle.  One suspects that Poncelet would
conceal any indebtedness of synthetic geometry to analysis; and this
impression is strengthened by all of his subsequent work.  He became the champion of the synthetic point of view in France and wrote
with polemic vehemence against rival analysts in general, and against
Gergonne in particular.  The controversy of the nineteenth century
between the proponents of synthesis and analysis reminds one of that
between the ancients and the moderns a hundred years before, or of
that between the formalists and intuitionists of today; but it en-
gendered a greater degree of bitterness.  The crux of the issue, apart
from the methodological question of the relative power of the two
methods, was the existence in the space of pure geometry of imaginary
elements.  As Bell has said, "this proves to be a pseudo question
without meaning";[19] but, like many another controversy, the results
were fruitful and unexpected.  Paradoxically, "perhaps nobody contributed more to the first development of [recent] Analytic Geometry
than Poncelet who by his destructive criticism, has achieved precisely
what he would prevent, the growing up of Analytic Geometry to the
level, and even to far above the level, of Synthetic Geometry."[20]

During the early stage of the conflict over methods, Poncelet and
Gergonne were earnest but friendly rivals, and the analyst made space
available in his *Annales* for the articles of the synthetist.  Poncelet's
paper of 1818 proposed some problems on polygons inscribed in conics
which he had solved by pure geometry and which he held would be difficult to handle by coordinate methods.  Gergonne in reply admitted
that he had exaggerated the advantages of analysis, carried away by
enthusiasm for his simple solutions of the problems of three circles and
four spheres, because his construction had not produced the sensation
he had expected.  Nevertheless, he again pointed to the generality
and uniformity of processes in analytic geometry, and he asserted that

[19] *Development of Mathematics*, p. 313–315.
[20] H. De Vries, "How Analytic Geometry Became a Science," *Scripta Mathematica*, XIV
(1948), p. 5–15.  See especially p. 6.

if one were to use more *adresse*, analysis would lead to solutions which need not yield to the geometry of the ancients in simplicity and in elegance.[21] After all, Gergonne pointed out, analytic geometry in the sense in which he used it (and which he ascribed to Monge), was very young and improvements were to be expected.

Gergonne's prediction of improvements to come in analytic geometry was indeed prophetic, for in the very same year there appeared a little volume by Gabriel Lamé (1795–1870), entitled *Examen des différentes méthodes employées pour résoudre les problèmes de géométrie*, which has been acclaimed (with perhaps pardonable exaggeration) as making the year 1818—the one in which Monge died—"the birthyear of Analytic Geometry as a science."[22] Lamé, also a graduate of the École Polytechnique, was primarily an engineer, but in his *Examen* he made two important contributions to the *adresse* which Gergonne wished for analytic geometry. The first consisted of the very simple expedient of representing whole equations by single letters, so that loci would appear as $E = 0$ or $E' = 0$. The second contribution was the very elementary yet important principle that if one combines the equations of two loci in any manner whatsoever, the resulting equation will represent a third locus through the points of intersection of the first two. Lamé and his successors limited themselves to the case where the combination is linear, the most useful one. Thus Lamé noted in particular that if $E = 0$ and $E' = 0$ are two loci of the same degree, then on linking them by parameters or "multipliers" in the form $mE + m'E' = 0$, the result is a curve or surface of like degree passing through the intersections of the two loci. This work marks the entrance into coordinate geometry of the systematic study of families of curves and surfaces. Monge had found envelopes of families of surfaces in differential geometry, and Lagrange had pointed out that a singular solution of a differential equation generally is an envelope of the integral curves. Lines of curvature and geodesics had been developed in differential geometry by Monge and Euler. The study of envelopes in the calculus goes back as far as Leibniz and is implicit in earlier work on evolutes. But before 1818 systems of curves were not a part of elementary analytic geometry. Even the geometry of the radical axis, which now appears in every textbook, was not included. It is odd to note that radical systems of circles, the theory of which is so simply presented in analytic form, arose first through synthetic geometry. There is some evidence that the Arabs of about the year 1000

[21] Poncelet, "Réflexions sur l'usage de l'analise algébrique dans la géométrie," *Annales de mathématiques*, VIII (1817–1818), p. 141–155; Gergonne, "Reflexions sur l'article précédent," *Ibid.*, p. 156–161. Poncelet's article is reprinted in his *Applications de l'analyse*, II, 466–76.
[22] De Vries, *op. cit.*, p. 9.

knew of the radical axis of two circles.[23]  Arabic sources have preserved (among other things) the Archimedean theorems on the shoemaker's knife.  One of the properties of this familiar figure is that the
two inscribed circles (Fig. 39) are equal.  The Arabs generalized this
theorem to include the equality of the circles in the cases in which the
two smallest semicircles are no longer tangent to each other.  If they
intersect each other, their common chord takes the place of the tangent;
if they do not intersect, one erects a perpendicular to the common base
line at the point on this line from which the tangents to the two semicircles are equal.  This line is, of course, the radical axis; but the
Arabs did not call it by this name, nor did they investigate further the
properties of this remarkable line.  Moreover, their work seems to
have been overlooked by their successors.

The properties of the radical axis were rediscovered in 1812–1813 by
Gaultier, again in connection with pure geometry.[24]  Take a circle
with center $A$ and radius $AK = AG$, and take a point $O$ on the line $GK$
(Fig. 40).  Then the circle with center $O$ and radius $OM = OG.OK$
is said to be radical to the circle with center $A$.  There are two cases
according as $O$ is outside or inside circle $A$; and in the latter case
circle $O$ is said to be a reciprocal radical circle.  Given a pair of circles,
their radical axis (known also as the "line of Gaultier" or the "power
line") is defined as the locus of the centers of all circles radical to the
pair of circles; given three circles, their radical center is the center of
the circle radical to the three circles.  The various cases are considered in turn, and the properties of a radical system are developed.
An analogous treatment is presented also for spheres.

Gaultier's paper contained virtually no analytic geometry, and so
the radical axis did not become part of the subject for another dozen
years or so, after Lamé's abridged notation began to be used systematically.  Lamé himself did not fully exploit his principle of "multipliers."  He did give the condition that three lines shall be concurrent,
expressing this in the equivalent of modern determinant notation and
also in the abridged notation in $mE + m'E' + m''E'' = 0$
[identically].  The corresponding condition on four planes is also given.
It is surprising that he did not develop systematically the idea of a
radical family of circles, $mC + m'C' = 0$; but this traditional portion

[23] See C. W. Merrifield, "On a Geometrical Proposition Indicating That the Property of
the Radical Axis Was Probably Discovered by the Arabs," London Math. Society, Proceedings, II (1866–1869), p. 175–177.  See also Apollonius Pergaeus, Conicorum libri v., vi, vii
(ed. by Borelli, Florentiae, 1661), p. 391–395.
[24] "Mémoire sur les moyens généraux de construire graphiquement un cercle déterminé
par trois conditions, et une sphère déterminée par quatre conditions," Journal de l'École
Polytechnique, cahier 16 (vol. IX, 1813), p. 124–214.

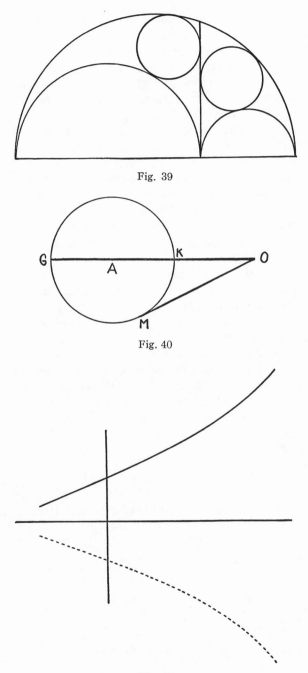

Fig. 39

Fig. 40

Fig. 41

of analytic geometry entered shortly afterwards, probably as the result of his influence.[25]

The importance of Lamé's ideas went unappreciated for almost a decade, and analytic geometry developed along certain unrelated lines. In the enthusiasm for the new analytic geometry of Monge, the older Eulerian aspect was all but forgotten.   When a paper on the graphical representation of functions did appear, it was concerned with pathological aspects of curves instead of with general principles.   For example, A. J. H. Vincent pointed out in Gergonne's *Annales* that the complete curve for the equation $y = e^x$ has a "branche pointillée," as well as a continuous branch (Fig. 41); and the equation $y = e^x + e^{-x}$ similarly represents not only the ordinary catenary but also (allowing for all possible combinations of signs for the roots of even index) three other discontinuous branches.[26]

The foremost French mathematician of the period, Augustin-Louis Cauchy (1789–1857), aided analytic extensively, but largely along the traditional lines.   His *Leçons sur le calcul* of 1826 made such effective use of the normal forms of the line and plane that these are often ascribed to him instead of to L'Huilier.[27]   To Cauchy also is ascribed the parametric form of the straight line in three dimensions[28], but anticipations of this are found in earlier works.   Jean Jacques Bret (b. 1781),[29] for example, had systematically used the form $x = \alpha + ar$, $y = \beta + br$, $z = \gamma + cr$, where $(\alpha, \beta, \gamma)$ are coordinates of a fixed point on the line, $a$, $b$, $c$ are constants determining the direction of the line, and the parameter $r$ is the distance between the points $(\alpha, \beta, \gamma)$ and $(x, y, z)$. The symmetric form of the line also is given by Cauchy in the modern

notation[30] $\dfrac{x - x_0}{\cos a} = \dfrac{y - y_0}{\cos b} = \dfrac{z - z_0}{\cos c} = \pm \sqrt{(x - x_0)^2 + (y - y_0)^2 + }$

$(z - z_0)^2$. Cauchy continued also the study and classification of quadric surfaces, giving a complete discussion of diametral planes and the problem of finding the center, in this respect rounding out the work

[25] Lamé invariably is credited with the introduction of abridged notation in 1818, but it should be noted that he had used it somewhat earlier in a paper, "Sur les intersections des lignes et des surfaces," *Annales de mathématiques*, VII (1816–1817), p. 229–240.   In the *Examen*, which appeared in 1818 when he was about twenty-three years old, he says that he had planned the work long before!   Mention should also be made of the fact that abridged notation was used (possibly independently) by Frégier in the *Annales* of 1818–1819.

[26] "Considérations nouvelles sur la nature des courbes logarithmiques et exponentielles," *Annales de mathématiques*, XV (1824–1825), p. 1–39.

[27] See his *Oeuvres* (2), V, p. 29.

[28] *Oeuvres* (2), V, p. 18–19.   Cf. *Encyclopédie des sciences mathématiques*, III (22), 9, III (17–18), 26.

[29] "Théorie analitique de la ligne droite et du plan," *Annales de mathématiques*, V (1814–1815), p. 329–341.   Cf. also p. 93 of volume IV.

[30] *Oeuvres* (2), V, p. 19.

of Euler, Monge, and Hachette.[31]   Where Euler had included only the proper quadrics, Cauchy gave essentially the definitive type of classification (in terms of the signs of the coefficients of the terms of even degree in the canonical forms) now found in all textbooks.   The continuing problem of orthogonal transformations attracted him as it did others of the nineteenth century.   Gergonne had tried his hand at improving on the formulas of Euler and Monge; and his *Annales* for 1824–1825 included similar work by a Swiss mathematician, C. Sturm (1803–1855).   Cauchy turned his attention to the question a score of years later,[32] using essentially the methods of Hachette; but he had earlier made use of the rotation of axes, in 1826, to complete Euler's study of the plane sections of a quadric.[33]   In 1826 Germinal Dandelin (1794–1847) considered the inverse question—to find a plane cutting a given quadric in a given conic.[34]   Dandelin a few years before had given the striking theorem, known by his name:   If two spheres are inscribed in a circular cone so that they are tangent also to a given plane cutting the cone in a conic, the points of contact of the spheres with the plane are foci of the conic and the intersections of the given plane with the planes of the circles in which the spheres touch the cone are directrices of the conic;[35] but this theorem had been anticipated synthetically (in slightly different form) in 1758 by Hugh Hamilton[36] (1729–1805).

The generatrices of the one-sheeted hyperboloid were also studied by Cauchy,[37] who wrote them in the form

$$\frac{y}{b} + \epsilon \frac{z}{c} = \lambda \left( 1 - \frac{x}{a} \right)$$

$$\frac{y}{b} - \epsilon \frac{z}{c} = \frac{1}{\lambda} \left( 1 + \frac{x}{a} \right)$$

where $\epsilon = \pm 1$.   Cauchy's work in determinants should also be mentioned because of the part these played later in analytic geometry. Apart from the work of Leibniz and Cramer, symmetric notations equivalent to determinants had appeared in the analytic studies of Lagrange and Monge.   Similar devices had been used also by Vander-

[31] See Cauchy, *Oeuvres* (2), V, p. 248; VIII, 12, p. 47.
[32] *Oeuvres* (1), IX, p. 253.
[33] *Oeuvres* (2), V, p. 273–280, XIII, p. 341.
[34] *Nouveaux mémoires de l'Académie Royale des sciences et belles-lettres de Bruxelles*, III (1826), p. 8.
[35] *Ibid.*, II (1822), p. 169–202 and Fig. 1.
[36] *Treatise of conic sections* (1758), Book II, theorem 37.   Cf. Taylor, *Ancient and modern geometry of conics*, p. 204–205.
[37] *Oeuvres* (2), V, p. 231.

monde and Laplace,[38] as well as by others in the early nineteenth century. In 1812–1813, for example, J. P. M. Binet (1786–1856) wrote a long and tedious paper in which he made use of analogous notations (which he called "resultants") in analytic theorems on volumes, areas, and lengths of rectangular configurations in three dimensions.[39] His work includes the equivalent of the multiplication of determinants; but in this early use of determinants the familiar square array was missing. This element, together with the double-subscript notation, was supplied by Cauchy at the very time of Binet's paper.[40] Cauchy applied his "resultants" in attributing signed values to angles, areas, and volumes in coordinate geometry.[41] The invariance of such expressions as $xx_1 + yy_1 + zz_1$ and $\begin{vmatrix} x & y & z \\ x_1 & y_1 & z_1 \\ x_2 & y_2 & z_2 \end{vmatrix}$ under an orthogonal transformation is easily justified by interpreting these geometrically, but Cauchy studied them from a purely arithmetic point of view. Other mathematicians, too, seem to have hesitated, before about 1830, in relating determinants and geometry.

Cauchy was violently opposed to the geometric principle of continuity, regarding it as ordinary induction; but Poncelet continued to lead synthetic geometry in a wave of progress. Gergonne did indeed continue his application of coordinate methods to classical problems, and in 1821 he gave an analytic proof of the theorem on the "Newton line"—the locus of the centers of conics tangent to four lines.[42] But Poncelet seems almost to have won the first skirmish in the battle of methodologies. Gergonne himself joined Poncelet in developing an important idea which at first appeared to be appropriate only to synthetic geometry—the principle of duality. In spherical trigonometry the reciprocity of figures had long been known through the polar triangle of a given spherical triangle. In the *Conics* of Apollonius the idea of poles and polars with respect to a conic is implied, but it was not until the nineteenth century that the general theory of polar reciprocals was brilliantly developed by Poncelet. Some aspects of this theory appeared in his paper of 1818, mentioned above, on polygons in-

[38] See *Mém. de l'Acad.*, 1772, part II.

[39] "Sur un système de formules analytiques, et leur application à des considérations géométriques," *Journal de l'École Polytechnique*, cahier 16 (v. IX, 1813), p. 280–354.

[40] See Cauchy, *Oeuvres* (2), I, p. 64 f., 90, 125 ff.

[41] See either Loria, "Perfectionnements . . .," *Mathematica*, XVIII (1942), or his article, "A. L. Cauchy in the history of analytic geometry," SCRIPTA MATHEMATICA, I (1932), p. 123–128.

[42] "Recherche du lieu des centres des sections coniques assujetties à quatre conditions," *Annales de mathématiques*, XI (1820–1821), p. 379–400. For Newton's work see *Principia*, I, lemma 25 and prop. 27.

and-circumscribed with respect to a conic. In fact, the theorems of Pascal and Brianchon are duals of each other and indicate the manner in which points are paired with lines with regard to a conic section. In three-space, it was known that in a similar way the quadric surfaces served to set up a one-to-one correspondence between points and planes. As early as 1806, the very year in which he gave the dual of Pascal's theorem, C. J. Brianchon (1783–1864) showed that the reciprocal polar of a surface of second order is another surface of the same kind.[43] In 1824 Poncelet presented to the Académie des Sciences his general theory of polar reciprocity, an extract of which was published in 1826. Meanwhile, Gergonne had noticed, as early as 1813, that in certain propositions of elementary plane geometry one arrives at new theorems through an interchange of the words "point" and "line." He introduced the word "duality" to indicate the relationship between the theorems in this case; and he saw that the idea could be applied to solid geometry by interchanging the words "point" and "plane." It appeared to Gergonne that through his principle of duality one can prove two theorems at once, and he proceeded to take advantage of this fact. In his *Annales* he began the ubiquitous practice of publishing geometrical theorems in double columns with "point" and "line" (or "point" and "plane") interchanged. Whereas previously theorems in a dual pair had each been proved independently, Gergonne became convinced by 1825–1826 that duality was a universal principle which could be invoked as a justification of the two theorems whenever one or the other had been proved.[44] Poncelet, meanwhile, had noted the similarity between the theory of poles and polars and the principle of duality; and he found in Gergonne's parallel columns some theorems which had been anticipated in his own study of the tangents to conics. He promptly and vehemently accused Gergonne of plagiarism. Duality, he insisted, was only a consequence of his own theory of reciprocal polars. Gergonne denied that duality depended upon polar reciprocity and pointed to the fact that his theory dispensed completely with the intermediary conic or quadric. He described his principle as a law of symmetry; and in 1827–1828 he attempted to justify it, but with no more success than Poncelet had had in connection with the law of continuity. Both principles depended, for their elucidation, upon analytic geometry; but duality at the time appeared— even to Gergonne—to be far removed from algebraic geometry. How suddenly the picture changed! After a comparative lull, coordinate geometry was about to burst forth in a period of expansion which was

---

[43] *Journal de l'École Polytechnique*, cahier 13 (v. VI, 1806), p. 297.
[44] See *Annales demathématiques*, XVI (1825–1826), p. 209–231.

unprecedented, both in scope and in rapidity; and the grand geo-metrical principles of Poncelet and Gergonne became merely two aspects of a new and powerful analysis.

There are many dates which are important in the rise of analytic geometry—1637, 1707, 1748, 1797–1798, 1818—but no triennium contributed more to the subject than did the years 1827–1829.[45]  As is to be expected, one finds a brilliant former student of the École Polytechnique, Etienne Bobillier (1797–1832), taking the lead in the new development. Bobillier began where Lamé had left off, for in Gergonne's *Annales* for 1827–1828 he gave the first extensive explanation and application of the method of abridged notation.[46]  Representing the three equations of the sides of a triangle, referred to Cartesian coordinates, by $A = 0$, $B = 0$, $C = 0$, he showed that the lines of the plane can be written as $aA + bB + cC = 0$, where the ratios of $a$, $b$, $c$ serve to determine a specific line.  In this way he avoided much of the tedious algebraic elimination with which analytic geometry had been overburdened.  With striking originality, he wrote in the form $aBC + bCA + cAB = 0$ the equation of all conics circumscribed about the triangle.  For given values of (or rather ratios of) $a$, $b$, $c$, the tangents to the conic at the vertices of the inscribed reference triangle, which thus form a triangle circumscribed about the conic, are then $bA + aB = 0$, $bC + cB = 0$, and $aC + cA = 0$.  The lines $aB = bA$, $bC = cB$, and $aC = cA$ connect the vertices of the inscribed triangle with the corresponding vertices of the circumscribed triangle. Bobillier showed that the sides of the circumscribed triangle intersect the corresponding sides of the inscribed triangle in three collinear points—a case of Desargues' theorem.  The proof of this is easily given in terms of his abridged notation and of linear dependence, anticipations of which had already appeared in Lamé's *Examen*.  Similarly for the tetrahedron $A = 0$, $B = 0$, $C = 0$, $D = 0$, the planes through the vertices tangent to a circumscribed quadric cut the opposite faces of the tetrahedron in four lines which are generatrices of a hyperboloid.  By similar methods he proved the theorems of Pascal and Brianchon,[47] as well as numerous other properties of conic sections. Bobillier contributed not only to Gergonne's *Annales*, but also to the newer *Correspondance mathématique et physique*.  The latter journal,

---

[45] It is interesting to note that these years virtually coincide with the crucial period in non-Euclidean geometry.

[46] "Essai sur un nouveau mode de recherche des propriétés de l'étendu," *Annales de mathématiques*, XVIII (1827–1828), p. 320.

[47] "Démonstrations nouvelles de quelques propriétés des lignes du second ordre," *Annales de mathématiques*, XVIII (1828), p. 359.  See also Loria, "Perfectionnements . . .," *Mathematica*, XVIII (1942), 136 ff.  An excellent and extensive account of Bobillier's work is available in Coolidge, *History of Conic Sections*, p. 84–85.

founded by L.-A.-J. Quetelet (1796–1874) and Garnier, carried numerous articles on analytic geometry. One of these, by Bobillier, extended to three dimensions the idea of foci for equations of second degree.[48]

Bobillier seems to have been the first one to study the straight line and the conic by referring them to a triangle. The quantities $a$, $b$, $c$, which in his treatment determined a line or a circumscribed conic, can be looked upon as coordinates, but this coordinate concept does not seem to have been clear in his work. Homogeneous coordinates are implied at every turn, but this system is not explicitly formulated. One of the most striking characteristics of the new upsurge in analytic geometry was the simultaneity of new discoveries, and homogeneous coordinates are an instance of this. They are not due to one individual, but to as many as four men—if one may include Bobillier among them. Had Bobillier not died at the early age of thirty-five, he might have become the leading analytic geometer of all times; but as it was, this honor did not fall, as one would have expected, to a Frenchman. The long predominance of France in the field of coordinate geometry—from Fermat and Descartes down to the time of Monge—finally was challenged by Germany, where homogeneous coordinates were invented independently (and nearly simultaneously) by three mathematicians.

The Napoleonic invasions seem to have affected German mathematics in somewhat the same way that the Revolution modified the French pattern. Schools of technology were developed as centers of research, in line with the feeling (which Bonaparte clearly shared) that the level of mathematical attainment was closely related to the welfare of the state. Whereas analytic geometry before 1827 had been very much a French science, it was destined to find its greatest representative in Germany, where French influence had been strongly felt. A German translation of Lacroix's textbook had appeared as early as 1805. Synthetic geometry, as well as analytic, received renewed impetus in Germany, and the battle between the analysts and purists continued on a second front. The *Annales* had been the focal point of geometrical research and controversy in France, and A. L. Crelle (1780–1855) became the "Gergonne of Germany" by the establishment in 1826 of the *Journal für die reine und angewandte Mathematik*. Although Crelle's *Journal* began under technological influences, it soon became so abstract in its emphasis that waggish references to it omitted one letter in the title, changing the "und angewandte" to "unangewandte."

Cauchy's determinants had not met with favor in France, but in

[48] *Corresp. math. et phys.*, IV (1828), p. 137 ,157, 216.

Germany they were most effectively employed by C. G. J. Jacobi (1804–1851). In Crelle's *Journal* for 1827 Jacobi called attention to Euler's formulas for the rotation of axes in three dimensions, and he showed how they could be improved by applying the determinant notation. However, Jacobi's interest in analytic geometry was quite transitory and he shortly became one of its bitterest opponents. This is greatly to be regretted inasmuch as determinants would have served greatly in the development of the subject. As it was, Jacobi applied them, together with the now universal double-subscript notation, to problems in the calculus where his name has been immortalized in the familiar "Jacobians."

The year 1827 is of considerable importance in the history of analytic geometry in Germany for reasons far removed from Jacobi's work. It is sometimes said that Descartes arithmetized geometry, but this is not strictly correct. For almost two hundred years after his time coordinates were in essence geometric. Cartesian coordinates were line segments, and polar coordinates were vectorial radii and circular arcs. Even the areal coordinates of Carnot were largely geometric. The arithmetization of coordinates took place not in 1637 but in the crucial years 1827–1829. Bobillier should be remembered as anticipating the new point of view to a certain extent, but otherwise the change came with a certain suddenness in 1827 with the *Barycentrische Calcul* of A. F. Möbius (1790–1860). Originally an astronomer, Möbius nevertheless seems to have studied carefully the works of the French geometers. In his highly original book, Möbius, like Bobillier, studied figures by means of a triangle of reference; but his coordinates were no longer lines. As the title indicates, the coordinates of a point, with respect to a triangle in the plane of which the point lies, are three numbers proportional to weights so chosen that if they are placed at the vertices of the triangle, the given point will be the center of gravity of the system. The barycentric coordinates of the centroid of the reference triangle $A$, $B$, $C$, with sides $a$, $b$, $c$, for example, are $(1, 1, 1)$—or, more generally, any three equal numbers; for the incenter the coordinates are $(a, b, c)$; for the orthocenter $(\tan A, \tan B, \tan C)$; for the circumcenter $(a \cos A, b \cos B, c \cos C)$. Möbius' work is striking not only for the use of three coordinates in two dimensions, but also for the subordination of the idea of length to numerical (or mechanical) considerations. One can, of course, reconcile his system with the division of lines in given ratios. For example, the point with barycentric coordinates $(a, b, c)$ with respect to the triangle $A$, $B$, $C$ can be located by finding the point $M$ which divides $BC$ in the ratio $c{:}b$, and then finding the point which divides $AM$ in the ratio $a + c{:}a$. In either case, the idea of number takes the place of that of geometry, for it is here (as in the work of Bobillier) only the *ratios* of the lines or weights which have significance.

One of the chief advantages of barycentric coordinates (or of other homogeneous systems) is that it gives analytic significance to the ideal elements which Poncelet had been using in pure geometry; and Möbius made use of this fact.   Where the coordinates of the vertices $A$, $B$, $C$ are respectively $(1, 0, 0)$, $(0, 1, 0)$, and $(0, 0, 1)$; and the midpoints of the sides are given by $(0, 1, 1)$, $(1, 0, 1)$, and $(1, 1, 0)$; the points at infinity on the medians are written as $(-1, 1, 1)$, $(1, -1, 1)$, and $(1, 1, -1)$.   The equations of the lines $BC$, $CA$, and $AB$ are respectively $a = 0$, $b = 0$, and $c = 0$; and the lines through the vertices parallel to the opposite sides are $b + c = 0$, $c + a = 0$, and $a + b = 0$; and from these equations it is obvious that the intersections of the above pairs of parallel lines satisfy the relationship $a + b + c = 0$, and hence this is the equation of the line at infinity in the plane.

A few great mathematicians, notably Gauss and Cauchy, recognized the *Barycentrische Calcul* for what it was—a work of great originality and significance; but the unusual language and notation which were adopted (and which Cauchy himself criticized) obstructed its success.   For the familiar phraseology in which $(a, b, c)$ would be called the coordinates of a point with respect to the triangle $ABC$, Möbius used the circumlocution $Aa + Bb + Cc$ is the "barycentric expression" of the point.   Moreover, although Möbius employed his method with striking success in the solution of elementary problems—such as finding necessary and sufficient conditions that four given points lie in a plane or on a circle—he did not stress its role as a general coordinate system applicable to the study of curves.   This had been true of Bobillier; and much the same thing can be said of a third independent discoverer of homogeneous coordinates, Karl Wilhelm Feuerbach (1800–1834).   The *Grundriss zu analytischen Untersuchungen der dreieckigen Pyramide* of Feuerbach did for three dimensions much the same thing that Möbius in the very same year (1827) had done in the plane, but his approach was geometrical instead of mechanical.   It is interesting to note that whereas in plane geometry he had adopted synthetic and trigonometric methods, in three-space he turned to the elegant analytic methods of Lagrange.   In doing so, Feuerbach ran across a striking generalization of the results of Lagrange: If five points are given (no. three of which are collinear and no. four of which are coplanar), and if the algebraic distance of each point from an arbitrary plane is multiplied by the signed volume of the tetrahedron determined by the other four points, then the algebraic sum of these products is zero.   Intrigued by this and other properties of the triangular pyramid, Feuerbach developed a new instrument for the study of "tetrahedrometry"— a set of coordinates which are numbers proportional to the volumes of the triangular pyramids formed by a given point and the faces of a tetrahedral frame of reference.  If units are properly chosen, his coordinates are quickly related to barycentric, but his work was not based upon the idea of weight or center of gravity.  In fact, the coordinate concept was not uppermost in his mind.  Feuerbach's *Untersuchungen* was only introductory to a longer work, *Analysis der dreieckigen Pyramide*, which was never published;[49] but in neither book were the new coordinates systematically developed.  The author was concerned instead with new theorems on the pyramid and with the determination of the forty-four elements of the tetrahedron, given any six independent parts.  In this respect he differed radically from a fourth independent discoverer of homogeneous coordinates, Julius Plücker (1801–1868), a man who approached the subject from an entirely new angle and one for whom the *results* were of little concern as compared with the *methods*.

No single person has contributed more to analytic geometry, both as to volume and power, than did Plücker.  No previous mathematician —not even Descartes, Fermat, Newton, Euler, or Monge—had been primarily an algebraic geometer.  Monge was, indeed, a geometer; but he was equally capable in, and concerned with, synthetic, coordinate, and differential geometry.  Plücker, on the other hand, was in a real sense the first specialist in analytic geometry.  Where his predecessors in each case were responsible for a few papers or a volume devoted to the subject, Plücker published half a dozen large quarto volumes, averaging over three hundred pages per volume, each one devoted entirely to analytic geometry.  Devoting the greater part of his life singlemindedly to coordinate methods, he contributed also scores of important papers (well over six hundred pages in all) to the learned periodicals of his time in Germany, France, England, and Italy.  In spite of the bulk of his work, Plücker's aim was not to amass results through an exploitation of existing principles; he sought instead to *rebuild* analytic geometry anew.  Each of his volumes bears a subtitle or carries a preface in which the author refers to a "new method," or even a "new geometry," which he is about to present.  As Descartes seems to have realized that he was blazing a new trail, so too Plücker had a clear conception of the transformation which he was working in analytic geometry.  And yet in all his work Plücker modestly felt

[49] See Albert Kiefer, *Die Einfuehrung der homogenen Koordinaten durch K. W. Feuerbach* (Strassburg, 1910).

that he was but building along the lines which Monge had suggested. Plücker had taken his doctorate at Bonn; and he studied also at Berlin and Heidelberg. In 1823 he spent some time at Paris,[50] attending lectures given by geometers of the school of Monge, and here he innocently walked into the crossfire between Gergonne and Poncelet. Plücker was not born an analyst. His first paper[51] was on a favorite topic of the time—the tangents to conics—and he treated the subject synthetically! Plücker sent the paper to Gergonne, and the latter exercised his editorial prerogative so freely that when the article appeared in the *Annales* for 1826, Plücker said that he recognized it only because it carried his name. Gergonne had adapted the material to his custom of publishing dual theorems in parallel columns. In addition, and without Plücker's knowledge, he had added, as though it were part of the original manuscript, a reference to Poncelet's *Traité* of 1822, a work which Plücker had not yet seen. Poncelet, believing naturally that Plücker was familiar with the material in the *Traité*, published a violent note charging the latter with plagiarism. Plücker defended himself, supported by Gergonne; but Poncelet renewed his accusations. Although the brunt of the attack was then turned against Gergonne, Plücker seems to have been deeply hurt; and it may be largely the ruthlessness of Poncelet's attack which drove into the enemy camp the greatest of all champions of analytic geometry—the man who during the next score of years "invented . . .as much (or more) new geometry as was created by all the Greek mathematicians in the two or three centuries of their greatest activity."[52]

Plücker reported that his introduction to analytic geometry had taken place in 1825 with the reading of the sixth edition of Biot's textbook. While drawing on paper three intersecting circles and their common chords, he noticed that the chords were concurrent. This theorem, which Plücker ascribed to Monge and which Gaultier had also given, he sought to prove analytically without recourse to the tedious algebraic elimination which was generally used at that time. In his search he was thus led to discover independently the method of abridged notation which Lamé had proposed. Writing the equations of the circles as $C = 0$, $C' = 0$, $C'' = 0$, and their common chords (real

[50] An excellent account of his life and a summary of his work is found in Wilhelm Ernst, *Julius Plücker* (Bonn, 1933). Plücker's mathematical and scientific papers have been collected in *Gesammelte wissenschaftliche Abhandlungen* (2 vols., Leipzig, 1895–1896), of which the first volume contains those on mathematics.

[51] "Théorèmes et problèmes sur les contacts des sections coniques," *Annales de mathématiques*, XVII (1826), p. 37–59. His doctoral thesis of 1823 had been on aspects of the calculus, especially Taylor's series. An analysis of the 1826 manuscript is found in A. Schoenflies, "Über den wissenschaftlichen Nachlass Julius Plückers." *Mathematische Annalen*, LVII (1904), p. 385–403.

[52] Bell *Development of Mathematics*. p. 15.

or ideal) as $C' - C'' = 0$, $C'' - C = 0$, $C - C' = 0$, Plücker saw that two of the equations of the chords imply the third, and hence the theorem is proved.[53] This is the ingeniously simple proof which is now found in most elementary analytic geometry textbooks.

Simultaneity of discovery is far from unusual in the history of mathematics, and there are numerous instances of this in the development of analytic geometry. The independent invention of the subject by Descartes and Fermat is but one of these. During the years 1827–1829 Plücker invented several important aspects of the subject, in each case apparently without knowledge of the work of at least two rivals. His use of abridged notation, is one of these, for it had been used earlier by Lamé and Frégier, and simultaneously by Bobillier and Gergonne. It is to Gergonne that the use of λ, instead of Lamé's multipliers $m$, and $m'$, is due. So widely has Gergonne's symbol been adopted that the formation of the equation $C_1 + \lambda C_2 = 0$ often is referred to as "lambdalizing."[54] But Plücker is nevertheless the real hero in this connection, for he made by far the widest and most effective use of such notations. It is therefore not without justice that it has become customary to speak of "Plücker's abridged notation," and to refer to the parameter, in combinations like $C_1 + \mu C_2 = 0$, as "Plücker's $\mu$."

One of the striking applications of abridged notation was made by Plücker in connection with the well-known "Cramer paradox." Between 1750 and 1827 little had been added to the subject of higher plane curves, but Plücker was here about to open a new era. Cramer had noticed, as had Euler at about the same time, that although a cubic curve generally is determined uniquely by $\frac{1}{2}n(n + 3) = 9$ points, nevertheless two cubic curves intersect in $n^2 = 9$ points. Why should nine points sometimes determine a cubic uniquely and sometimes not? Cramer and Euler realized that somehow the interdependence of points was involved. Plücker gave a clearer answer to the problem[55] by proving that if all but one of the $\frac{1}{2}(n + 1)(n + 2) - 1 = \frac{1}{2}n(n + 3)$ points which determine a curve of order $n = 3$ are given, then all of the one-parameter family of curves of order $n$ through these points pass also through a group of $\frac{1}{2}(n - 1)(n - 2)$ points which are determined by the given points. Thus if eight points are specified, a ninth point is thereby determined such that all cubics through the eight

[53] "Mémoire sur les contacts et sur les intersections des cercles," *Annales de mathématiques*, XVIII (1827), p. 29–47.
[54] See De Vries, *op. cit.*, p. 10.
[55] "Recherches sur les courbes algébriques de tous les degrés," *Annales de mathématiques*, XIX (1828), p. 97–106. Cf. also A. Brill and M. Noether, "Die Entwickelung der Theorie der algebraischen Funktionen in älterer und neuerer Zeit," *Jahresbericht der Deutschen Mathematiker-Vereinigung*, III (1892–1893), p. 107–566.

points have also the ninth point in common. Plücker easily proved the theorem as follows: Let $M = 0$ and $M' = 0$ be two distinct curves of $n$th order passing through the $1/_2 (n + 1)(n + 2) - 2 = 1/_2 n(n + 3) - 1$ given points. Then, according to Lamé's principle, $M + \mu M' = 0$ is the equation of curves of degree $n$ through the given points, a particular member of the family being specified by a value assigned to $\mu$. But all of these curves intersect each other in the same points, $n^2$ in number; and hence the $1/_2 n(n + 3) - 1$ points initially given determine a concomitant set of $n^2 - [1/_2 n(n + 3) - 1] = 1/_2 (n - 1)(n - 2)$ additional points which necessarily lie on any curve through the given points. The curve of fourth order, for example, has an equation containing fifteen coefficients or, to use Plücker's expression, fourteen "necessary constants." If, then, fourteen points are given, the quartic curve through these can be written as $M + \mu M' = 0$, where $M = 0$ and $M' = 0$ are two distinct quartics through thirteen of the given points and where $\mu$ is so chosen that the coordinates of the fourteenth point satisfy the equation $M + \mu M' = 0$. This quartic—and also other quartics through the thirteen points—will pass also through the other three points in which $M = 0$ and $M' = 0$ intersect; so that the thirteen points determine an additional three points associated with, or dependent upon, the thirteen. No set of fourteen or more points selected from the combined set of sixteen points would determine a unique quartic curve.

Somewhat similar explanations of the dependence of points were given at about the same time by Gergonne, Jacobi, and Lamé. Gergonne, for example, had announced that if two curves of order $m = p + q$ have $p(p + q)$ of their points of intersection on a curve of order $p$, the remaining $q(p + q)$ points will lie on a curve of order $q$. As an example of this he gave a beautifully simple analytic proof of Pascal's theorem: Let the three odd-numbered sides of the hexagon inscribed in a conic be taken as a (composite) cubic curve and the other three sides as a second such curve. Now six of the nine points of intersection of the two cubics lie on a curve of order $p = 2$; and hence, by Gergonne's theorem, the other three intersections lie on a curve of order $3 - 2 = 1$, a straight line.[56] Plücker's proof of the same theorem so well illustrates his use of abridged notation that it may be appropriately repeated here: Let the equations of the sides of the hexagon be $p = 0$, $q = 0$, $r = 0$ and $p' = 0$, $q' = 0$, $r' = 0$. Then the equation $pqr + \mu p'q'r' = 0$ represents all cubics through the nine points of intersection of the lines. Six of the points of these cubics necessarily lie on the circumscribed conic and, by a suitable choice of $\mu$,

[56] See De Vries, *op. cit.*, p. 11.

a cubic can be determined which has a seventh point in common with the conic. But a proper cubic intersects a conic in at most six points; and hence the cubic in question must be composite, consisting of the conic and a straight line. Consequently the three remaining intersections—the intersections of opposite sides of the hexagon—are collinear.[57]

Following Gergonne's custom, Plücker published, in parallel columns, both the theorem on Cramer's paradox and its dual. He stated the latter as follows: All the curves of $m$th class which touch the same $\dfrac{m+1}{1} \cdot \dfrac{m+2}{2} - 2$ fixed lines also touch $m^2 - \dfrac{m+1}{1} \cdot \dfrac{m+2}{2} + 2$ other fixed lines. In the same year (1828) he presented also, in the *Annales*, analogous theorems for surfaces: All surfaces of $m$th order [class] which pass through [touch] $\dfrac{m+1}{1} \cdot \dfrac{m+2}{2} \cdot \dfrac{m+3}{3} - 3$ given points [lines] also pass through [touch] $m^3 - \dfrac{m+1}{1} \cdot \dfrac{m+2}{2} \cdot \dfrac{m+3}{3} - 3$ common fixed points [lines].[58] Although Plücker's work was most influential in clarifying Cramer's "veritable paradox," it should be noted that more rigorous formulations of the problem continued to appear for almost another hundred years.[59]

It was Plücker, more than anyone else, who elevated abridged notation to the status of a principle. In 1828 he built the first volume (consisting of 270 pages) of his important *Analytisch-geometrische Entwicklungen* about this one idea. Emphasizing (in the preface) that this was a new way of handling analytic geometry—in which all elimination is eliminated—Plücker remarked that the results are only the details of a general method. His manner of treatment is purely analytic, he wrote, in the sense in which the term was used since Monge; and by this he means that there is an exact correspondence between analytic expressions and geometric constructions.[60] Plücker's guiding star was the firm conviction that what synthetic geometry has accomplished can be done as well—or better—by means of coordinates. He was determined to win back the territory which Poncelet had won for synthesis through the principles of continuity and duality; and this he accomplished in the next year or two.

[57] See Felix Klein, *Vorlesungen über die Entwicklung der Mathematik im 19 Jahrhundert* (2 vols., Berlin, 1926–1927), I, p. 122.

[58] "Recherches sur les surfaces algébriques de tous les degrés," *Annales de mathématiques*, XIX (1828), p. 129–137.

[59] Coolidge, *History of Geometric Methods*, p. 133. ascribes to Luigi Berzolari in 1914 the first "satisfactory" answer to the paradox.

[60] Cf. *Wissenschaftliche Abhandlungen*, I, p. 61.

In his contributions to Gergonne's *Annales*, Plücker had used only Cartesian coordinates; but in 1829 he wrote an article for Crelle's *Journal*, "Über ein neues Coordinatensystem." Homogeneous coordinates, which he here announced, were not so new as he believed; they had been invented three times previously—by Möbius, Feuerbach, and Bobillier. But where his predecessors had made very limited use of the new system, Plücker had a clearer view of the methodological principles involved. He did for homogeneous coordinates what he had done for abridged notation—he applied them systematically to the study of curves in general. The paper of Plücker opens with a statement that is reminiscent of Carnot's multitude of types of coordinates: "Any particular procedure for fixing the position of a point, with respect to points or lines considered as known in position, corresponds to a system of coordinates." Plücker took as his coordinate frame three lines, no two of which are parallel; and he chose as the "triangular coordinates" of a point $M$ the signed distances $(p, q, r)$ of $M$ from the three reference lines, measured along lines making given angles with the reference lines. Later he systematically adopted perpendicular distances, in which form his coordinates correspond to those now known as trilinear. To convert these coordinates to barycentric coordinates, one divides them respectively by $a$, $b$, and $c$, the lengths of the sides of the reference triangle. Whereas the equation of the line at infinity in Möbius' coordinates was $a + b + c = 0$, in Plücker's system it was $ap + bq + cr = 0$.

Plücker, the analyst, noted with particular satisfaction that by means of his new coordinate system he arrived at two of Poncelet's spectacular synthetic discoveries—that all the points at infinity in a plane lie on a line; and that concentric circles have double imaginary contact at infinity. For concentric similar conics he noted agreement with Poncelet's observation that the double contact at infinity is real or ideal according as the curves are hyperbolas or ellipses. Yet of more importance to analytic geometry as a whole was Plücker's introduction of the homogeneous equation, $f(p, q, r) = 0$, of a plane curve.[61] He studied the properties of the conics, for example, through the equation $Ap^2 + 2Bpq + 2Cpr + Dq^2 + 2Eqr + Fr^2 = 0$. This made the study of the behavior of the curves at infinity quite analogous to the investigation for ordinary points. He showed how to transform the equation of a curve from Cartesian to homogeneous coordinates, and vice versa. The special case $X = \dfrac{x}{t}$, $Y = \dfrac{y}{t}$, where

$(X, Y)$ are Cartesian coordinates and $(x, y, t)$ are homogeneous, appeared in 1831 in volume II of Plücker's *Analytisch-geometrische Abhandlungen*. Here, too, the system is generalized to multilinear coordinates $(p, q, r, s, t, \ldots)$ where $s, t, \ldots$ are linearly related to $p, q, r$.

Homogeneous coordinates, especially those of Möbius, had done much to arithmetize geometry; but in 1829 Plücker introduced a still more revolutionary point of view which broke completely from the idea of coordinates as line segments. Again Plücker appreciated the significance of the change, for in the titles of his papers in Crelle's *Journal* he used the phrases "Ueber ein neues Princip" and "Ueber ein neues Art."[62] In the early days of analytic geometry the parameters $a, b, c$ in the equation $ax + by = c^2$ were understood to designate line segments, in keeping with the idea of dimensionality. Gradually, however, the coefficients came to have more and more the status of pure numbers. Just as geometrical homogeneity was disappearing, homogeneous coordinates were introduced; but these led, not to a return to geometric concepts, but to complete arithmetization. Beginning with the equation of the line $Ay + Bx + C = 0$, Plücker wrote it in the homogeneous form $aA + bB + cC = 0$. The three coefficients $(A, B, C)$ determine a straight line, just as the homogeneous coordinates $(a, b, c)$ determine a point. There is therefore an analogy between the two sets of quantities; if $(a, b, c)$ are called coordinates, the same phraseology may be applied to $(A, B, C)$. Plücker took advantage of this situation and called the latter "line coordinates." The coordinates of the line at infinity, for example, are $(0, 0, C)$, and coordinates of lines through the origin are of the form $(A, B, 0)$. To conform to the Cartesian convention that unknowns (i. e., variables) are denoted by letters near the end of the alphabet, Plücker rewrote his equation as $au + bv + cw = 0$. If $(a, b, c)$ are the coordinates of a varying point and $u, v, w$ are fixed, the equation represents the line common to all of the points; if $(u, v, w)$ are coordinates of a varying line and $a, b, c$ are fixed, the equation represents the point common to all the lines. Just as a first-degree equation in point coordinates represents a line, so in line coordinates such an equation represents a point. Here Plücker discovered an immediate analytic counterpart of the geometric principle of duality, about which Gergonne and Poncelet had quarreled; and it now became clear that the justification which pure geometry had sought in vain was supplied at once by the powerful methods which analysis had at hand. The interchange of the words "point" and "line" merely corresponds to an interchange of the words "constant" and "variable" with respect to the quantities $a, b, c$ and

[62] Crelle's *Journal*, V (1829), p. 268–286; VI (1829), p. 107–146.

$u, v, w$. But the algebraic processes remain the same; and hence every theorem appears immediately in two forms, one the dual of the other.

Duality for space was justified by Plücker in 1831, in a paper in Crelle's *Journal* with the characteristic title, "Note sur un théorie générale et nouvelle des surfaces courbes."[63] Here the equation $tz + uy + vx + w = 0$ is thought of as representing a point or a plane according as $(t, u, v)$ are regarded as "plane coordinates" or $(z, y, x)$ are thought of as point coordinates. It is odd to note that although Plücker was largely responsible for homogeneous coordinates, he seems here to have preferred non-homogeneous coordinates. In two dimensions also he frequently used the ratios $\dfrac{u}{w}$ and $\dfrac{v}{w}$ instead of $(u, v, w)$, and hence these two quantities (the negative reciprocals of the intercepts of the line) have come to bear the name "Plücker coordinates."

A natural consequence of Plücker's invention of line coordinates was his analytic development of the idea of the class of a curve and its tangential equation. The notion of a curve as the envelope of its tangent lines was not new; it had been proposed by DeBeaune, and in 1692 Leibniz had given his rule for finding envelopes. Brianchon and Poncelet had noted the advantage of developing point and line conceptions of a curve simultaneously. Monge had given the crucial theorem that through a given point there are $n(n - 1)$ tangents to a given curve of order $n;$ but his theorem was overlooked.[64] Gergonne in 1826 applied the principle of duality to curves and introduced the word "class" of a curve to indicate the number of possible tangent lines; but he made the mistake of assuming that the order and class of a curve were the same. Möbius in 1827 had determined the condition $\phi(u, v, w) = 0$ that a line $ux + vy + wz = 0$ should be tangent to a curve $f(x, y, z) = 0$; and this condition is equivalent to the line equation of the curve. Möbius, however, did not express the idea that $(u, v, w)$ are coordinates of a line and that the degree of $\phi$ determines the class in the same sense that the degree of $f$ indicates the order. This clear general principle is due essentially to Plücker in 1830.[65] A point has only a line equation, and a line has only a point equation; but all other curves have both point and line equations. That is, Plücker developed analytically the notion of a curve as the locus generated by a point and enveloped by a line; the point moves continuously along the line while the line rotates continuously about the point.

---

[63] Or see *Wissenschaftliche Abhandlungen*, I, p. 224–234.

[64] See De Vries, *op. cit.*, p. 13.

[65] See Crelle's *Journal*, VI (1830), p. 107. The geometer Michel Chasles, however, in 1829 wrote to Quetelet that he had had the idea of line and plane coordinates independently of Plücker. See Plücker's *Wissenschaftliche Abhandlungen*, I, 600; or A. Schoenflies and M. Dehn, *Einführung in die analytische Geometrie* (2nd ed., Berlin, 1931), p. 58.

The elaboration of this idea forms the basis of volume II of his *Entwicklungen* of 1831. He noted that the conic sections are always of class two, because the equation $\phi = 0$ in this case takes the form $au^2 + buv + cv^2 + duv + evw + fw^2 = 0$; and he studied this equation in the way that previous writers had considered the general point-equation of the conic sections. Combining abridged notation and tangential coordinates, Plücker wrote $A = 0$ and $A' = 0$ as equations of two curves of class two, and he noted that $A + \mu A' = 0$ represents loci of the same class which have the same four common tangents. Similarly the proof of the Pascal theorem (given above) is easily converted by tangential coordinates into a proof of the dual (Brianchon) theorem.

The next book-length contribution of Plücker was his *System der analytischen Geometrie* of 1835. The author here again emphasizes the Mongian idea of the concordance of analytic and synthetic forms, but he approaches the subject from a different point of view. The book therefore carries a typical supplementary phrase in the title, *auf neue Betrachtungsweisen gegründet*. The new principle this time is that known as the "enumeration of constants," the basis of "enumerative geometry." The application of duality to the singularities of curves had indicated that these come in couples—double point and bitangent, cusp and stationary tangent (point of inflection); and Plücker's discovery of line coordinates led him to the study of line singularities. Point singularities had been studied in the preceding two centuries, and it had long been known that, for a given curve, the number of such points is limited by the degree of the equation. Maclaurin long before, in *Geometria organica*, had shown that a curve of degree $n$ has at most $1/2(n - 1)(n - 2)$ double points; Plücker showed that it has at most $1/2n(n - 2)(n^2 - 9)$ double tangents.[66] Poncelet had found that singularities and the class of a curve are also related; whereas a curve of order $n$ generally is of class $n(n - 1)$, a double point causes a reduction by two in the class. But Plücker, in Crelle's *Journal* for 1834, had gone on to make a discovery which Cayley considered "the most important one beyond all comparison in the entire subject of modern geometry." This discovery made it possible to set up not only upper bounds for the number of point and line singularities of a curve, but to write down equations relating the actual number of singularities to the order and class of the curve. These equations are the famous "Plücker equations." Those most commonly used are

$$m = n(n - 1) - 2\delta - 3\kappa$$
$$\iota = 3n(n - 2) - 6\delta - 8\kappa$$

[66] *Wissenschaftliche Abhandlungen*, I, p. 298.

and the two correlative equations

$$n = m(m - 1) - 2\tau - 3\iota$$
$$\kappa = 3m(m - 2) - 6\tau - 8\iota$$

where $m$ is the class, $n$ the order, $\delta$ the number of nodes, $\kappa$ the number of cusps, $\iota$ the number of stationary tangents, and $\tau$ the number of bitangents. From these equations, for example, it becomes obvious that a curve of order two can have no singularities and hence must be of class two. A cubic can have not more than one cusp or one node; a quartic without nodes can have as many as four cusps, and one without cusps can have three double points.

In his *System* of 1835 Plücker used his equations to give a new classification of cubics and quartics, a phase of analytic geometry which had to some extent been forgotten since Euler and Cramer. In resuming this type of investigation, Plücker said that he wished to complete the work of Newton and Euler on cubics by making their study as systematic as that of conics. In this connection he took advantage of abridged notation to write cubics in the form $pqr + \mu s = 0$, where $p = 0, q = 0, r = 0$, and $s = 0$ are straight lines.[67] Of the 146 possible quartic curves Plücker listed 135. In another volume, *Theorie der algebraischen Curven*, published four years later, he carried his results further, especially through the use of imaginary elements. After Plücker's time, operations on imaginaries came to be regarded as a necessary part of algebraic geometry. The Plücker equations hold, of course, only if all ideal elements—imaginary, infinite, and infinite imaginary—are included together with the real ones. For example, after stating the century-old theorem that a line through two points of inflection of a cubic passes also through a third point of inflection, Plücker added the observation that of the nine possible points of inflection of a cubic, only three are real. In such work the principle of continuity of Kepler, Desargues, and Poncelet had reached analytic maturity.

With the help of imaginary coordinates Plücker was able to generalize some of the properties of conics to higher plane curves. Poncelet as early as 1818 had deduced (by algebraic methods) that conic sections have four foci, two real on the principal axis and two imaginary on the transverse or conjugate axis.[68] Plücker, in his *Entwicklungen* of 1831 and in Crelle's *Journal* for 1832, continued and extended this work.[69] The foci of the conics, for example, have the property that the tangents from these points to the curve have slopes of $\pm i$—that is,

[67] See *Wissenschaftliche Abhandlungen*, I, p. 586–590.

[68] *Annales de mathématiques*, VIII (1817–1818), 222–223.

[69] See also *Wissenschaftliche Abhandlungen*, I, 290 f.

they pass through the circular points of Poncelet. Plücker therefore defined a focus of a higher plane curve as a point with this property. A directrix similarly is a chord of contact of the two circular lines through a focus. Plücker showed that a curve of class $m$ has, in general, $m^2$ foci, of which only $m$ are real.

A number of Plücker's papers in Gergonne's *Annales* and Crelle's *Journal* had dealt with three dimensions, but his first four volumes had been limited to plane analytic geometry. In 1846, however, he published his *System der Geometrie des Raumes* in which he applied his methods to surfaces and skew curves. The phrase *in neuer analytischer Behandlungsweise*, which was characteristically included in the title, refers largely to the extension to space of principles previously applied in the plane: abridged notation and Plücker's $\mu$; homogeneous (tetrahedral) coordinates; duality (or "reciprocity," as he called it) and the tangential equation of a surface in plane coordinates; and the enumeration of constants. Following out work begun by Lamé in 1818, Plücker studied the ensemble of quadrics $f + \mu g = 0$ through the curve of intersection of two quadrics $f = 0$ and $g = 0$, using Cartesian and tetrahedral coordinates. As in the plane case of the Cramer paradox, he considered the conditions under which nine given points in space are independent in the sense that they determine a unique quadric surface. Plücker classified quadrics in a manner similar to that of Cauchy; and, looking upon them as envelopes of their tangent planes, he studied the general tangential equation of second degree. He studied also the rectilinear generators of the lined quadrics and the properties of skew curves drawn upon quadric surfaces. An essentially newer idea, contained in the work of 1846, is the view that the four conditions determining a line in space correspond to four coordinates of a line. Here there is an implication of an analytic geometry of four dimensions, a notion which Plücker anticipated but did not develop. Recurring to a favorite theme, he pointed out that every geometric relation is to be regarded as a pictorial representation of an analytic relation which nevertheless has its own independent value. The principle of reciprocity is no exception, and hence, conceived of purely analytically, it is not bound to the dimensions of space. As in the transition from the plane to three dimensions one introduces another variable, so it is possible to extend the discussion of duality analytically to a greater number of variables.[70] For four variables [or dimensions] the first-degree equation becomes $pP + qQ + rR + sS + tT = 0$, a form quite analogous to the dual homogeneous form in two or three dimensions.

[70] *System der Geometrie des Raumes*, p. 322.

The *Vorrede* of the *System der Geometrie des Raumes* again emphasizes that it is the *method* which is important, and that it is this aspect of his work which will remain in science. Following this hope—eminently justified—for a measure of mathematical immortality, Plücker cryptically added that he was laying his pen down, after twenty years of work of this type, and that he would not again take up such research. What was it that caused the most original and prolific analytic geometer of all times to abandon the subject to which he had devoted himself exclusively for so long a time? What led him to cut himself off from the ranks of mathematicians for the next twenty years? It is not possible to give a categorical answer, but one can discern several possibilities. Plücker was aware of the significance of what he was doing; but in his own country—as well as in Italy, and even to some extent in France—there was a reluctance on the part of his contemporaries to grant him the recognition he had earned. Jakob Steiner (1796–1863), the "greatest geometer since Apollonius," ruled the hearts and minds of his day; and he took an intense dislike to analytic methods. It is not easy to define "analysis" as applied to geometry, but in any case the term would seem to connote a certain amount of technique or "machinery." Analysis sometimes is referred to as a tool, a term never applied to synthesis. Steiner, however, felt that geometry could best be learned by concentrated thought, and he objected even to such "props" as the models and diagrams which synthetic geometers employed. Calculating, he said, replaces, while geometry stimulates, thinking.[71] So antagonistic was Steiner to the analytic point of view that he is said to have threatened to give up contributing to Crelle's *Journal* if it continued to publish material by Plücker.[72] Möbius seems to have remained neutral with respect to the controversy, for he was both synthesist and analyst. Jacobi, however, aligned himself with the synthetic camp and polemically opposed Plücker. If his "Zwistigkeit" with Poncelet in 1826 was influential in turning him from an early interest in pure geometry, it is equally possible that his conflict with Steiner was a factor in Plücker's abandonment of analytic geometry. There is, however, another explanation available which would appear more plausible. From 1825 until 1846 Plücker had taught mathematics—first at Bonn, then at Berlin, and finally at Halle. In 1847 he became professor of physics at Bonn; and it is said that there was some criticism of the fact that a chair in physics should be held by a pure mathematician. Whatever the reason, Plücker abandoned geometrical research for

[71] See Struik, *op. cit.*, II, p. 246.
[72] See Cajori, *History of Mathematics*, p. 311.

experimental investigation.   Beginning with 1847 he published a long series of papers, mostly in Poggendorff's *Annalen*,[73] devoted to discoveries—some made in collaboration with Hittorf—in magnetism and spectroscopy.   Among other contributions, he announced that chemical substances are identifiable by the characteristic spectral lines which they emit, an anticipation of the work of Bunsen and Kirchhoff.

The direction of his physical research may have been suggested by the electrical discoveries of Faraday, for Plücker had been in touch with British scientists and mathematicians, among whom he found ardent admirers.   It is paradoxical to note that England, the stronghold of synthetic methods throughout the eighteenth century, should have seized the initiative in continuing the analytic geometry of Plücker. While Monge and Lacroix were engineering the analytical revolution in France, coordinate geometry in England had scarcely developed beyond the work of Newton and Maclaurin.   Wallis' *Conics* had fallen out of use at Cambridge; and while analytic geometry was always present to some extent from 1800 to 1820, it was largely through its relationship to problems on mensuration.[74]   The only work on coordinate geometry commonly read there at the beginning of the nineteenth century was a thirty-page section on "The application of algebra to geometry" appended to James Wood, *The Elements of Algebra*.[75] This brief account presents the subject about as it was in the days of L'Hospital.   Coordinates are defined as geometrical lines, and the rotation transformation is described without the use of trigonometric symbolism.   Simple examples on conics are given, together with the characteristic of the equation of second degree.   A few other curves are included, with reference to Euler and Waring.   A section "On the construction of equations" is highly reminiscent of Descartes' *Géométrie*.   The sharp contrast between British and Continental methods in the calculus, about 1800, is well known; and evidently the situation with respect to analytic geometry was much the same.   It was probably the "Analytical Society," formed to promote (in the calculus) the "principles of pure *d*'ism in opposition to the dot-age of the university," which indirectly brought about a change in analytical geometry.   In 1816 the Society translated Lacroix's *Elements of the Calculus*, and it was not long before the Leibnizian differential methods superseded the fluxions of Newton.   But Lacroix's *Calculus* presupposed some of Lacroix's analytic geometry; and it was not long

---

[73] See *Wissenschaftliche Abhandlungen*, v. II.
[74] W. W. R. Ball, *A History of the Study of Mathematics at Cambridge* (Cambridge, 1889), p. 129.
[75] The 6th and 9th editions (Cambridge, 1815 and 1830, respectively), which I have used, are virtually identical.  See p. 276–305.

before textbooks appeared to fill the need.  One of the first of these was *A System of Algebraic Geometry* (London, 1823) by Dionysius Lardner (1793–1859).  The unfortunate state of the subject at the time in England is evidenced by the author's remark that "Hitherto, no treatise whatever on Algebraic Geometry has appeared in Great Britain."[76]  Lardner called attention also to the fact that Sir John Leslie (1766–1832) was at the time seeking to lead a counter revolution back to the ancients.  Leslie had written a book with the title *Geometrical Analysis and Geometry of Curve Lines* (Edinburg, 1821); but the word "analysis" here is used in the sense of Plato and Pappus to indicate "an inverted form of solution."  The works of Lardner and Leslie are but two of those which betray that Britain, too, shared in the widespread controversy between analysts and synthesists.

The material in Lardner's *Algebraic Geometry* is much like that in the earlier texts of Lacroix and Biot, showing that England in 1823 no longer was a century behind the times.  This is confirmed by a number of similar books which appeared within the next few years: *Principles of Analytical Geometry*, by H. P. Hamilton in 1826; *Analytical Geometry of Three Dimensions*, by John Hymers in 1830;[77] and *A Treatise on Algebraical Geometry*, by S. W. Waud in 1835.  All of these[78] resemble Continental textbooks of the early part of the century; and the book by Waud in particular is an excellent and thorough treatment which would be acceptable in elementary classes of today.  But while Great Britain finally could point to satisfactory textbook material, she had produced no outstanding analytic geometer since Waring.  Yet when Plücker abandoned the field in 1846, his mantle descended upon an Englishman, Arthur Cayley (1821–1895), who, despite time devoted to legal practice, rivals Euler and Cauchy in volume of output.

Cayley had a convenient medium of publication in the *Cambridge Mathematical Journal*.  This periodical, established in 1837 and later appearing under the title *Cambridge and Dublin Mathematical Journal*, did for Great Britain somewhat the same thing as Gergonne's *Annales* had done for France and Crelle's *Journal* was doing for Germany.

Cayley was not a specialist in analytic geometry.  Most of his 900-odd mathematical papers are on the algebra of invariants; but it was just on the algebraic side that Plücker was weakest.  One notes with surprise that Plücker failed to take advantage of determinants; and

[76] Preface, p. liii.
[77] I have not seen these works and cite them on the basis of Ball's *Mathematics at Cambridge*.  However, I have used Hamilton's *An Analytical System of Conic Sections* (3rd ed., Cambridge, 1834), a work which resembles the treatment on the Continent.
[78] Mention might also be made of an English translation of a French work: L. B. Francoeur, *A Complete Course of Pure Mathematics* (transl. by R. Blakelock), 2 vols., Cambridge, 1829–1830.

the beautifully symmetric formulas of Lagrange and Monge did not stir him to generalizations in this direction. In the period of Plücker's greatest activity, following 1826, Jacobi used determinants effectively in analysis, and one wonders if their feud soured the geometer on this subject. Whatever the reason, it probably was the lack of the determinant notation which kept Plücker from developing the analytic geometry of $n$ dimensions. The introduction of multidimensional geometry is still another example of simultaneity of discovery, for three men took this step independently of each other at about the same time.[79] The quaternions of William Rowan Hamilton (1805–1865) and the *Ausdehnungslehre* of Hermann Grassmann (1809–1877), both dating from 1844, were actually a part of vector and tensor analysis. Hamilton wished to build up, without Cartesian coordinates, a calculus of vectors in ordinary space; and so he fixed his attention upon the four-parameter operation which transforms one vector into another. The view of Grassmann was less restricted, for his basic elements, or "extensive magnitudes," involved an indefinite number of dimensions. The *Ausdehnungslehre*, however, resembled the *Barycentrische Calcul* of Möbius, both in its great originality of conception and a forbiddingly novel terminology. Grassmann's ideas, too, were slow to receive recognition. Cayley, on the other hand, in 1843 approached the question of higher dimensionality from the point of view of algebraic geometry.[80] Possessed of a strong aesthetic feeling with regard to mathematics, Cayley took pleasure in solving in new and ever more elegant ways the elementary problems connected with points, lines, and planes; and determinants afforded him an excellent means of extending the symmetric formulas of Lagrange. His "Chapters in the analytical geometry of $(n)$ dimensions" opens with the statement, "I take for granted all the ordinary formulas relating to determinants." Applying the square array of Cauchy (enclosed within double vertical bars) to the symmetric results of Monge, he wrote the area of a triangle and the two-point equation of the straight line in the forms, now usually given in textbooks, of determinants of order three. In the same way the volume of a tetrahedron and the equation of the plane through three points are written as determinants of order four. In an exactly analogous manner one can extend this work to $n$ dimensions by means

[79] Ludwig Schläfli (1814–1895) also seems to have developed the idea independently, but his work was almost ten years later than that of the others. See H. S. M. Coxeter, *Regular Polytopes* (London, 1948), p. 141. Multi-dimensional algebraic geometry was also developed in 1854 in the well-known *Habilitationschrift* of G. F. B. Riemann (1826–1866). Schläfli contributed also to the classification of cubic surfaces, e. g., in *Philosophical Transactions*, CLIII (1863), p. 193–241.

[80] See *Cambridge Mathematics Journal*, IV (1843–1845), p. 119–127. The articles of Cayley are conveniently referred to also in his *Collected Mathematical Papers* (Cambridge, 1889–1897). See v. I, p. 55–62. Cf. also VI, 456.

of determinants of order $n + 1$.   Cayley put in determinant form also
Plücker's final idea before his desertion, the four coordinates of a line
in three-space: If ($\alpha$, $\beta$, $\gamma$, $\delta$) and ($\alpha'$, $\beta'$, $\gamma'$, $\delta'$) are the homogeneous
point-coordinates of two points determining the line, Cayley expressed
the coordinates of the line through the points as the determinants of
the matrix

$$\begin{vmatrix} \alpha & \beta & \gamma & \delta \\ \alpha' & \beta' & \gamma' & \delta' \end{vmatrix}$$

Cayley has been compared to Euler and Cauchy in range and
analytical power, and for the prolific production of new views and
theories.   There is "hardly a subject in the whole of pure mathematics
at which he has not worked."[81]   It will not be possible here to give a
systematic account of these contributions, even of those limited to
analytic geometry.   One of his discoveries, however, is so striking as
to deserve mention here:  Cayley noted in 1849 that whereas quadrics
contain either no straight lines or an infinite number, there is a definite finite number of lines upon a cubic surface.   George Salmon
(1819–1904), a man who was deeply influenced by the work of Cayley,
later determined that there were just twenty-seven lines.[82]   These
are not necessarily all real; but since the time of Plücker imaginary
elements had become an integral part of algebraic geometry.   Cayley
wrote, "The notion, which is really the fundamental one (and I
cannot too strongly emphasize the assertion) underlying and pervading
the whole of modern analysis and geometry, is that of imaginary
magnitude in analysis and of imaginary space in geometry."[83]   Trilinear coordinates, too, enjoyed a great popularity in England at the
time, and this is reflected in Cayley's attitude toward elements at
infinity.   Writing for the ninth edition of the *Encyclopaedia Britannica*,
he said (in the article on "Geometry") that, "The whole tendency
[in modern methods] is toward generalization...   The treatment of
the infinite is in fact another fundamental difference between the two
methods.   Euclid avoids it, in modern mathematics it is systematically
introduced, for only thus is generality obtained."   Cayley's article
on "Curve" (in the eleventh edition) contains similar views, as well
as an extensive historical account of the development of the principle
of duality and the "Plückerian dual generation of a curve."   In the
latter article the Plücker equations are treated at great length.   So

[81] See A. R. Forsyth, "Obituary notices" in *Proceedings of the London Royal Society*,
LVIII (1895), i–xliii, esp. p. xxi.   Cf. also Ch. Hermite, *Comptes rendus*, CXX (1895), p.
234.
[82] See Archibald Henderson, *The Twenty-Seven Lines Upon the Cubic Surface* (Cambridge, 1911).
[83] *Collected Mathematical Papers*, XI, p. 434.

fascinated was Cayley by this subject that he attempted to extend them to surfaces and skew curves, as well as to the higher singularities of plane curves.

Simultaneously with Cayley in England, the German mathematician L. O. Hesse (1811–1874) also used determinants effectively in geometry —as well as in analysis, where his name is celebrated in the well-known Hessians. Plücker through abridged notation had avoided algebraic elimination; but Hesse through determinants showed how to make elimination simpler. Except for the use of double indices, adopted from Cauchy and Jacobi, the work of Hesse resembles that of Lagrange in its emphasis on elegant symmetries of calculation. In Crelle's *Journal* for 1848 he adopted the now familiar form $(x_1, x_2, x_3)$ for homogeneous coordinates in a plane and the convenient double index notation for the coefficients in the general equation of second degree, notations which lend themselves readily to work with determinants. Hesse also was largely responsible for the adoption of determinants in textbooks. His two popular texts, *Vorlesungen über die analytische Geometrie des Raumes* of 1861, and *Vorlesungen aus der analytische Geometrie der geraden Linien, des Punktes und des Kreises* of 1865, except for the wide use of determinants, might almost be said to have done for analytic geometry in the sense of Plücker what the texts of Lacroix and Biot had accomplished for analytic geometry in the sense of Monge. Here one finds abridged notation, homogeneous coordinates, quadric surfaces in point and plane coordinates, polar theory —all in best modern form. Numerous textbooks of similar character appeared also in other countries, especially England. Salmon, for example, published *A Treatise on the Conic Sections* (1848), *Higher Plane Curves* (1852), and *A Treatise on Analytical Geometry of Three Dimensions* (1862), all of which have passed through numerous editions and are still widely used. In the latter work Salmon called attention to an aspect of analytic geometry which had developed inconspicuously over a long period—the idea of spherical coordinates.

In a broad sense the analytic geometry of the sphere goes back to the geography of Hipparchus and the Greek theory of spherics; but the first methodical treatment was that of Christof Gudermann (1798–1882) in 1830. In this year Gudermann published a paper, "Ueber die analytische Sphärik," in Crelle's *Journal*, and a book, *Grundriss der analytischen Sphärik*, both dealing with coordinates on a sphere.[84] He took as a frame of reference two quadrants $VX$ and $VY$ which cut each other at $V$ in any angle (corresponding to oblique coordinates in a

---

[84] See Loria, "Perfectionnements . . .," *Mathematica* XXI (1945), p. 66–83, whose account I here follow.

plane). To find the coordinate of a point $M$ of the sphere, he joined $M$ to $X$ and $Y$ by arcs of great circles and found the intersections $P$ and $Q$ of these arcs with $VX$ and $VY$. Then the "axial coordinates" $x = VP$, $y = VQ$ are uniquely determined, except for points on the arc $XY$. Gudermann suggested also a system of "central coordinates" on the sphere—analogous to polar coordinates in the plane—in which the position of $M$ is determined by the great-circle arc $VM$ and angle $MVX$. He solved the fundamental problems relating to points and great circles, and even considered some simple curves drawn on the sphere. Evidently inspired by the work of Plücker, he later generalized his scheme (in Crelle's *Journal* for 1838) by defining trilinear coordinates, using numbers proportional to the sines of the arcs drawn through a point perpendicular to the sides of a spherical triangle of reference. Möbius in 1846 also published a paper on spherical analytic geometry in which he applied his barycentric calculus to the surface of a sphere.

In Great Britain, T. S. Davies (1794–1851) in 1833 also presented a thorough treatment of spherical coordinates, especially in polar form.[85] Davies traces attempts to give spherical coordinates back to one by James Skene in Aberdeen in the *Gentlemen's Diary* for 1795; but he makes no mention of Gudermann and recognizes no *general* development before his own. Using $\theta$ as polar angle and $\phi$ as radius vector (or polar distance), Davies gave the equation of the circle with center $(\lambda, \kappa)$ and radius $\rho$ as $\cos \rho = \cos \lambda \cos \phi + \sin \lambda \sin \phi \cos(\theta - \kappa)$. If the circle is a great circle, the equation becomes $\cot \phi = -\tan \lambda \cos (\theta - \kappa)$; and if the center is the pole of the equator, the equation is $\cos \rho = \cos \phi$. Similarly he found the great circle through two points; the intersections of two great circles, the angle between them; the great circle through a point and making a given angle with a given great circle; and the circle through three points. He gave formulas for the transformation of coordinates, and for various projections: orthographic, stereographic, and gnomonic; and he studied a variety of curves, including the spherical ellipse, hyperbola, and parabola, spherical epicycloids, and various spirals, especially the loxodrome. In the next decade C. Graves (1812–1899) presented an extensive treatment of rectangular spherical coordinates.[86] He gave such formulas as distance between two points; the normal distance from a point to a great circle; the angle between two great circles; the equation of the great circle through a given point perpendicular to a given great

[85] "On the equations of loci traced upon the surface of the sphere, as expressed by spherical coordinates," *Transactions of the Royal Society of Edinburgh*, XII (1833–1834), 259–362, 379–428.
[86] I have not seen this work but cite it on the basis of Loria, "Perfectionnements .. "

circle; and the equation of the great circle tangent to a given spherical curve. He included transformations of rectangular spherical coordinates and also transformations from polar to rectangular coordinates on the sphere. Salmon noticed that in Graves's rectangular system on a unit sphere, the projection upon the plane tangent to the sphere at the origin leads to a corresponding system of Cartesian coordinates. Thus the Gudermannian equation of a spherical curve corresponds to the Cartesian equation of its central projection on the tangent plane at the origin. Spherical and Cartesian coordinates can also be related through stereographic projection. Cayley, too, studied spherical analytic geometry, using a system of tripolar coordinates; and long after (1895) he considered the nine-point circle of a spherical triangle.[87]

Spherical coordinates are a special case of analytic geometry on surfaces in general, but the development of the latter was quite different. Euler had represented plane curves parametrically, and the extension to space is so immediate that it is difficult to ascribe it to any one individual. Lagrange's transformation from spherical or polar coordinates $(p, q, r)$ to rectangular coordinates $(x, y, z)$ becomes, for $r$ constant, a parametric representation of the spherical surface. In this sense the parameters $p$ and $q$ may be called the spherical coordinates of a point on the sphere. It is this approach which was generalized by Carl Friedrich Gauss (1777–1855), the greatest mathematician of modern times, to lead to general curvilinear coordinates. In his classic work of 1827, *Disquisitiones generales circa superficies curvas*, he used three differential equations in two parameters to define a surface, and he referred to the parameters $p$ and $q$ which determine two systems of geodesic lines on the surface as "curvilinear coordinates." This work, however, was a part of differential, rather than algebraic, geometry.

Gauss used the name "curvilinear coordinates" to designate a system on an arbitrarily given surface; but the phrase can be used as well to indicate various systems of coordinates either in a plane or in space of three dimensions. Two equations $x = f(p, q)$ and $y = g(p, q)$ suffice to set up a correspondence of values between $p$ and $q$ on the one hand and $x$ and $y$ on the other; so that $p$ and $q$ may be thought of as curvilinear coordinates of a point $(x, y)$ in the plane. Polar coordinates are a special case in which $f$ is $p \cos q$ and $g$ is $p \sin q$. In 1874 C. A. Laisant (1841–1920) gave another example, $x = r \cosh w$, $y = r \sinh w$—a system which he called hyperbolic polar coordinates.[88] Similarly, in space of three dimensions, three equations $x = f(u, v, w)$, $y = g(u,$

[87] *Collected Mathematical Papers*, XIII, p. 548.
[88] *Essai sur les fonctions hyperboliques* (Paris, 1874), p. 71–83.

$v, w)$, $z = h(u, v, w)$ set up a correspondence between the "curvilinear" coordinates $(u, v, w)$ and the Cartesian coordinates $(x, y, z)$. Lagrange's equations furnish but a special instance of this, and the types of curvilinear coordinates are limitless in number. Possibly the most important single form of curvilinear coordinates is that proposed by Lamé in 1837 and named by him "elliptic coordinates."[89] For space of three dimensions these are defined as follows: The equation $\dfrac{x^2}{a^2 + \lambda}$

$+ \dfrac{y^2}{b^2 + \lambda} + \dfrac{z^2}{c^2 + \lambda} = 1$ represents an ellipsoid, an hyperboloid of one nappe, or an hyperboloid of two nappes according as $\lambda$ is a value chosen between $-a$ and $-b$, or between $-b$ and $-c$, or between $-c$ and $+\infty$. If one chooses three values of $\lambda$, one in each of the above intervals, these three numbers determine three quadrics which intersect each other (orthogonally) in eight points, one in each octant and symmetric in pairs with respect to the principal planes of the system. These numbers, $\lambda_1$, $\lambda_2$, $\lambda_3$, may therefore be regarded as coordinates of the points so determined. Similarly, with respect to the equation $\dfrac{y^2}{p + \lambda} + \dfrac{z^2}{q + \lambda} = 2x + \lambda$, one can choose three values of $\lambda$, one in each of the intervals $-\infty$ to $-p$, $-p$ to $-q$, and $-q$ to $+\infty$, to obtain three paraboloids which intersect in four points; and the three values $\lambda_1$, $\lambda_2$, $\lambda_3$ are known as parabolic coordinates of the points. Similar systems of curvilinear coordinates can be applied in two dimensions through the use of confocal conics in place of the quadrics. In such a system the idea of coordinate is divorced from geometric significance as thoroughly as in the work of Möbius and Plücker, where also coordinates were mere numbers. Analytic geometry more and more was being arithmetized.

Lamé was a civil and railroad engineer who had been led to curvilinear coordinates through his work on the conduction of heat in ellipsoids; and in 1859 he published a volume on the role such coordinates can play in mechanics, heat, and electricity. His forecast (in *Leçons sur les coordonnées curvilignes et leurs applications*) of their significance is worth quoting, for it is an inspiring tribute to those men who, through their analytic geometry, have participated in the march of science:

Should anyone find it singular that we have been able to found a Course of Mathematics on the sole concept of a system of coordinates, he may be

[89] "Sur les surfaces isothermes," *Journal des mathématiques*, II (1837), p. 156. Cf. also IV (1839), p. 134; VIII (1843), p. 397.

reminded that it is precisely these systems which characterize the phases and stages of science. Without the invention of rectangular coordinates, algebra might still be where Diophantus and his commentators left it, and we should lack both the infinitesimal calculus and analytic mechanics. Without the introduction of spherical coordinates, celestial mechanics would be absolutely impossible; and without elliptic coordinates, illustrious mathematicians would have been unable to solve several important problems of this theory . . . Subsequently the reign of general curvilinear coordinates supervened, and they alone are capable of attacking the new problems [of mathematical physics] in all their generality. Yes, this definitive epoch will arrive, but tardily: those who first recognized these new implements will have ceased to exist and will be completely forgotten—unless some archaeological mathematician revives their names. Well, what of it, provided science has advanced?[90]

At the time that Lamé penned these words in praise of coordinate geometry, the foremost champion of the subject was sulking in his tent. It is true that Plücker was contributing to science through his work in physics; but one wonders whether he did not now and then cast a fond eye in the direction of the field of his earlier triumphs. Was he roused once more to activity by the challenge of Lamé? Or was he encouraged to return to his first love by the ardor with which Cayley pursued his ideas? Or was he perhaps induced by the death (in 1863) of his arch opponent, Steiner, to give up his self-imposed role of mathematical exile? Whatever the cause may have been, Plücker in 1865 turned back to the work he had broken off so abruptly in 1846.

During the period of Plücker's geometrical retirement, the notion of space of more than three dimensions had been developed (especially by Cayley) from a formal point of view, with little regard for geometric interpretation.[91] In 1846 Plücker himself had hinted at such a purely algebraic generalization. In 1865, however, he returned to his faith that analytic operations and geometric constructions run parallel to each other, and he shattered the naive notion that it is impossible to imagine a space of more than three dimensions. When considering the stuff of which a space is composed, one is inclined to think first of points. In spite of duality, it seems to be more natural to look upon a curve as a locus of points rather than as an envelope of tangent lines—even though the curve may be a caustic, actually generated by rays of light instead of by a moving point. In space of two dimensions this does

[90] Quoted from Bell, *Development of Mathematics*, p. 487.
[91] In a provocative little book, *Art and Geometry* (Cambridge, Mass., 1946), p. 121, Wm. M. Ivins has written, "Cayley and Grassmann invented conceptual spaces of more than three dimensions"; but this is misleading. Even as late as 1883 Cayley insisted that multi-dimensional geometry is a part of pure mathematics only and not of the realm of conception. See Forsyth, *op. cit.*, p. xxxii.

not much matter.   There are as many points as there are lines.   For ordinary space of three dimensions, however, the situation is quite different.   A point in this case has three independent coordinates, and there are as many planes as points; but a line is determined by *four* conditions.   Before his retirement Plücker had referred to the four parameters of a line as "coordinates," but not until 1865 did he seize upon this idea as the basis for a "new geometry of space."   He developed the idea that space need not be thought of as a totality of infinitely many points; it can equally well be visualized as composed of infinitely many straight lines.   As Bell has vividly expressed the change, "Instead of our familiar solid space looking like an agglomeration of infinitely fine birdshot it now resembles a cosmic haystack of infinitely thin, infinitely long straight straws."[92]   That is, the dimensionality of a space depends upon the type of element out of which, in our mind's eye, it is visualized as built.   Any figure which formerly had been regarded as a *locus* of points can instead be taken as a space *element*, and the number of dimensions is indicated by the number of parameters determining such a locus.

The cordial relations between Plücker and the British geometers is evidenced both by the fact that his new geometry of space first appeared in the publications of the Royal Society of London,[93] and by the response in England to his ideas.   Cayley, for example, in 1868 developed analytically the notion of a plane as a space of five dimensions, the elements of which are conics.[94]   In this same year appeared also the first volume of Plücker's last work, the *Neue Geometrie des Raumes gegründet auf die Betrachtung der geraden Linie als Raumelement*. In this he built up his line geometry as one ordinarily does point geometry.   A single equation in the coordinates of ordinary point-space is called a surface; one equation in the four coordinates of his line-space Plücker called a "complex."   Two equations in ordinary space determine a curve; and in his new space Plücker called the locus corresponding to two equations a "congruence."   Three equations in point-geometry lead to a single element of the space, a point; but in line-geometry there is still another intermediate configuration possible, a one-parameter family of lines being known as a "range."   Plücker set himself the task of building up the properties of the new space in a manner corresponding to his treatment of ordinary space.   The quadratic line complex, for example, has properties resembling those

---

[92] *Men of Mathematics*, p. 400.

[93] *Proceedings*, XIV (1865), p. 53–58; and *Philosophical Transactions*, CLV (1865), p. 725–791.   Or see *Wissenschaftliche Abhandlungen*, I, p. 462–545.

[94] "On the curves which satisfy given conditions," *Philosophical Transactions*, CLVIII (1868), p. 75–143; or see *Collected Mathematical Papers*, VI, p. 191–291.

of the quadric surface, and he undertook to study this in detail.   He did not live to complete this work, but he had gone over the ground with his students, especially Felix Klein (1849–1925) and R. F. A. Clebsch (1833–1872), who saw to the final publication.

The death of Plücker did not bring an end to the development of analytic geometry; for, like many great men, he had enthusiastic disciples.   Plücker had begun his study under the students of Monge who, upon the death of the great French geometer in 1818, had continued to develop the subject along lines their teacher had suggested. Half a century later, in 1868, Plücker in turn passed away, leaving analytic geometry again transformed; and succeeding generations of students have added enormously to the growth of the subject. It has been estimated[95] that from 1870 to 1890 the rate of development in geometry was doubled, with analytic methods predominating. Nor is there any likelihood that an end has yet been reached.   As Wieleitner has conservatively estimated, perhaps in one or two hundred years from now, analytic geometry will be as different from ours as is ours from that of Descartes and Fermat.[96]   However, specialization from Plücker onward has increased to such an extent that a general account of the history of elementary analytic geometry may reasonably stop short of anything since his death.   While "history shows that the general tendency [in geometry as a whole, and in analytic geometry as well] has been more and ever more generalization,"[97] the contributions of individuals have tended in the opposite direction. To follow the work of even a few of Plücker's students—such as Klein and Sophus Lie (1842–1899) in the theory of groups and invariance or Clebsch in the invariants of algebraic forms—would lead to fields far beyond the scope of a survey.[98]   But it may not be inappropriate to mention here one of the implications of analytic geometry with respect to the foundations of mathematics.   The train of thought leading to analytic geometry arose when Greek mathematicians insisted on seeking for geometrical, rather than arithmetical, solutions of equations such as $x^3 = 2$.   In the work of Monge and Plücker, however, coordinate geometry was emancipated from the constructions of pure geometry—it was arithmetized.   Auguste Comte (1798–1857) was much impressed by this arithmetizing tendency and its bearing upon positivistic philosophy.   He went so far as to place analysis alone in

[95] See Cajori, *History of Mathematics*, p. 278–279.
[96] "Zur Erfindung . . .," p. 426.
[97] Coolidge, *History of Geometric Methods*, p. 422–423.   Cf. Loria, "Perfectionnements . .," *Mathematica*, XXI (1945), p. 63–83.
[98] The reader who wishes to pursue further such lines of development should consult the admirable works of Bell (especially his *Development of Mathematics*) and Coolidge (in particular the *History of Geometrical Methods*).

the category of abstract mathematics, pure geometry being put (along with mechanics), under the heading of concrete mathematics; and he called analytic geometry the most decisive step in mathematical education.[99] Yet although Comte emphasized the parallels between analytic expressions and geometric constructions, he gave no serious consideration to the fundamental assumption upon which such an association must be based—the postulate, tacitly accepted since the days of Descartes, that to each point of a line there corresponds a real number, and conversely.[100] This problem had been considered by Riemann in 1854, but it was especially in 1872 that the Cantor-Dedekind axiom placed the arithmetization of analytic geometry upon a sound logical foundation. The work of Richard Dedekind (1831–1916) and Georg Cantor (1845–1918), however, was essentially a part of the development of the calculus,[101] which continued, in many respects, to overshadow coordinate geometry. Even in the case of the theory of curves many of the later contributions were also far removed from algebraic geometry in the strict sense. Such pathological examples as space-filling curves, and curves everywhere continuous but without a tangent anywhere, brought much of the subject under the more general heading of the theory of functions. Except in the case of elementary mathematics, it has become increasingly difficult to distinguish clearly between the various fields, such as algebra and geometry. The association of number and magnitude, out of which analytic geometry arose, is now on a sounder basis than ever before. To trace the history of the subject beyond 1872 would carry one far from the field of elementary Cartesian geometry, and hence this year, only four years after the death of Plücker, may be taken as a sort of terminal date; but in closing it is appropriate to add a word on the historians of analytic geometry.

Chemistry has been claimed, with excessive arrogation, as a French science, largely because of the "chemical revolution" of Lavoisier. If there is indeed some basis for this Gallic claim, there is far more obvious justification for an assertion that analytic geometry is a French contribution to mathematics. The closest medieval and early modern anticipations were due to two Frenchmen, Oresme and Viète; the inventors, Descartes and Fermat, were both French; and French

[99] See *The Philosophy of Mathematics* (transl. by W. M. Gillespie, New York, 1851), p. 202–203, and *Traité élémentaire de géométrie analytique à deux et à trois dimensions* (Paris, 1843), p. 9.
[100] See, e. g., Tobias Dantzig, *Number, the Language of Science* (3rd ed., New York, 1939), p. 178.
[101] See C. B. Boyer, *The Concepts of the Calculus* (New York, 1939), p. 285–290. The work of Cantor and Dedekind was paralleled at about the same time by that of Méray and Weierstrass, providing another striking instance of the simultaneity of discoveries.

also were the central figures who participated in the "analytical revolution" of Monge. Although the foremost specialist in the subject was German, nevertheless Plücker was introduced to the subject by French teachers and textbooks. Yet nothing illustrates the international character of mathematics so well as the fact that, although the subject is largely French in origin, the historiography of it is due mostly to other lands. Brief summaries are found in nearly every language, but the most extensive general accounts of the development of Cartesian geometry are by an Italian, two German scholars, and an American. Much of the inspiration and the material for this volume has been drawn from the works (cited in the bibliography) of Loria, Tropfke, Wieleitner, and Coolidge; and so in conclusion the author wishes to refer the reader to these sources and to express his admiration for the work of these and other men who have enriched not only mathematics itself but also the story of its development.

# ANALYTICAL BIBLIOGRAPHY

The works listed here are selected for their importance or their relevance. For purposes of convenience the collected works of an author, where available, are cited in preference to scattered papers. Further bibliographic references will be found in the footnotes of the text above.

## I. Some Important Primary Sources Illustrating the History of Analytic Geometry (arranged roughly in chronological order)

[Neugebauer, O., and A. Sachs], *Mathematical cuneform texts* (American Oriental Series, vol. XXIX). New Haven, Conn., 1945.
  Illustrates the earliest applications of algebra to geometry.
[Chase, A. B., L. S. Bull, H. P. Manning, and R. C. Archibald], *The Rhind mathematical papyrus.* 2 vols., Oberlin, 1927–1929.
  Illustrates the application of number to geometry in Egypt.
[Heath, T. L.], Apollonius of Perga, *Treatise on conic sections.* Cambridge, 1896.
  This excellent English edition of the *Conics* includes an introduction which presents an extensive history of the conic sections before Apollonius.
[Heath, T. L.], *The works of Archimedes.* Cambridge, 1897.
  Important for the knowledge of the *Conics* in antiquity, including Archimedes' graphical solution of cubic equations.
[Thomas, Ivor], *Selections illustrating the history of Greek mathematics.* 2 vols., Cambridge, Mass., 1939–1941.
  A convenient source book of fragments pertaining to the Greek equivalent of analytic geometry.
[Ver Eecke, Paul], Pappus of Alexandria, *La collection mathématique.* 2 vols. Paris and Bruges, 1933.
  A readily available edition of an important source of inspiration in the development of analytic geometry.
[Wieleitner, Heinrich], "Der 'Tractatus de latitudinibus formarum' des Oresme," *Bibliotheca Mathematica* (3), XIII (1913), 115–145.
  Includes valuable commentary and analysis of the work of the most important medieval precursor of analytic geometry. See also the article on Oresme by Wieleitner listed in part II of this bibliography.
Viète, François, *Opera mathematica* (ed. by van Schooten, Lugduni batavorum, 1646).
  Contains applications of algebra to geometry by the most important early modern precursor of Descartes and Fermat. For French translations of parts of his work see *Bullettino di Bibliografia e di Storia delle Scienze Matematiche e Fisiche*, I (1868), 223–276.
Fermat, Pierre de, *Oeuvres.* Ed. by Paul Tannery and Charles Henry, 4 vols. and supp., Paris, 1891–1922.
  Contains the Latin, and also French translation, of the *Introduction to loci*, as well as of other works bearing upon analytic geometry. The Latin of the *Introduction to loci* is found also in the *Varia opera mathematica* of Fermat (Tolosae, 1679).
Descartes, René, *The geometry.* Transl. by D. E. Smith and Marcia L. Latham, with a facsimile of the first edition, 1637. Chicago and London, 1925.
  A convenient edition and translation, with notes. For other aspects of the geometrical work of Descartes, see his *Oeuvres* (ed. by Charles Adam and Paul Tannery, 12 vols. and supp., Paris, 1897–1913).

Roberval, G. P. de, "Divers ouvrages," *Mémoires de l'Académie Royale des Sciences depuis 1666 jusqu'à 1699*, VI (Paris, 1730), 1–478.
> Pages 94–246 contain work on the derivation of the equations of loci and on the graphical solution of cubics and quartics.

Van Schooten, Frans, *Geometria a Renato Des Cartes*. 2nd ed., 2 vols., Amstelaedami, 1659–1661.
> One of the most important works in the history of analytic geometry. Besides very extensive commentary on Cartesian geometry by Van Schooten, the 2nd edition contained the *Elements of curved lines* of Jan de Witt, sometimes called the first textbook of analytic geometry. Van Schooten's work included also significant additions to analytic geometry by Debeaune and others. A third edition appeared in 1683. The first edition of Leyden, 1649, did not contain the work of de Witt.

Sluse, René de, *Mesolabum*. 2nd ed., Leodii Eburonum, 1668.
> This "book of means" was an important link in the Cartesian graphical solution of equations. Cubics and quartics are solved by intersecting conic sections. The first edition appeared in 1659.

Wallis, John, *Opera*. 3 vols., Oxonii, 1693–1699.
> This is the best edition of Wallis' works. The important *Treatise on conic sections* is also found in the more readily accessible *Opera mathematica* (2 vols., Oxonii, 1656–1657).

Huygens, Christiaan, *Ouevres complètes*. 22 vols., La Haye, 1888–1950.
> Important for correspondence with mathematicians of the time, especially Sluse. Huygens was one of the early Continental writers to understand negative coordinates.

Lahire, Philippe de, *Nouveaux élémens des sections coniques, les lieux géo métriques, la construction ou effection des équations*. Paris, 1679.
> The most important analytic work of a man otherwise known as a great synthetic geometer. It is in the strict Cartesian tradition and appeared again in 1701.

Ozanam, Jacques, *Traité des lignes du premier genre; traité des lieux géométriques; traité de la construction des équations*. Paris, 1687.
> This rather uninspired work is in harmony with the ideas of Descartes and closely resembles the work of Lahire.

Craig, John, *Tractatus mathematicus de figurarum curvilinearum quadraturis et locis geometricis*. Londini, 1693.
> Contains the important *Nova methodus determinandi loca geometrica* with the equivalent of the modern characteristic for determining the nature of a conic section.

Leibniz, G. W., *Mathematische Schriften*. Ed. by C. I. Gerhardt. *Gesammelte Werke*. Ed. by G. H. Pertz. Third series, *Mathematik*, 7 vols., Halle, 1849–1863.
> Especially useful for Leibniz' correspondence with the Bernoulli brothers.

Bernoulli, Jacques, *Opera*. 2 vols., Genevae, 1744.
> Important for the graphical solution of equations and for the use of polar coordinates.

Bernoulli, Jean, *Opera omnia*. 4 vols., Lausannae and Genevae, 1742.
> Contains anticipations of solid analytic geometry.

Guisnée, N., *Application de l'algebre a la geometrie*. Paris, 1705.
> A popular analytic geometry of the first half of the eighteenth century.

Newton, Sir Isaac, *Opera quae exstant omnia*. Ed. by Samuel Horsley, 5 vols., Londini, 1779–1785.
> This edition, and the *Opuscula mathematica, philosophica et philologica* (3 vols., Lausannae and Genevae) are convenient for those who read Latin. Many of the treatises of importance in the history of analytic geometry are available also in English: *The method of fluxions* (London, 1736), for polar coordinates; *Universal arithmetick* (London, 1769 etc.), for the graphical solution of equations; *Enumeration of lines of the 3rd order* (London, 1760), on graphical representation. An extensive account of the last-mentioned is found also in W. W. R. Ball's article, "On Newton's classification of cubic curves," *London Mathematical Society, Proceedings*, XXII (1890), 104–143.

L'Hospital, G. F. A. de, *Traité analytique des sections coniques*. Paris, 1707.

The most popular text on analytic geometry of the eighteenth century, this work appeared in numerous editions.

Varignon, Pierre, "Nouvelle formation de spirales," *Mémoires de l'Académie des Sciences*, 1704, pp. 69–131.

An early instance of the use of polar coordinates.

Rolle, Michel, "De l'evanoüissement des quantitez inconnuës dans la géométrie analytique," *Mémoires de l'Académie des Sciences*, 1709, pp. 419–450.

One of the first uses in print of the name analytic geometry. The article is on the graphical solution of equations.

Reyneau, Ch. R., *Analyse démontrée.* 2 vols., Paris, 1708.

A work resembling that of L'Hospital, but not so well known.

Parent, Antoine, *Essais et recherches de mathématique et de physique.* 2nd ed., 3 vols., Paris, 1713.

One of the first instances of the systematic use of solid analytic geometry. This work first appeared in 1705.

Maclaurin, Colin, *A treatise of algebra.* London, 1748.

Important for the "use of algebra in the resolution of geometrical problems." This posthumous work was planned as early as 1729. For a twenty-page account of the life and work of Maclaurin see his *Account of Sir Isaac Newton's philosophical discoveries* (London, 1748).

Stirling, James, *Lineae tertii ordinis Neutonianae.* Londini, 1717.

Stirling added so much new material to the little work of Newton that this is virtually a new book.

Rabuel, Claude, *Commentaires sur la géométrie de M. Descartes.* Paris, 1730.

A prolix traditional treatment.

Clairaut, A. C., *Recherches sur les courbes a double courbure.* Paris, 1731.

A classic, composed when the author was only sixteen years old. It is the first book devoted entirely to solid analytic geometry.

Hermann, Jacob, "De superficiebus ad aequationes locales revocatis," *Commentarii Academiae Petropolitanae*, VI (1732–1733), 36–67.

An important contribution to the early history of solid analytic geometry, composed about the same time as the first works on the subject by Clairaut and Euler. On plane analytic geometry see an article by Hermann, "De locis solidis ad mentem Cartesii concinne construendis," in the same journal, IV (1729), 15–25.

Wolff, Christian, *A treatise of algebra; with the application of it to a variety of problems in arithmetic, to geometry, trigonometry, and conic sections.* Transl. from the Latin, London, 1739.

An excellent example of the nature of Cartesian geometry at that time.

De Gua de Malves, J. P., *Usages de l'analyse de Descartes.* Paris, 1740.

One of the most important works on higher plane curves, following the work of Newton and Stirling.

Caraccioli, J. B., *De lineis curvis.* Pisis, 1740.

One of the best treatises of its day on higher plane curves, both algebraic and transcendental. Treatment is in part analytic, in part synthetic.

Chelucci, Paolino, *Institutiones analyticae earumque usus in geometria.* Romae, 1738.

Shows how slowly analytic geometry developed in Italy at first. The geometric uses consist of the graphical solutions of cubic equations by means of conics.

Agnesi, Maria Gaetana, *Instituzioni analitiche.* Milano, 1748.

An analytic geometry quite characteristic of its time, and well known on the Continent. An English edition appeared at London in 1801

Euler, Leonard, *Opera omnia.* Ed. by Ferdinand Rudio, 22 vols. in 23, Lipsiae and Berolini, 1911–1936.

The contributions of Euler to analytic geometry covered close to half a century. They began in the *Commentarii Academiae Petropolitanae* for 1728 with a paper on solid analytic geometry; and in the *Novi Commentarii* for 1775–1776 there is an article by Euler on the transformation of coordinates in three dimensions. The

272 BIBLIOGRAPHY

most important single work of his on analytic geometry is the *Introductio in analysin infinitorum* (2 vols., Lausannae, 1748), of which translations are available in French and German. For full bibliographic information on Euler see Gustaf Eneström, "Verzeichnis der Schriften Leonhard Eulers," *Jahresbericht der Deutschen Mathematiker-Vereinigung, Ergänzungsbände*, IV, 2 parts, 1910–1913, an impressive list including 866 entries, not counting multiple editions.

Cramer, Gabriel, *Introduction a l'analyse des lignes courbes algébriques*. Geneve, 1750.
    This large volume of 680 pages was definitive in its field for over half a century.

La Chapelle, L'Abbé, *Traité des sections coniques et autres courbes anciennes*. Paris, 1750.
    This book is Cartesian in spirit, but not at all like modern works on analytic geometry. Equations of curves are seldom given.

Gallimard, J. E., *Les sections coniques et autres courbes anciennes*. Paris, 1752.
    Like the work of La Chapelle, this book is based upon the geometry of Descartes, but does not resemble modern texts. The language of proportions is used instead of equations.

Goudin, M. B., and A. P. Dionis du Sejour, *Traité des courbes algébriques*. Paris, 1756.
    Important for the emphasis upon the transformation of coordinates and for the analytic treatment of the straight line. See also Goudin's *Traité des propriétés communes à toutes les courbes* (Paris, 1778), a 3rd edition of which appeared in 1803.

Waring, Edward, *Miscellanea analytica, de aequationibus algebraicis, et curvarum proprietatibus*. Cantabrigiae, 1762.
    Probably the most important English work on analytic geometry in the second half of the eighteenth century.

Riccati, Vincenzo, and Girolamo Saladini, *Institutiones analyticae*. 2 vols. in 3, Bononiae, 1765–1767.
    Includes a good account of analytic geometry characteristic of the period.

Sauri, L'Abbé, *Cours complet de mathématiques*, 5 vols., Paris, 1774.
    Includes a treatment of analytic geometry of the time.

Frisi, Paolo, *Operum*. 3 vols., Mediolani, 1782–1785.
    Title page of vol. I carries the modern name "Analytic geometry."

Bézout, Etienne, *Cours de mathématiques*. Part III, with notes by A. A. L. Reynaud, Paris, 1812.
    The compendia by Bézout enjoyed quite a vogue during the last quarter of the eighteenth century and influenced the American textbooks of the early nineteenth. The treatment of analytic geometry is typical of the time about 1775.

Lagrange, J. L., *Oeuvres*. 14 vols., Paris, 1867–1892.
    The contributions to analytic geometry are found especially in vol. III, pp. 617–692. Important for arithmetizing coordinate geometry. Not a single diagram is used in his classic treatment of the tetrahedron.

Monge, Gaspard, *Feuilles d'analyse*. Paris, 1795.
    Monge is the most important contributor to analytic geometry whose works have not been collected. His *Feuilles d'analyse*, which appeared in five editions to 1850, is the most important single contribution he made, but by no means the only one. Like Euler, he added to the subject over a long period of time. His classic paper on developable surfaces was delivered in 1771 but went through several editions, *Application de l'algèbre a la géométrie; traité des surfaces du second degré* (3rd ed., Paris, 1813).

Lacroix, S. F., *Cours de mathématiques*. Vol. IV, *Traité élémentaire de trigonométrie rectiligne et sphérique et application de l'algèbre à la géométrie*, Paris, 1798–1799.
    The first really modern textbook treatment of plane analytic geometry. This book

set the pattern for a host of elementary textbooks which appeared in the following century. Much of the material in this book had appeared about a year before in the same author's *Traité du calcul* (vol. I, Paris, 1797). The 25th edition of the textbook on trigonometry and analytic geometry appeared in 1897! Lacroix undoubtedly was the greatest textbook writer of modern times, if one allows for multiple editions. See also his *Essais sur l'enseignement en général, et sur celui des mathématiques en particulier* (Paris, 1805).

Lefrançais, F. L. [or Français], *Essai sur la ligne droite et les courbes du second degré*. Paris, 1801.
    A good textbook along the lines set by Lacroix. A 2nd edition appeared in 1804 with the title *Essais de géométrie analytique*.

Puissant, Louis, *Recueil de diverses propositions de géométrie, résolues ou démontrées par l'analyse algébrique*. Paris, 1801.
    An excellent textbook following the principles of Monge and Lacroix. Much-enlarged editions appeared in 1809 and 1824.

Biot, J. B., *Essai de géométrie analytique*. Paris, 1802.
    A textbook that rivalled Lacroix's in popularity. Plane and solid analytic geometry are integrated. This book was translated into many languages, and exerted a wide influence. It was used for many years at West Point. The 6th edition of 1823 consists of about half again as many pages (some 450) as the 2nd edition of 1805.

Carnot, L. N. M., *Géométrie de position*. Paris, 1803.
    Only a portion of this important book is devoted to analytic geometry, but these pages (about 425–475) are significant for the imagination with which coordinates are used. Various systems are suggested, including polar, bipolar, and intrinsic coordinates.

Lhuilier, S. A. J., *Eléméns d'analyse géométrique et d'analyse algébrique, appliquées a la recherche des lieux geometriques*. Paris, 1809.
    Noteworthy for the systematic use of the normal form of the line and the plane.

Hachette, J. N. P., *Eléments de géométrie à trois dimensions*. Paris, 1817.
    The author is to be remembered also for his editing of the *Correspondance sur l'Ecole Impériale Polytechnique*, (3 vols., 1813–1816).

Ampère, A. M., "Sur les avantages qu'on peut retirer dans la théorie des courbes, de la considération des paraboles osculatrices," *Journal de l'Ecole Polytechnique*, cah. XIV (vol. 7, 1808), 159–181.
    One of the early suggestions for the use of intrinsic coordinates.

Garnier, J. G., *Géométrie analytique ou application de l'algèbre à la géométrie*, 2nd ed., Paris, 1813.
    A good textbook, characteristic of its day. The 1st edition appeared in 1808. The author is to be remembered also as co-editor, with Quételet, of *Correspondance Mathématique et Physique*.

Gergonne, J. D., "Essai sur l'expression analytique des courbes indépendamment de leur situation sur un plan," *Annales de Mathématiques pures et appliquées*, IV (1813–1814), 42–55.
    On intrinsic coordinates. Gergonne, founder and editor of the *Annales*, displayed almost unrivalled enthusiasm for analytic geometry, and the numbers of his journal are full of excellent articles on the subject both by himself and by others.

Gaultier, L., "Mémoire sur les moyens généraux de construire graphiquement un cercle déterminé par trois conditions, et une sphère déterminée par quatre conditions," *Journal de l'Ecole Polytechnique*, cah. XVI (vol. 9, 1813), 124–214.
    The first systematic presentation, largely synthetic, of the radical axis and the radical plane.

Bret, J. J., "Théorie analitique de la ligne droite et du plan," *Annales de Mathématiques*, V (1814–1815), 329–341.
    Early systematic use of the parametric form of the line in three dimensions.

Wood, James, *The elements of algebra*. (Part IV, *The application of algebra to geometry*, pp. 276–305.) 6th ed., Cambridge, 1815.
> A brief introduction to analytic geometry along the lines given by Descartes and Fermat. This account, one of the first in England since Wallis, shows how far behind developments on the Continent the country had fallen.

Lamé, Gabriel, *Examen des différentes méthodes emplyées pour resoudre les problèmes de géométrie*. Paris, 1818.
> Important for the use of abridged notation in connection with linear combinations of equations of curves.

Cauchy, Augustin, *Oeuvres complètes*. 25 vols., Paris, 1882–1932.
> Important for the classification of quadrics, the use of determinants, and for certain forms of the line in three dimensions.

Lardner, Dionysius, *A system of algebraic geometry*. London, 1823.
> Just about the first textbook on algebraic geometry, in the sense of Monge and Lacroix, to appear in Great Britain. Includes a fifty-page historical introduction. See also his *Treatise on algebraic geometry* (London, 1831).

Bobillier, Etienne, "Essai sur un nouveau mode de démonstration des propriétés de l'étendue," *Annales de mathématiques*, XVIII (1827–1828), 320–339, 359–367.
> One of the most original early users of abridged notation. Important also for the adumbration of homogeneous coordinates.

Poncelet, J. V., *Applications d'analyse et de géométrie, qui ont servi de principal fondement au Traité des propriétés projectives des figures*. 2 vols., Paris, 1862–1864.
> Overshadowed by the author's famous projective geometry of 1822, the present work often is overlooked by historians. Composed half a century before it was published, it shows the important part that analytic considerations played in Poncelet's early thought.

Plücker, Julius, *Gesammelte wissenschaftliche Abhandlungen*. Ed. by Arthur Schoenflies, 2 vols., Leipzig, 1895–1896.
> Vol. I, containing Plücker's *Gesammelte mathematische Abhandlungen*, is one of the most important primary sources on the history of analytic geometry. Here are collected the papers of the one who was perhaps the greatest algebraic geometer of all times. The articles, many of which appeared originally in Gergonne's *Annales* and Crelle's *Journal*, are full of important new ideas on abridged notation, homogeneous equations, line coordinates, and the singularities of algebraic curves. Plücker was also the most prolific analytic geometer of all time; besides the score of papers, totaling over 600 pages in the above volume, he published the following works, averaging over 300 pages per volume:
> *Analytisch-geometrische Entwicklungen* (2 vols. in 1, Essen, 1828–1831).
> *System der analytischen Geometrie* (Berlin, 1835).
> *Theorie der algebraischen Curven* (Bonn, 1839).
> *System der Geometrie des Raumes* (Düsseldorf, 1846).
> *Neue Geometrie des Raumes* (2 vols. in 1, Leipzig, 1868–1869).

Davies, T. S., "On the equations of loci traced upon the surface of the spheres as expressed by spherical coordinates," *Transactions of the Royal Society of Edinburgh*, XII (1833–1834), 259–362, 379–428.
> One of the most extensive developments of spherical coordinates of the time, including a wide variety of formulas and equations.

Comte, Auguste, *Traité élémentaire de géométrie analytique a deux et a trois dimensions*. Paris, 1843.
> Much more discursive than the usual textbook, but written by a philosopher who was profoundly influenced by analytic geometry. See also his *Philosophy of mathematics* (transl. by W. M. Gillespie, New York, 1851).

Cayley, Arthur, *Collected mathematical papers*. 13 vols., Cambridge, 1889–1897.
> Most of Cayley's 900-odd papers are on the algebra of invariants, but some are de-

voted to important aspects of analytic geometry, especially the theory of curves and surfaces, the geometry of n dimensions, and the use of determinants. See also his articles for the *Encyclopaedia Britannica* on "Geometry" (9th ed.) and "Curve" (11th ed.).

Hesse, Otto, *Vorlesungen aus der analytischen Geometrie der geraden Linie, des Punktes und des Kreises in der Ebene.* 3rd ed., Leipzig, 1881.
This book, the 1st edition of which appeared in 1865, may be said to have done for analytic geometry in the sense of Plücker what the texts of Lacroix and Biot did for analytic geometry in the sense of Monge. It represents the strictly formal type of approach, with extensive use of determinants. See also his *Vorlesungen über die analytische Geometrie des Raumes* of 1861.

Laisant, C. A., *Essai sur les fonctions hyperboliques.* Paris, 1874.
See pp. 71–83 for the use of elliptic and hyperbolic polar coordinates.

Baltzer, Richard, *Analytische Geometrie.* Leipzig, 1882.
One of the important textbooks of the 19th century. It is especially good for historical notes, particularly on the material of the "golden age." A thoroughly modern treatment.

Briot, C. A. A., and J. C. Bouquet, *Elements of analytical geometry of two dimensions,* 14th ed., transl. by J. H. Boyd, Chicago and New York, 1896.
One of the most popular of the more extensive textbooks of the 19th century. Still a useful work.

## II. Secondary Works on the History of Analytic Geometry (arranged alphabetically).

Archibald, R. C., *Outline of the history of mathematics,* 6th ed., Mathematical Association of America, 1949.
Excessive brevity limits its usefulness, but bibliographic references are a good feature.

Bell, E. T., *The development of mathematics,* New York, 1940.
Especially good on the development of ideas in the nineteenth century.

Berenguer, P. A., "Un geómetra español del siglo XVII," *El Progreso Matemático,* V (1895), 116–121.
Claims Antonio Hugo de Omerique as a precursor of modern analytic geometry through his *Analysis geometrica;* but the claim can not be substantiated because the second, and more important, part of this work has been lost.

Bopp, Karl, "Die Kegelschnitte des Gregorius a St. Vincentio," *Abhandlungen zur Geschichte der mathematischen Wissenschaften,* XX (1907), 87–314.
Includes comparative remarks on the synthetic and analytic methods in the seventeenth century.

Bortolotti, Ettore, "L'algebra nella storia e nella preistoria della scienza," *Osiris,* I (1936), 184–230.
On the role of Bombelli in the representation and construction of quantities geometrically by line segments.

Bortolotti, Ettore, *L'algebra, opera di Rafael Bombelli da Bologna.* Bologna, 1929.
Points out the use by Bombelli of coordinates and of the combination of algebra and geometry. This book is extensively reviewed in *Scripta Mathematica,* IV (1936), 166–169.

Bortolotti Ettore, *Lezioni di geometria analitica.* 2 vols., Bologna, 1923.
Vol. I contains a long "Introduzione storica" (pp. ix-xxxix) which is especially good on the Italian precursors of Descartes.

Bortolotti, Ettore, *Studi e ricerche sulla storia della matematica in Italia nei secoli XVI e XVII,* Bologna, 1928.
Especially useful for the analysis of the "Algebra geometrica" of Paolo Bonasoni, a precursor of Viète.

Bosmans, Henri, "La première édition de la 'Clavis mathematica' d'Ought-

red, son influence sur la 'Géométrie' de Descartes," *Annales de la Société Scientifique de Bruxelles*, XXXV (1910–1911), 24–78.
    Portrays Oughtred as a link from Viète to Descartes.
Bosmans, Henri, "Pour une historie de la géométrie analytique, d'après G. Loria," *Mathesis* (3), VI (1906), 260–264.
    Essentially a summary of Loria's paper with similar title.
Boyer, C. B., "Analytic geometry: the discovery of Fermat and Descartes," *The Mathematics Teacher*, XXXVII (1944), 99–105.
    Emphasizes that they discovered, not graphs, coordinates, or the analytic view, but the fundamental principle of analytic geometry.
Boyer, C. B., "Cartesian geometry from Fermat to Lacroix," *Scripta Mathematica*, XIII (1947), 133–153.
    Especially on the changes in attitude and purpose.
Boyer, C. B., "Historical stages in the definition of curves," *National Mathematics Magazine*, XIX (1945), 294–310.
    Origins of some of the familiar curves in ancient and modern times.
Boyer, C. B., "Note on an early graph of statistical data," *Isis*, XXXVII (1947), 148–149.
    On the slow penetration of graphical methods into fields other than geometry.
Brill, A., and Noether, M., "Die Entwickelung der Theorie der algebraischen Funktionen in älterer und neuerer Zeit," *Jahresbericht der Deutschen Mathematiker-Vereinigung*, III (1892–1893), 107–566.
    Touches frequently upon the history of analytic geometry and is an important secondary source.
Brunschvicg, Léon, *Les étapes de la philosophie mathématique*, Paris, 1912.
    Chap. VII is an analysis of the geometry of Descartes, but it should be read with care because it attributes to Descartes a degree of arithmetization which is absent from *La géométrie*.
Cajori, Florian, "Generalizations in geometry as seen in the history of developable surfaces," *American Mathematical Monthly*, XXXVI (1929), 431–437.
    Especially on the contributions of Euler and Monge.
Cajori, Florian, *A history of mathematics*. 2nd ed., New York, 1931.
    Noteworthy for a summary of contributions to analytic geometry in the nineteenth century (pp. 309–328).
Cajori, Florian, "Origins of fourth dimension concepts," *American Mathematical Monthly*, XXXIII (1926), 397–406.
    Both analytic and synthetic aspects are included.
Cantor, Moritz, *Vorlesungen über Geschichte der Mathematik*. 4 vols., Leipzig, 1900–1908.
    The most extensive history of mathematics before 1800.
Carrus, S., see Fano, G.
Chasles, Michel, *Aperçu historique sur l'origine et le développement des méthodes en géométrie*. 2nd ed., Paris, 1875.
    A valuable, although somewhat outdated, work by a great geometer. Contains too strong a claim for Descartes as sole inventor (pp. 94–95).
Chasles, Michel, "Sur la doctrine des porismes d'Euclide," *Correspondance mathématique et physique*, X (new series IV, 1838), 1–20.
    Sees, in the porisms "véritablement une Géométrie analytique."
Coddington, Emily, *A brief account of the historical development of pseudospherical surfaces from 1827–1887*. Lancaster, 1905.
    Mostly on differential geometry, but especially useful for the bibliography on surfaces.

Coolidge, J. L., *A history of the conic sections and quadric surfaces*. Oxford, 1945.
> A valuable and attractive work. Includes analytic treatment from ancient to modern times.

Coolidge, J. L., *A history of geometrical methods*. Oxford, 1940.
> A worthy successor to Chasles' *Aperçu*. Analytic geometry is not always separately handled, but the treatment is admirable.

Coolidge, J. L., "The beginnings of analytic geometry in three dimensions," *American Mathematical Monthly*, LV (1948), 76–86.
> Excellent summary from Plato to Euler.

Coolidge, J. L., "The origin of analytic geometry," *Osiris*, I (1936), 231–250,
> Defends the thesis that analytic geometry was an invention of the Greeks, perhaps of Menaechmus. This appears also as a part of his *History of geometrical methods*.

Coolidge, J. L., "The origin of polar coordinates," *American Mathematical Monthly*, LIX (1952), 78–85.
> Attention is called to the work of Cavalieri.

Cournot, A. A., *De l'origine et des limites de la correspondance entre l'algèbre et la géométrie*. Paris, 1847.
> More philosophical than historical, but contains some relevant material.

Darboux, Gaston, *Principes de géométrie analytique*. Paris, 1917.
> Contains numerous historical allusions.

Darboux, Gaston, "A survey of the development of geometric methods," translated by H. D. Thompson, *Bulletin, American Mathematical Society*, XI (1905), 517–543.
> A good summary of geometrical work from Lagrange to Chasles, but leans heavily on the account by Fano, q. v.

Dehn, Max, see Schoenflies.

Delambre, J. B., *Rapport historique sur les progrès des sciences mathématiques depuis 1789 et sur leur état actuel*. Paris, 1810.
> First-hand information on the figures who influenced the development of analytic geometry during a crucial period, notably Monge and Lacroix.

DeVries, Hk., "How analytic geometry became a science," *Scripta Mathematica*, XIV (1948), 5–15.
> One of the most important accounts of the history of analytic geometry during the early nineteenth century.

Dingeldey, F., and E. Fabry, "Coniques" and "Systèmes de coniques," *Encyclopédie des sciences mathématiques*, III, 17 and 18, 1–256.
> These two articles are full of historical material. There are close to a thousand footnotes with innumerable bibliographic references.

Duhem, Pierre, *Etudes sur Léonard de Vinci*. 3 vols., Paris, 1906–1913.
> Extensive account of the medieval precursors of Descartes.

Duhem, Pierre, "Oresme," *Catholic Encyclopedia*, XI (1911), 296–297.
> Holds that Oresme "forestalls Descartes in the invention of analytic geometry."

Dupin, Charles, *Essai historique sur les services et les travaux scientifiques de Gaspard Monge*. Paris, 1819.
> Generally good account, but inadequate on his place in analytic geometry.

Eneström, Gustav, "Auf welche Weise hat Viète die analytische Geometrie vorbereitet?," *Bibliotheca Mathematica* (3), XIV (1914), 354.
> Objects to the claim that Viète gave the equation of the line.

Eneström, Gustav, "Die Briefwechsel zwischen Leonhard Euler and Johann I. Bernoulli," *Bibliotheca Mathematica* (3), IV (1903), 344–388.
> Includes, on pp. 354f, correspondence on space coordinates.

Eneström, Gustav, "Kleine Mitteilungen," *Bibliotheca Mathematica* (3), XI (1911), 241–243.
> Indicates that the object of Descartes' *Geometry*, the geometric construction of the

solutions of algebraic equations, makes any dependence upon the work of Oresme appear improbable.

Eneström, Gustav, "Ueber das angebliche Vorkommen krummliniger Koordinaten bei Leibniz," *Bibliotheca Mathematica* (3), X (1909–1910), 43–47.
Points out that the history of curvilinear coordinates really begins with Gauss.

Eneström, Gustav, "Ueber die Bedeutung von Quellenstudien bei mathematischer Geschichtsschreibung," *Bibliotheca Mathematica* (3), XII (1911–1912), 1–20.
Shows how errors are spread by failure to examine sources, citing cases from Viète and Descartes.

Eneström, Gustav, "Ueber die verschiedenen Auflagen und Uebersetzungen von Descartes' 'Géométrie'," *Bibliotheca Mathematica* (3), IV (1903), 211.
Lists about a dozen editions.

Ernst, Wilhelm, *Julius Plücker*. Bonn, 1933.
A well-rounded account of his life and work, including his analytic geometry.

Fabry, E., see Dingeldey, F.

Fano, G., and S. Carrus, "Exposé parallèle du développement de la géométrie synthetique et de la géométrie analytique pendant le 19ième siècle," *Encyclopédie des sciences mathématiques*, III, 3, 185–259.
A good comparative study, including the work of Monge, Gergonne, Lamé, Bobillier, Möbius, Plücker, and others.

Forsyth, A. R., "Obituary notices," *Proceedings of the London Royal Society*, LVIII (1895), i–xliii.
Excellent summary of the life and work of Cayley, with numerous references.

Funkhouser, H. G., "Historical development of the graphical representation of statistical data," *Osiris*, III (1937), 269–404.
Includes reference to the work of Oresme. See also his "Note on a tenth century graph," *Osiris*, I (1936), 260–262.

Gelcich, E., "Eine Studie über die Entdeckung der analytischen Geometrie mit Berücksichtigung eines Werkes des Marino Ghetaldi Patrizier Ragusaer aus dem Jahre 1630," *Abhandlungen zur Geschichte der Mathematik*, IV (1882), 191–231.
Excellent critical account of relations between algebra and geometry at that time. Concludes that Ghetaldi lacked the principle of coordinates.

Gibson, Boyce, "La 'Géométrie' de Descartes au point de vue de sa méthode," *Revue de Métaphysique et de Morale*, IV (1896), 386–398.
Emphasizes Descartes' reduction of problems to geometrical constructions.

Gomes Teixeira, F., *Traité des courbes spéciales remarquables planes et gauches* Transl. from the Spanish, 2 vols., Coïmbre, 1908–1909.
An extensive catalog of curves with numerous historical notes.

Greenwood, Thomas, "Origines de la géométrie analytique," *Revue Trimestrielle Canadienne*, XXXIV (1948), 166–179.
Philosophical account of the elementary principles of Fermat and Descartes.

Grévy, A., see Staude, O.

Günther, Sigismund, "Le origini ed i gradi di sviluppo del principio delle coordinate," *Bullettino di Bibliografia e di Storia delle Scienze Mathematiche e, Fisiche*, X (1877), 363–406.
Shows that the use of coordinates goes back to Greek times, and that in some respects the work of Oresme and Kepler resembles analytic geometry.

Heath, T. L., *A history of Greek mathematics*. 2 vols., Oxford, 1921.
The chief secondary source on Greek mathematics. See also Heath's editions of Archimedes and Apollonius.

Heiberg, J. L., "Die Kentnisse des Archimedes über die Kegelschnitte,"

*Zeitschrift für Mathematik und Physik*, XXV (1880), *Historisch-Literarische Abtheilung*, pp. 41–67.
A technical analysis, with frequent reference to the Greek text.

Hill, J. E., "Bibliography of surfaces and twisted curves," *Bulletin, American Mathematical Society*, III (1896–1897), 133–146.
Includes reference to some of the work of the 18th century.

Kaestner, A. G., *Geschichte der Mathematik.* 4 vols., Goettingen, 1796–1800.
Old, but still useful for some aspects, especially from Cardan to Galileo.

Karpinski, L. C., "Is there progress in mathematical discovery and did the Greeks have analytic geometry?" *Isis*, XXVII (1937), 46–52.
Rejects the thesis of Coolidge that the Greeks had analytic geometry. Emphasizes the idea of progress and the development of algebraic notations.

Karpinski, L. C., "The origin of the mathematics taught to Freshmen," *Scripta Mathematica*, VII (1939), 133–140.
Emphasizes the importance for analytic geometry of the notation of Viète.

Kiefer, Albert, *Die Einfuehrung der homogenen Koordinaten durch K. W. Feuerbach.* Strassburg i. E., 1910.
Analysis of Feuerbach's *Untersuchungen der dreieckigen Pyramide* (1827) and his unpublished *Analysis der dreieckigen Pyramide*.

Klein, Felix, *Vorlesungen über die Entwicklung der Mathematik im 19. Jahrhundert.* 2 vols., Berlin, 1926–1927.
Especially good for an account of the work of Plücker written by one of his students.

Klein, Felix, *Vorlesungen über höhere Geometrie.* 3rd ed., Berlin, 1926.
Contains frequent historical allusions to the analytic geometry of the golden age.

Kötter, Ernst, "Die Entwickelung der synthetischen Geometrie von Monge bis auf Staudt (1847)," *Jahresbericht der Deutschen Mathematiker-Vereinigung*, vol. 5, part 2, 1896 (pub. 1901).
This valuable work of almost 500 pp. contains so many allusions and references to analytic geometry that it constitutes an important secondary source.

Kommerell, V., "Analytische Geometrie der Ebene and des Raumes," in Moritz Cantor, *Vorlesungen über Geschichte der Mathematik*, IV (1908), 451–576.
Especially useful for the account of the work of Euler, Lagrange, and Monge, although much of the emphasis is upon differential, rather than analytic, geometry.

Krazer, Adolf, *Zur Geschichte der graphischen Darstellung von Funktionen.* Karlsruhe, 1915.
This 31-page *Festschrift* contains an excellent critical account of the work of Oresme, although not so complete as those of Wieleitner in *Bibliotheca Mathematica*, q. v.

Lattin, Harriet P., "The eleventh century ms Munich 14436: its contribution to the history of coordinates, of logic, of German studies in France," *Isis*, XXXVIII (1948), 205–225.
Describing a graphic representation of the course of the planets through the zodiac, following the text of Pliny.

Lehmann, Ernst, "De La Hire und seine Sectiones Conicae," *Jahresbericht des Königlichen Gymnasiums zu Leipzig*, 1887–1888, 1–28.
Emphasis is upon synthetic methods, but analytic contributions are included.

Libri, Guillaume, *Histoire des sciences mathématiques in Italie, depuis la renaissance des lettres jusqu'à la fin du dixseptième siècle.* 4 vols., Paris, 1838–1841.
Emphasizes Benedetti (III, 124) and Cataldi (IV, 95).

Loria, Gino, "A. L. Cauchy in the history of analytic geometry," transl. by Evelyn Walker, *Scripta Mathematica*, I (1932), 123–128.
The substance of this article appears also in Loria's later paper, "Perfectionnements...."

Loria, Gino, "Aperçu sur le développement historique de la théorie des courbes planes," *Verhandlungen des ersten internationalen Mathematiker-Kongresses in Zürich, von 9. bis 11. August 1897,* ed. by F. Rudio, Leipzig, 1898, pp. 289–298.
A good summary from ancient to modern times.

Loria, Gino, "Da Descartes e Fermat a Monge e Lagrange. Contributo alla storia della geometria analitica," *Reale Accademia dei Lincei. Atti. Memorie della classe di scienze fisiche, matematiche e naturali* (5), XIV (1923), 777–845.
Together with Loria's "Perfectionnements . . . ," this constitutes one of the most important works on the history of analytic geometry.

Loria, Gino, "Descartes géomètre," in *Etudes sur Descartes, Revue de Métaphysique et de Morale,* XLIV (1937), 199–220.
Summary and analysis of *La géométrie.*

Loria, Gino, "Gli 'Acta Eruditorum' durante gli anni 1682–1740 e la storia delle matematiche," *Archeion,* XXIII (1941), 1–35.
This summary of contents shows how thoroughly analytic geometry was overshadowed at the time by the calculus.

Loria, Gino, *Il passato e il presente delle principali teorie geometriche. Storia e bibliografia.* 4th ed., Podova, 1931.
Ample account of the development of geometry in modern times, mostly on the 19th century.

Loria, Gino, "Le rôle de la représentation géométrique des grandeurs aux différentes époques de l'historie des mathématiques," *Revue de Métaphysique et de Morale,* XLVI (1939), 57–64.
On the association of number and geometry, but not specifically on analytic geometry.

Loria, Gino, "Michele Chasles e la teoria delle coniche," *Osiris,* I (1936), 421–450.
Contains a brief historical introduction on conics, as well as a bibliography of 28 items by Chasles on conics.

Loria, Gino, "Perfectionnements, évolution, métamorphoses du concept de 'coordonnées.' Contribution a l'histoire de la géométrie analytique," *Mathematica,* XVIII (1942), 125–145; XX (1944), 1–22; XXI (1945), 66–83.
This work without doubt constitutes, together with Loria's "Da Descartes e Fermat . . . ," one of the most important histories of analytic geometry. This article appeared also in *Osiris,* VIII (1948), 218–288.

Loria, Gino, "Phases de développement de la géométrie analytique," *Assoc. Fr. Grenoble,* 1925, pp. 102–150.
I have not seen this work.

Loria, Gino, "Pour une histore de la géométrie analytique," *Verhandlungen des dritten internationalen Mathematiker-Kongresses in Heidelberg von 8. bis 13. August 1904* (Leipzig, 1905), pp. 562–574.
Critical analysis of Fermat and Descartes, and a summary of subsequent developments.

Loria, Gino, "Qu'est-ce que la géométrie analytique?" *L'Enseignement Mathématique,* XIII (1923), 142–147.
Emphasizes Euler's *Introductio* for its use of formulas in the solution of geometrical problems.

Loria, Gino, "Sketch of the origin and development of geometry prior to 1850," Transl. by G. B. Halsted, *Monist,* XIII (1902–1903), 80–102, 218–234.
Good for general survey, but the emphasis is more upon synthesis than upon analysis.

Loria, Gino, *Spezielle algebraische und transcendente ebene Kurven.* Leipzig, 1902.
 Altogether excellent collection of curves with extensive historical material. This valuable work appeared also in two-volume editions in German (Leipzig, 1910–1911) and Italian (Milan, 1930).
Loria, Gino, *Storia delle mathematiche.* 3 vols., Turin, 1929–1933.
 A characteristically excellent work.
Mangoldt, H. von, and L. Zoretti, "Les notions de ligne et de surface," *Encyclopédié des sciences mathématiques*, III, 2, 152–184.
 Contains numerous historical references.
Marie, Maximillien, *Histoire des sciences mathématiques et physiques.* 12 vols., Paris, 1883–1888.
 Gives convenient summaries of the work of many individuals, but should be used with care.
Merrifield, C. W., "On a geometrical proposition indicating that the property of the radical axis was probably discovered by the Arabs," *Proceedings, London Mathematical Society*, II (1866–1869), 175–177.
 Concerns a property of the "shoemaker's knife" of Archimedes.
Milhaud, Gaston, "Descartes et la géométrie analytique," *Nouvelles études sur l'histoire de la pensée scientifique*, Paris, 1911, pp. 155–176.
 Emphasizes that the works of Descartes and Fermat were a natural continuation of the works of antiquity.
Milhaud, Gaston, *Descartes savant.* Paris, 1921.
 Critical analysis of the development of the scientific and mathematical thought of Descartes.
Milne, J. J., *An elementary treatise on cross-ratio geometry with historical notes.* Cambridge, 1911.
 See especially pp. 146–149 on the problem of Pappus.
Milne, J. J., "Note on Cartesian geometry," *Mathematical Gazette*, XIV (1928–1929), 413–414.
 On Descartes and the problem of Pappus.
Montucla, Etienne, *Histoire des mathématiques.* 2nd ed., 4 vols., Paris, 1799–1802.
 Old but still somewhat useful. Too close in time to Descartes for proper perspective.
Morley, F. V., "Thomas Hariot," *Scientific Monthly*, XIV (1922), 60–66.
 Claims analytic geometry for Hariot, but this has been refuted. See Smith, *op. cit.*, II, 322.
Müller, Felix, "Zur Literatur der analytischen Geometrie und Infinitesimalrechnung vor Euler," *Jahresbericht der Deutschen Mathematiker-Vereinigung*, XIII (1904), 247–253.
 Emphasizes the numerous analytic works which appeared between Fermat and Euler.
Neugebauer, Otto, "Apollonius-Studien," *Quellen und Studien zur Geschichte der Mathematik, Astronomie und Physik*, Part B, Studien, II (1932–1933), 215–254.
 Presents new arguments to show that Apollonius knew of the focus (but not necessarily the directrix) of the parabola.
Neugebauer, Otto, "The astronomical origin of the theory of conic sections," *Proceedings of the American Philosophical Society*, XCII (1948), 136–138.
 Suggests an origin in the theory of sundials.
Noether, M., see Brill A.
Pierce, J. M., "References in analytic geometry," *Harvard University Library Bulletin*, I (1875–1879), nos. 8, 10, 11 (1878–1879), pp. 157–158, 246–250, 289–290.

Excellent analysis of the works of Viète and Descartes.

Prag, A., "John Wallis," *Quellen und Studien zur Geschichte der Mathematik. Astronomie und Physik*, Part B, *Studien*, I (1931), 381–412.
On Wallis' arithmetization of geometry and the calculus.

Saltykow, N., " 'La géométrie' de Descartes. 300° anniversaire de géométrie analytique," *Bulletin des Sciences Mathématiques* (2), LXII (1938), 83–96, 110–123.
Emphasizes the originality of Descartes in combining many elements previously known.

Saltykow, N., "Souvenirs concernant le géomètre Yougoslave Marinus Ghetaldi," *Isis*, XXIX (1938), 20–23.
On the relationship between Viète and Ghetaldi.

Sauerbeck, Paul, "Einleitung in die analytische Geometrie der höheren algebraischen Kurven nach den Methoden von Jean Paul de Gua de Malves," *Abhandlungen zur Geschichte der Mathematischen Wissenschaften*, XV (1902), 1–166.
An extensive history of higher plane curves from Newton to De Gua.

Schoenflies, A., and M. Dehn, *Einführung in die analytische Geometrie der Ebene und des Raumes*. 2nd ed., Berlin, 1931.
An excellent "Historische Uebersicht" (pp. 379–393) traces the history from antiquity to the early 19th century.

Scott, Charlotte A., "On the intersections of plane curves," *Bulletin, American Mathematical Society*, IV (1897–1898), 260–273.
The best general reference on the history of the Cramer paradox.

Scott, J. F., *The mathematical work of John Wallis*. London, 1938.
Contains a summary, somewhat inadequate, of Wallis' important analytic work on conic section.

Sergescu, Petru, "Les mathématiques dans le 'Journal dels Savants' 1665–1701, *Archeion*, XVIII (1936), 140–145.
Illustrates the way in which the calculus overshadowed analytic geometry of the time.

Simon, Max, *Ueber die Entwicklung der Elementargeometrie im XIX. Jahrhundert. Jahresbericht der Deutschen Mathematiker-Vereiningung, Ergänzungsbände*, I (1906).
An unusually extensive bibliographical work, but somewhat lacking in critical discrimination.

Smith, D. E., *History of mathematics*. New York, ca 1923–1925.
The section on analytic geometry (II, 316–331) is inadequate but includes further references.

Staude, O., and A. Grévy, "Quadriques," *Encyclopédie des sciences mathématiques*, III, 22, 164 pp.
Full of historical notes containing thousands of bibliographic references. Mostly on the 19th century.

Steele, A. D., "Ueber die Rolle von Zirkel & Lineal in der griechischen Mathematik," *Quellen und Studien zur Geschichte der Mathematik, Astronomie und Physik*, Part B. Studien, III (1934–1936), 287–369.
Doubts that Plato intended the rigid limitation to the line and circle as found in Aristotle and Euclid.

Struik, D. J., *A concise history of mathematics*. 2 vols., New York, 1948.
Although very brief, it contains appropriate reference to developments in analytic geometry.

Tannery, Paul, "Notions historiques," in Jules Tannery, *Notions de mathématiques* (Paris, 1903), pp. 327–348.
These notes are almost all relevant to the history of analytic geometry, including

curves in antiquity, origin of coordinates, and the use of the words analysis and synthesis.

Tannery, Paul, "Pour l'histoire des lignes et surfaces courbes dans l'antiquité," *Mémoires scientifiques*, II (1912), 1–47.
   A summary of the curves and surfaces known to the ancients.

Taton, René, "Monge, créateur des coordonnées axiales de la droite, dites de Plücker," *Elemente der Mathematik*, VII (1952), 1–5.
   Shows that Monge in 1771 (pub. 1785) anticipated Plücker and Cayley.

Taton, René, *L'oeuvre scientifique de Monge*. Paris, 1951.
   Contains a good chapter (pp. 101–147) on "la géométrie analytique."

Taylor, Charles, "The geometry of Kepler and Newton," *Cambridge Philosophical Society, Transactions*, XVIII (1900), 197–219.
   Especially on the conics in Kepler's *Ad Vitellionem* and in Newton's *Principia*.

Taylor, Charles, *An introduction to the ancient and modern geometry of conics*. Cambridge, 1881.
   Contains an historical "Prolegomena" (pp. xvii–lxxxviii) on geometry from the beginnings to the 19th century, as well as historical footnotes throughout the body of the work.

Tropfke, Johannes, *Geschichte der Elementar-Mathematik*. 2nd ed., vol. VI, Berlin and Leipzig, 1924.
   One of the most extensive histories of analytic geometry is included in pp. 92–169.

Turnbull, H. W., *The mathematical discoveries of Newton*. London and Glasgow, 1945.
   Contains some relevant material, especially on the mss which Horsley collated as the *Geometria analytica*.

Tweedie, C., *James Stirling, a sketch of his life and works, along with scientific correspondence*. Oxford, 1922.
   Includes material on Stirling's important edition of Newton's *Enumeration of cubic curves*.

Weaver, J. H., "On foci of conics," *Bulletin, American Mathematica Society*, XXIII (1916–1917), 357–365.
   Historical survey from Euclid to Poncelet.

Weaver, J. H., "The duplication problem," *American Mathematical Monthly*, XXIII (1916), 106–113.
   Historical survey from the Pythagoreans to Descartes.

Wieleitner, Heinrich, *Geschichte der Mathematik*. Part II, 2 vols., Leipzig, 1911–1921.
   This is one of the most important histories of mathematics. The second volume of part II includes a valuable history of plane analytic geometry (pp. 1–46), of solid analytic geometry (pp. 47–60), and of curves (pp. 61–92). It is based in part on a ms left by Braunmühl. Part I was written by Günther.

Wieleitner, Heinrich, Geschichte der Mathematik. 2nd ed., 2 vols., Berlin, 1939.
   Good, but very brief. Not to be confused with the more important work listed above.

Wieleitner, Heinrich, "Die Anfänge der analytischen Raumgeometrie," *Zeitschrift für mathematischen und naturwissenschaftlichen Unterricht*, XLIX (1918), 73–79.
   Good summary from Fermat to Clairaut.

Wieleitner, Heinrich, *Die Geburt der modernen Mathematik*. Vol. I, *Analytische Geometrie*. Karlsruhe, 1924.
   Especially good on the work of Fermat and Descartes.

Wieleitner, Heinrich, "Marino Ghetaldi und die Anfänge der Koordinatengeometrie," *Bibliotheca Mathematica* (3), XIII (1912–1913), 242–247.
   Sees, in Ghetaldi, a combination of algebra and geometry, but not analytic geometry.

Wieleitner, Heinrich, "Ueber den Funktionsbegriff und die graphische Darstellung bei Orseme," *Bibliotheca Mathematica* (3), XIV (1914), 193–243.
    An excellent analysis. Rejects the claim that Oresme invented analytic geometry.
    Emphasizes that the coordinate concept of Oresme is different from that of Des-
    cartes.
Wieleitner, Heinrich, "Zur Entstehung der analytischen Raumgeometrie,"
*Zeitschrift für mathematischen und naturwissenschaftlichen Unterricht*, LIX
(1928), 357–358.
    Corrections, on Descartes and the Greeks, to his article in the same journal ten years
    earlier.
Wieleitner, Heinrich, "Zur Erfindung der analytischen Geometrie," *Zeitschrift
für mathematischen und naturwissenschaftlichen Unterricht*, XLVII (1916),
414–426.
    Excellent summary and analysis of the work of Fermat and Descartes. This ar-
    ticle is reprinted under the same title in W. Dieck, *Mathematisches Lesebuch* (5 vols.,
    Sterkrade, 1920–1921), IV, 60–71.
Wieleitner, Heinrich, "Zur Frühgeschichte der Räume von mehr als drei Di-
mensionen," *Isis*, VII (1925), 486–489.
    Especially good for the ideas of Oresme.
Wieleitner, Heinrich, "Zwei Bemerkungen zu Stirlings 'Linea tertii ordinis
Neutonianae'," *Bibliotheca Mathematica* (3), XIV (1914), 55–62.
    Thorough account of the place of Stirling in the history of analytic geometry.
Wölffing, E., "Bericht über den gegenwärtigen Stand der Lehre von den
natürlichen Koordinaten," *Bibliotheca Mathematica* (3), I (1900), 142–159.
    The most important secondary source on intrinsic coordinates.
Wolff, Georg, "Leone Battista Alberti als Mathematiker," *Scientia*, LX
(1936), 353–359; and supp., pp. 142–147.
    Holds that Alberti's use of a coordinate system makes him a precursor of Descartes.
Woolard, E. W., "The historical development of celestial co-ordinate systems,"
*Publ., Astronomical Society of the Pacific*, LIV (1942), 77–90.
    Survey of the use of coordinates in ancient astronomy.
Zuethen, H. G., *Die Lehre von den Kegelschnitten im Altertum* (Kopenhagen,
1886).
    Extensive, but rather discursive, account of the early history of the conics.
Zeuthen, H. G., *Geschichte der Mathematik im XVI. und XVII. Jahrhundert*.
German ed. by Raphael Meyer, Leipzig, 1903.
    A good summary of the geometry of Fermat and Descartes is given on pp. 192–233.
Zeuthen, H. G., "Sur les rapports entre les anciens et les modernes principes
de la géométrie," *Atti del IV Congresso Internazionale dei Matematici* (Roma,
1908), III (1909), 422–427.
    Not specifically on analytic geometry, but touches upon it.
Zeuthen, H. G., "Sur l'usage des coordonnées dans l'antiquité," *Kongelige
Danske Videnskabernes Selskabs, Forhandlingen. Oversigt*, (1888), 127–144.
    Insists that the geometric algebra of antiquity closely resembles the use of coordin-
    ates, and that Fermat was an almost immediate disciple of Apollonius and Pappus.
    His views here conflict with those of Günther in "Die Anfänge . . . ," q. v.
Zoretti, L., see Mangoldt, H. von.

# Index

Abridged notation, 233f, 236, 240, 245ff, 249, 252f, 254
Abscissa, 76, 111, 118, 120f, 127, 150f, 194
Acta Eruditorum, 225
Agnesi, Maria, 177–180; 175
  Witch of Agnesi, 179f; 81, 142
Ahmes papyrus, 5
Ahrens, J. T., 226
Albert of Saxony, 45
Alberti, L. B., 52
Algebra, 2f, 7f, 10, 19, 22, 28f, 35f, 41ff, 48, 51f, 54f, 57ff, 84ff, 96, 101, 109, 122ff, 147, 212, 215f, 223
Alhazen, 42f
  Problem of Alhazen, 43, 117
Al-Khowarizmi, 41f
Ampère, A. M., 228
Analysis, 14, 37; 10, 17, 23, 57, 65, 71, 111, 180f, 190, 215ff, 223, 232, 248, 255
Analytic geometry
  Fundamental principle, 30, 75, 81, 87, 99, 101f, 181, 211, 214
  Golden age, chapter IX
  Invention of, 74ff, 82ff, 101, 217
  Of $n$ dimensions, 258f, 264f
  Of three dimensions (see Solid analytic geometry)
  Name, 142, 155, 211, 215ff, 220
  Origins, 1ff, 39, 48
  Textbooks, 191, 210f, 213ff, 218ff
Analytic method, 14, 17, 23, 37, 64ff, 71, 75, 80, 85, 107, 182f, 211, 266
Analytical revolution, 218, 256, 268
Analytical Society, 256
Anaxagoras, 9
Anderson, Alexander, 66
Angle bisectors, 218
*Annales de mathématiques pures et appliquées*, 226f, 230ff, 236f, 239, 245f, 249 254, 257
Anthemius of Tralles, 41, 68
Apollonius of Perga, 23–31; 17, 20ff, 33, 35ff, 42, 46, 51, 66, 74, 78, 106, 108, 110, 112, 120, 159, 222
  Circle of Apollonius, 31
  Problem of Apollonius, 74, 226, 232, 234
Application of areas, 8f; 22, 24, 34, 58, 66
Arabs, 42f, 52, 55, 57, 233f
Arc length, 228ff, 170
Archibald, R. C., 1, 9, 31
Archimedes, 31–34; 10, 15, 21, 23ff, 36f, 42f, 63, 66, 78, 114, 125
  Spiral of Archimedes, 32, 127, 144f, 154, 186
Archytas, 12f, 17f, 33

Aristaeus, 23f, 30
Aristotle, 7, 15, 31, 41, 44, 57, 134
Arithmetization, 7, 84, 111f, 202, 218, 242, 250, 263, 266f
Atomism, 9f, 15

Babylonians, 1ff, 8, 22, 35f, 41f, 70, 169
Bacon, Francis, 54
Bacon, Roger, 40
Baker, Thomas, 122
Ball, W. W. R., 138, 256f
Barrow, Isaac, 112, 135
Bartholinus, Erasmus, 109
Becker, Oskar, 16
Beekmann, Isaac, 82
Bell, E. T., 16, 59, 93, 99, 108, 118, 224, 245, 264ff
Benedetti, G. B., 67
Berenguer, P. A., 133, 155
Bernoulli, Jacques, 127ff; 33, 72, 130, 133, 145f, 163, 172, 194
  Lemniscate of Bernoulli, 33, 133
Bernoulli, Jean, 129f; 133f, 146, 153, 158, 165, 185, 188
Beumer, M. G., 218
Bézout, Étienne, 193; 199, 208
  Theorem of Bézout, 163, 184
Billy, Jacques de, 69
Binet, J. P. M., 238
Biot, J. B., 220–221; 211, 213, 223, 245, 257, 260
Bobillier, Étienne, 240f; 242f, 246, 249
Boethius, 40, 51f, 55
Bombelli, Rafael, 57f, 69
Bonasoni, Paolo, 58f
Bopp, Karl, 119
Borelli, G. A., 120
Bortolotti, Ettore, 44, 56, 58ff
Bosmans, Henri, 60, 70, 119
Bovelles, Ch. de, 73
Boyle, Robert, 74
  Boyle's law, 162
Bradwardine, Thomas, 51ff, 55
Bragelogne, C. B. de, 164
Braikenridge, William, 163
Bret, Jean Jacques, 236
Brianchon, C. J., 239, 251
  Brianchon's theorem, 252
Brill, A., 246
Brocard, H., 136
Brunschvicg, Léon, 104, 204
Bunsen, R. W., 256
Burley, Walter, 45
Byzantine, 41, 43

Cajori, Florian, 6, 49, 58, 70, 141, 166, 255, 266
Calculator (see Suiseth)
Cambridge and Dublin Mathematical Journal, 257f
Campanus, Johannes, 50, 55
Cantor, Moritz, 51, 68, 158, 166, 174, 198
Cantor, Georg, 267
  Cantor-Dedekind axiom, 267
Caraccioli, J. B., 175
Cardan, Jerome, 56ff, 67, 73
Carnot, L. N. M., 228f; 249
Cartesian parabola, 88, 90, 94, 142
Cassirer, Ernst, 231
Cataldi, P. A., 67f
Cauchy, A. L., 236–238; 223, 241, 243, 259f
Cavalieri, Bonaventura, 125, 127
Cayley, Arthur, 257–260; 262, 264f
Cesàro, Ernesto, 229–230; 171
Chapelle, (see LaChapelle)
Characteristic of a conic, 92, 111, 116, 131, 153f, 172, 183, 214
Chasles, Michele, 22f, 65, 100, 103, 163, 226, 251
Chelucci, Paolino, 175
Chuquet, N., 44, 54f, 57f, 61, 69
Circle of Apollonius, 31
Circle, equation of, 2, 77ff, 104f, 150, 195, 213, 261
Circles, systems of, 233f
Cissoid of Diocles, 32
Clairaut, A. C., 167–170; 130, 173f, 177, 201f, 215
Clebsch, R. F. A., 266
Colson, John, 141
Comte, Auguste, 266f; 100, 103
Conchoid, 33, 105, 143, 164, 185
Condamine, C. M. de la, 170
Cone, equation of, 166ff, 170f, 189
Conics, conic sections, 17f, 20, 22ff, 30ff, 34, 68, 72, 77ff, 97f, 110ff, 116, 119, 126, 143, 148, 151ff, 178, 182f, 192, 195, 216
Discriminant of conic, 220
General equation of conic, 91, 111, 116, 126, 131, 153, 156, 159ff, 172, 183, 219, 249, 252, 260
  Polar equation of conic, 173, 214, 220
Conoid, 125, 172, 207
Constructions, 8, 34, 36, 41, 43, 58, 62ff, 66f, 78, 82ff, 85ff, 91, 95ff, 110, 115, 117, 122ff, 126, 147, 149, 176ff, 190, 192, 199
Continuity, principle of, 231f, 238f, 248, 253
Continuum, 6f, 10, 15f, 46, 267
Coolidge, J. L., 16ff, 27, 68, 70, 81, 84, 99f, 119, 125, 127f, 132, 148, 151, 158, 169, 185, 215, 222, 224, 240, 248, 268
Coordinates, 26ff, 46, 227ff, 249ff; 32, 35, 40, 52, 75ff, 83ff, 111, 120f, 133f, 142ff, 150, 178

Barycentric coordinates, 242f
Bipolar coordinates, 142f, 227
Coordinates of a line, 259
Curvilinear coordinates, 262f
Elliptic coordinates, 263
Homogeneous coordinates, 241–244, 249–252, 254, 260
Intrinsic coordinates, 227–229
Line coordinates, 250ff
Negative coordinates, 86, 111f, 115, 118, 133, 139, 150f, 154, 164, 176, 186
Parabolic coordinates, 228, 263
Plane coordinates, 251, 254
Plücker coordinates, 251
Polar coordinates, 32, 52, 127, 133, 142ff, 145f, 154, 168, 172ff, 175, 180, 185ff, 199f, 227ff, 242, 262
Spherical coordinates, 202f, 213, 260ff
Tangential coordinates, 251f
See also Transformation of coordinates
Copernicus, N., 72f
Corancez, L. A. O. de, 212
Correspondance Mathématique et Physique, 240f
Cotes, Roger, 136
Cournot, A. A., 186, 229
Coxeter, H. S. M., 258
Craig, John, 130ff; 152ff, 156, 159, 172, 183
Cramer, Gabriel, 194–197; 146, 175, 200, 208, 213, 217, 223, 237, 253
  Cramer's paradox, 196, 246ff; 163, 255
  Cramer's rule, 195f
Crelle, A. L., 241
  Crelle's Journal (see Journal fur die reine und angewandte Mathematik)
Curtze, M., 51
Curvature, radius of, 144, 187, 199, 227ff
Curve
  Cartesian curve, 89f, 140
  Class of a curve, 251f
  Classification of curves, 32, 36, 38, 90f, 96, 133f, 138ff, 147, 149, 164, 178, 181f, 194
  Curve tracing, 181; 142f, 161f, 178
  Definition of curve, 12, 30, 32f, 38, 72f, 88ff, 99, 134ff, 138, 188
  Equation of curve, 2, 18f, 23, 26ff, 38, 49; chapters V, VI, and VII, passim, 251f
  Higher plane curves, 79, 139, 164, 182, 194
  Order of curve, 198
  Organic description of a curve, 140, 163
  Parametric representation of a curve, 180, 187
  Skew curves, 166ff, 23, 32, 93, 188, 190, 254
  Tangential equation of curve, 252, 254
  Transcendental curves, 184f; 133ff, 139, 156, 182ff, 194
Cycloid, 33, 72f, 134, 154, 175, 187f, 227f
Cylinder, equation of, 165f, 171, 189

D'Alembert, Jean-le-Rond, 187, 192, 200, 215f, 222, 224
Dandelin, Germinal, 237
Dandelin's theorem, 237
Dantzig, Tobias, 267
Davies, T. S., 261
Debeaune, Florimond, 107ff; 114, 122, 136, 172, 192, 252
Dedekind, Richard, 267
De Gua de Malves, J. P., 174f; 146, 184, 194, 207
De Gua's triangle, 174, 194f
Dehn, Max, 35, 251
Delambre, J. B., 211
Delian problem, 12f; 17ff, 56, 99
Democritus, 5, 9f, 15, 17
Desargues, Gérard, 118ff; 68, 135, 232, 253
Descartes, René, chapter V, especially pp. 82–102, and passim
Folium of Descartes, 86f, 92, 98f, 112, 133, 139, 174
Ovals of Descartes, 89, 109, 142
Determinants, 237f, 257ff; 195f, 241f
De Vries, H., 232f, 246f, 251
De Witt, Jan, 114ff; 118f, 130f, 149, 153, 156, 159, 172, 176, 178
Dingeldey, F., 17, 31, 120
Dinostratus, 11f, 17
Diogenes Laertius, 14
Dionis du Sejour, A. P., 197f; 206
Diophantus, 35ff, 39, 41f, 55, 61, 264
Distance, formula for, 169f, 182, 201ff, 205, 209, 213
Duality, 238f; 230, 250ff, 259
Dürer, Albrecht, 72, 135
Duhamel, J. M. C., 155
Duhem, Pierre, 48ff
Duns Scotus, 45
Dupin, Ch., 204
Duplication of the cube, 12, 19, 99
Dupre, Huntley, 228

Ecole Central, 215
École Normale, 210f
École Polytechnique, 208, 218, 222, 225ff, 230, 233
Correspondance sur l'École Polytechnique, 221, 223
Journal de l'École Polytechnique, 208 221ff, 225f, 234
Egyptians, 1ff, 28, 36, 41
Ellipse, 17, 24ff, 30f, 33, 41, 68, 78, 89, 92, 105, 108, 110, 114, 116, 119, 132, 149, 151f, 160, 183
Generalized ellipses, 175
Ellipsoid, 189
Eneström, G., 64, 101, 155, 207
Enriques, Federigo, 8
Enumerative geometry, 252f
Equation, cubic, 3, 12f, 34, 43, 56f, 63f, 78, 82, 86, 91, 96ff, 105, 109, 140, 146f, 159, 163, 174, 253

Equation, quadratic, 8f, 22, 101, 112, 125f, 178, 214
Equation of curve (see Curve, equation of)
Equation of surface (see Surface, equation of)
Eratosthenes, 64
Ernst, Wilhelm, 245
Estienne de la Roche, 55
Euclid, 7f, 16, 21ff, 33f, 36f, 42, 51f, 66, 71, 110
Eudoxus, 10, 14ff, 21f
Euler, Leonhard, 164–167, 180–185; chapters VII and VIII passim, 130, 132, 134, 136, 153, 227f, 233, 237, 244, 253
Euripides, 1
Eutocius, 17, 25, 31, 34, 41

Faraday, Michael, 256
Faulhaber, J., 207
Favaro, A., 57, 122
Fermat, Pierre de, chapter V, especially pp. 74–82; 31, 50f, 61, 68, 103ff, 108, 112
Parabolas and hyperbolas of Fermat, 80, 111, 133, 136, 142, 172
Ferrari, Lodovico, 57
Ferro, Scipione del, 56f
Feuerbach, K. W., 243f, 249
Fibonacci, 40, 44, 55, 58
Fontana, Gregorio, 187; 174
Fontenelle, Bernard le Bovier de, 164; 146
Forsyth, A. R., 259, 264
Francoeur, L. B., 257
Frézier, A. F., 236, 246
Frisi, A. F., 177
Frisi, Paolo, 216
Function, 46; 5, 44, 49f, 65, 80, 130, 142, 180ff, 188, 190
Funkhouser, H. G., 43
Fuss, N., 216

Galileo, 50f, 68, 73, 134
Gallimard, J. E., 192
Garnier, J.-G., 221; 220, 223, 241
Gaultier, L., 226, 234, 245
Line of Gaultier (see Radical axis)
Gauss, C. F., 262f; 166, 243
Gelcich, E., 51, 67
Geminus, 33
Genty, Abbé Louis, 82
Gergonne, J.-D., 225–233, 238–240; 196, 237, 245ff, 251
Gergonne construction, 226
Gergonne's Annales (see Annales de mathématiques pures et appliquées)
Germaine, Sophie, 212
Ghetaldi, M., 66ff; 71, 74
Gibson, Boyce, 83
Girard, Albert, 69f
Goethe, Wolfgang, 192
Goudin, M. P., 197ff; 206
Gourieff, S., 199
Grandi, Guido, 185f; 179
Roses of Grandi, 185f

Graphical representation, 27f, 35, 43, 46ff, 64, 72, 75, 98, 104, 128f, 133, 136, 140, 147, 161f, 177ff, 194ff, 235f
Grassmann, Hermann, 258, 264
Graves, C., 261
Greeks, chapters I and II; 41, 43, 56f, 68, 88, 99, 147, 212
Gregory, James, 117, 128, 154
Gregory of St. Vincent, 118f; 127
Grünbaum, Adolph, 6
Gudermann, Christof, 260f
Günther, Siegmund, 7, 27, 46
Guidubaldo del Monte, 68, 112, 114
Guisnée, N., 149f, 153, 192f, 215

Hachette, J. N. P., 222; 210, 219, 221, 223, 226, 237
Haestrecht, Godefroy, 107
Halley, Edmund, 66, 181
Halstead, G. B., 27, 99
Hamilton, H. P., 237, 257
Hamilton, William Rowan, 258
Hankel, Hermann, 51, 190
Harpedonaptae, 1
Harriot, Thomas, 69f; 97, 112, 122
Heath, Sir T. L., 3, 7f, 17f, 21, 23f, 27, 31f, 35f
Heiberg, J. L., 31
Henderson, Archibald, 259
Hérigone, P., 70
Hermann, Jacob, 170–174; 163, 183, 185, 197, 200, 216
Hermite, Ch., 259
Heron of Alexandria, 35, 37
Hesse, L. O., 223, 260
Heuraet, H. van, 114
Hindus, 7, 40ff, 55, 57
Hindu-Arabic numerals, 41, 44, 59
Hipparchus, 28, 35
Hippasus of Metapontum, 5ff
Hippias of Elis, 5f, 9, 11f, 19, 32, 134
Hippocrates of Chios, 5, 9f, 12, 16, 19, 34
Hittorf, J. W., 256
Hobbes, Thomas, 112
Homogeneity, 61, 84f, 106, 109, 140, 162
Horsley, Samuel, 145, 216
Hospital (see L'Hospital)
Hube, Mich., 198f, 216f
Hudde, Johann, 113f, 169
Huilier (see L'Huilier)
Hultsch, F., 37
Hume, James, 65
Huygens, Christiaan, 51, 117f, 122, 127, 133, 139, 142, 162
Hymers, John, 257
Hyperbola, 18f, 24ff, 30f, 34, 68, 77f, 92, 105, 108, 110, 113, 116, 119, 126, 132, 151, 183, 263
Generalized hyperbolas, 175
Hyperboloid, 81, 125, 156, 158, 189, 237, 240

Iamblichus, 33, 52

Incommensurability, 7, 14, 16f, 20, 60
Infinitesimal, 10, 15, 17, 80, 126, 183
Irrational, 7, 20, 41, 267
Isidore of Miletus, 41, 68
Ivins, W. M., 264

Jacobi, C. G. J., 242, 247, 255, 258, 260
Jacobians, 242
James of Forli, 45
Jones, Wm., 146
Jordanus Nemorarius, 50
Journal für die reine und angewandte Mathematik, 241f, 249ff, 252, 254f, 257, 260

Kaestner, A. G., 68, 193, 199, 216f
Kant, Immanuel, 37
Karpinski, L. C., 27, 60
Karpis of Antioch, 33
Kasir, D. S., 43
Kaye, G. R., 41
Kepler, Johann, 68f, 232, 254
Kiefer, Albert, 244
Kirchhoff, G. R., 256
Klein, Felix, 9, 248, 266
Klügel, G. S., 216f
Kötter, E., 125
Kokomoor, F. W., 66
Kowalewski, G., 230
Krafft, G. W., 171
Krause, K. C. F., 229
Krazer, Adolf, 46, 48

LaChapelle, Abbé de, 192ff
Lacroix, S. F., 210–218; 153, 177, 193, 204f, 208, 223, 228, 256f, 260
Lagny, T. F. de, 136
Lagrange, J. L., 200–204; 170, 205ff, 233, 237, 258, 260, 263
Lahire, Philippe de, 118ff; 126, 139, 149f, 153, 156, 164f, 167
Laisant, C. A., 262
Lambo, Ch., 55
Lamé, Gabriel, 233–236, 263f; 240, 245f
Langsdorf, K. C., 198
Laplace, P. S., 208, 238
Lardner, Dionysius, 257
Latitude of forms, 44ff, 50, 53, 60, 72
Lattin, Harriet, 43
Lavoisier, Antoine, 218, 267
Lefrançais, F. L., 218–223
Legendre, A.-M., 211f
Lehmann, Ernst, 119
Leibniz, G. W. von, 15, 100f, 127, 129f, 133f, 149, 158, 185, 196, 233, 251
Leonardo da Vinci, 54f, 135
Leslie, Sir John, 257
Lexell, A. I., 207f
L'Hospital, G. F. A. de, 150–154; 127, 132ff, 139, 156, 159, 162f, 172, 176, 183, 192, 194, 196, 214, 223, 256
L'Huilier, S. A. J., 223, 236
Libri, G., 67

Lie, Sophus, 266
Limaçon, 185
Livet, J.-J., 222
Loci, 9, 19f, 22ff, 29ff, 36ff, 67, 72, 75ff, 81, 83, 87f, 95ff, 101, 104, 106, 112, 119, 154, 265
Locke, John, 103
Logarithmic spiral, 72, 199, 227f
Loria, Gino, 19, 27, 67, 86, 92, 99, 111, 118, 126, 132, 134, 136, 179f, 186, 190, 197, 223f, 228f, 238, 240, 260f, 266, 268

Maclaurin, Colin, 162f, 175ff; 140, 202, 252, 256
Magnus, H. G., 223
Marliani, Giovanni, 53
Maroger, A., 37
Marre, Aristide, 55
Matthiessen, Ludwig, 9, 57, 122
Maupertuis, P. L. M., 164
Maxima and minima, 80, 82, 86, 142, 171
Menaechmus, 17ff, 21, 24, 26, 42f, 56, 63f, 78, 110
    Menaechmian triads, 18, 32, 34
Méray, Ch., 267
Merrifield, C. W., 234
Mersenne, Marin, 73, 104
Method of exhaustion, 11
Meusnier, J. B., 207f; 189, 203
Milhaud, Gaston, 83, 103
Milne, J. J., 37, 83
Mitzcherling, Arthur, 9
Möbius, A. F., 242f; 249, 255, 258, 261, 263
Monge, Gaspard, 204ff, 209f; 137, 168, 198, 211ff, chapter IX passim
    Circle of Monge, 221
    Theorem of Monge, 222
Montucla, Étienne, 86, 108
Moore, Jonas, 70
Morley, F. V., 70
Moscow papyrus, 5
Müller, E., 185
Müller, Felix, 154
Mydorge, C., 69f; 112, 114

Napier, John, 134
Nasir Eddin, 73
Neugebauer, Otto, 1, 3, 8, 20, 30, 35
Newton, Sir Isaac, 138–145; 15, 36, 102, 112, 127, 130, 134, 137, chapters VII and VIII passim, 228, 244, 253, 256
    Newton's parallelogram, 141, 174, 195
Nicholas of Cusa, 72
Nicole, François, 164; 146, 170, 174
Noether, M., 246
Normal line, 93f, 205, 210
Normal plane, 205
Number, 1ff
    Real number 16, 267
Núñez, P., 72

Ockhamites, 45

Omar Khayyam, 43, 56, 63, 78
Omerique, A. H. de, 133, 155
Ordinate, 76, 111, 120f, 127, 130, 150, 194
Oresme, Nicole, 45ff; 50f, 54f, 61, 86, 101, 106, 119, 267
Oughtred, Wm., 69ff, 84
Ozanam, Jacques, 126f; 149f

Pacioli, 54ff
Pappus, 36ff; 22ff, 31ff, 41, 51, 61, 65, 95, 217, 223
    Problem of Pappus, 37ff, 74, 83, 87f, 91ff, 104, 109, 140
Parabola, 19, 24ff, 30f, 34, 41, 68, 77ff, 92, 105, 108, 110, 115f, 119, 126, 130ff, 144f, 151, 178, 183, 214
Paraboloid, 189, 207
Parameter, 59, 61, 69
Parent, Antoine, 156ff; 165, 167, 171, 207
Parmenides, 6
Pascal, Blaise, 68, 117, 127, 135
    Pascal's theorem, 247f, 252
Pascal, Étienne, 50
Peirce, J. M., 57
Perseus, 33
Peters, A., 229
Pitot, Henri, 166f
Plane, equation of, 158, 168, 171, 189, 201, 204, 209
    Normal form of, 201, 223
    Determinant form of, 258
Plato, 13ff; 34, 37, 65, 156, 217
Playfair, J., 162
Pliny, 43
Plot, R., 162
Plücker, Julius, 244–257, 264ff; chapter IX passim
    Plücker coordinates, 251
    Plücker's equations, 252f, 259
    Plücker's $\mu$, 246, 254
Poe, Edgar Allen, 225
Poisson, S. D., 226
Polynomial curve, 128f, 147, 153f, 155, 161f, 177f, 194f
Poncelet, Jean-Victor, 230ff, 238ff; 198, 225, 233, 245, 248f, 251ff
Porism, 22f
Prag, A., 110
Proclus, 14, 17, 33, 36, 41, 114
Proney, G. C. F. de, 212
Proportion, 16; 5, 12, 17, 19f, 24, 52f, 63, 107, 110, 117
Ptolemy, 35
Puissant, L., 218ff; 211, 223
Pythagoras, Pythagoreans, 4ff, 10, 21f, 24, 34, 51, 92
    Pythagorean triads, 3
    Pythagorean theorem, generalizations of, 207, 219

Quadratrix, 11f, 19, 32, 134, 143, 175, 187
Quadrature, 11, 31, 142

290

Quadric surface (*see* Surface, quadric)
Quetelet, L.-A.-J., 241, 251

Rabuel, Claude, 164
Radical axis, 233f
Randall, J. H., Jr., 50
Recorde, Robert, 70f, 130
Renaldini, Carlo, 133
Reyneau, Ch. René, 154, 175, 216
Riccati, Vincenzo, 199; 119, 187
Riemann, G. F. B., 258, 267
Roberval, G. P., 104ff; 92f, 125, 127, 134ff, 213
Rolle, Michel, 155, 215
Roomen, A. van, 60

Sachs, A., 35
Saladini, Girolamo, 199; 187, 193
Salmon, George, 260, 262
Saltykow, N., 67, 103
Sauerbeck, Paul, 175
Sauri, Abbé, 199; 193
Schenmark, Niels, 216
Schläfli, Ludwig, 258
Schmidt, M. C. P., 18
Schoenflies, A., 245, 251
Scholastics, 44ff
Scott, Charlotte A., 196
Scott, J. F., 109
Segner, Andreas, 162, 195
Seneca, 21
Shields, M. C., 162
Shoemaker's knife, 234
Simons, L. G., 221
Simplicius, 41
Skene, James, 261
Sluse, René de, 117ff; 127, 156
  Pearls of Sluse, 118
Smith, David Eugene, 33, 40, 49, 70, 174, 187, 210, 224
Snell, Willebrord, 66, 74
Solid analytic geometry, 165–172, 188ff, 200–210; 49, 81f, 93, 121, 125f, 156, 215, 219ff, 254ff
Spencer, R. C., 179
Sphere, director sphere, 222
Spiral (*see* Logarithmic spiral, and Archimedes, spiral of)
Stamm, Edward, 51
Steele, D. A., 13
Steiner, Jakob, 255; 226, 264
Stevin, Simon, 57f; 69, 112, 114
Stirling, James, 159ff; 146, 163, 174, 177, 183f, 194f
Stone, Edmund, 119, 154
Straight line, equation of, 197f; 76f, 92, 101, 108, 112, 114, 126, 150, 176, 178, 182, 195, 205f, 209, 213, 240, 243, 246f, 249ff, 258
  Determinant form, 258
  Normal form, 213, 223
  Parametric form, 236

Symmetric form, 236
Strong, E. W., 55
Struik, D. J., 255
Sturm, C., 237
Suiseth, Richard, 45, 47, 50
Sulvasutras, 7
Surfaces, 188ff; 23, 49, 81, 93, 121, 165ff, 198, 254
  Developable, 166f, 204, 206
  Equation of, 13, 49, 93, 121, 156, 158, 165, 189, 254, 265
  Parametric representation, 262f
  Quadric, 189; 125, 172, 221f, 236f, 254
Suter, H., 42
Symptomae, 29
Synthesis, 10, 14, 17, 37, 71, 216f, 232, 248, 255

Tangent line, 28, 82, 93, 95, 100, 130, 136, 142ff, 178, 219, 251f
Tangent plane, 157f, 254
Tannery, Paul, 14, 23, 29, 127
Tartaglia, Nicolo, 56f, 67
Taylor, Charles, 17, 19, 38, 68, 237
Teixeira, Gomes, 136
Tetrahedron, 201f, 221, 240, 243f, 258
Thales, 4
Theaetetus, 16
Thomas, Ivor, 38
Thorndike, Lynn, 48
Three famous problems, 12f, 33, 99
Tinseau, Ch., 207f; 206, 219
Torricelli, Evangelista, 134f
Transformation of coordinates, 28, 30, 78, 92, 113, 116, 139, 173, 182ff, 189, 197, 213, 219, 227f
Transformations in three dimensions, 189, 203, 207f, 215, 222, 242, 261f
Tropfke, Johannes, 120, 132, 153, 169, 173, 205, 214, 220, 224, 268
Turnbull, H. W., 117, 141, 162
Tweedy, C. T., 161

Ursus, Raimarus, 65

Vandermonde, C. A., 238f
Van Schooten, Frans, 112ff; 61, 66, 101, 108f, 128, 130, 164, 192
Vargas y Aguirre, D. J. de, 136
Variable, 6f, 29, 35, 44, 60f, 84, 142, 190
Varignon, Pierre, 146, 154, 172, 185
Vera, Francisco, 93
Ver Eecke, Paul, 17, 24, 37f
Viète, 59–65; 41, 56f, 66ff, 74, 76, 78, 80, 83, 88, 97, 99f, 104, 130, 133, 190, 223, 267
Vincent, A. J. H., 236
Vogt, Heinrich, 4, 7
Voltaire, 138
Von Fritz, Kurt, 5, 7

Walker, Evelyn, 134
Wallis, 109ff, 122ff; 71, 114, 116f, 126, 128, 136, 139, 147, 156, 158, 183, 222, 256
Wallner, C. R., 51
Waring, Edward, 198; 256f
Watt, James, 162
Waud, S. W., 257
Weaver, J. H., 9, 36
Weierstrass, Karl, 267
Werner, Johann, 66
Whewell, William, 229
Wieleitner, Heinrich, 46, 49, 51, 67, 100, 106, 114, 125, 132, 160, 169, 189, 198, 203, 205f, 208, 213, 220, 224, 266, 268

Wölfing, E., 228
Wolff, Christian von, 155f, 193
Wolff, Georg, 52
Wood, James, 256
Woolard, E. W., 1
Wren, Sir Christopher, 125, 156, 158

Yates, R. C., 136

Zeno of Elea, 5ff
    Paradoxes of, 6f, 15, 35
Zeuthen H. G., 7f, 17ff, 27, 29, 31, 34, 56, 145